OPEN CHANNEL FLOW

OPEN CHANNEL FLOW

SAIFUL ISLAM

KHANNA BOOK PUBLISHING CO. (P) LTD.

::: *Publisher of Engineering and Computer Books* :::

Head Office : 4C/4344, Ansari Road, Darya Ganj,
New Delhi-110002
Phone: 011-23244447/48 Fax: 011-23244447

Regd. Office : 1694-95, Nai Sarak, Delhi-110006
Phone: 011-23283211 Fax: 011-23256658

E-Mail : contact@khannabooks.com
Website : www.khannabooks.com

OPEN CHANNEL FLOW
SAIFUL ISLAM

ISBN: 978-93-81068-99-1
First Edition: 2013

Published by :
Khanna Book Publishing Co. (P) Ltd.
4C/4344, Ansari Road, Darya Ganj, New Delhi-110002
Phone : 011-23244448 Telefax : 011-23244447
E-mail : contact@khannabooks.com
Website : www.khannabooks.com

Regd. Off :1694-95, 1st Floor, Nai Sarak, Delhi-110006 (India)
Phone : 011-23283211 Telefax : 011-23256658

Printed in India by:
S.P.S. Printers & Binders, Delhi

Preface

Open channel flow finds application in the solution of problem related to several aspects of development of surface water resources. Problems in wide variety of fields, such as design of hydraulic structures, dispersion of pollutants, overland flow and sediment transport in rivers require the use of principles of open channel. The importance of this subject has grown in recent years to the extent that most civil engineering curriculam include a full semester course on the subject at the undergraduate and graduate levels.

This book has been written keeping in mind the recent syllabus of GBTU & MTU Universities. An attempt has been made to enrich the book with simple language, lucid expression, and make it available at the low price.

The author would feel obliged if an error in this book are brought to their notice. Constructive suggestions from the readers to further improve the book are always welcome

Saiful Islam

Acknowledgement

I express my sincere gratitude to my father Dr. Fakhrul Islam Azmi, HOD, Department of Arabic, Shibli National PG college, Azamgarh, for his constant inspiration and encouragement.

I have the benefits of valuable advice and suggestions from Prof. Shahiq Khan, HOD, Mr Sabir Khan and Mr. Piyush Gupta, Assistant Professor, Department of Civil Engineering, RGEC, Meerut and Mozaffar Imam, Assistant Professor, Department of Civil Engineering, Integral University, Lucknow.

I express my indebtness to Books, Literatures, Journals, which have been referred to during the preparation of this book. Mr Puneet Khanna, Managing Director, Khanna Publications deserves special thanks who has taken special pains in bringing out this book in this form.

I wish to thanks the production team member of Khanna Publication, Darya Ganj, New Delhi-110002, for their unfailing cooperation they executed during the course of publication of this book.

Saiful Islam

Contents

Unit 1

Basic Principles of Open Channel Flow

1.1 Introduction

The term open channel flow or Free Surface Flow represents flows through channels that are open to the atmosphere and liquid flows with a free surface. The free surface is the interface between the moving liquid and an overlying fluid medium and will have constant pressure. For example, water is the most common liquid with air at atmospheric pressure as the overlying fluid. The most important force behind open channel flow is gravity.

Examples of open channel flow include flow in rivers, canals, laboratory flumes, flow over weirs and spillways and overland runoff. It is necessary to remember that Froude number is the prime non-dimensional number governing the flow in open channels.

1.2 Applications of Open Channel Flow

Some of the important applications of the principles of open channel flow are as follows:

1. Measuring the discharge in a river or canal.
2. Developing a relationship between the depth of flow and discharge in a channel.
3. Designing a canal to carry given amount of water.
4. Estimating the area of submergence due to construction of a dam on a river.

5. Preventing very high velocity flows from damaging the channel.
6. Estimating the change in the flow condition due to a change in the bed width or bed elevation.
7. Estimating change in the flow depth due to the over-land runoff.
8. Estimating the time taken by the flood wave to pass through a given length of a river.
9. Estimating the amount of sediment carried by a channel.
10. Studying the spread of pollutants in a river.

1.3 DIFFERENCE BETWEEN OPEN CHANNEL FLOW AND PIPE FLOW

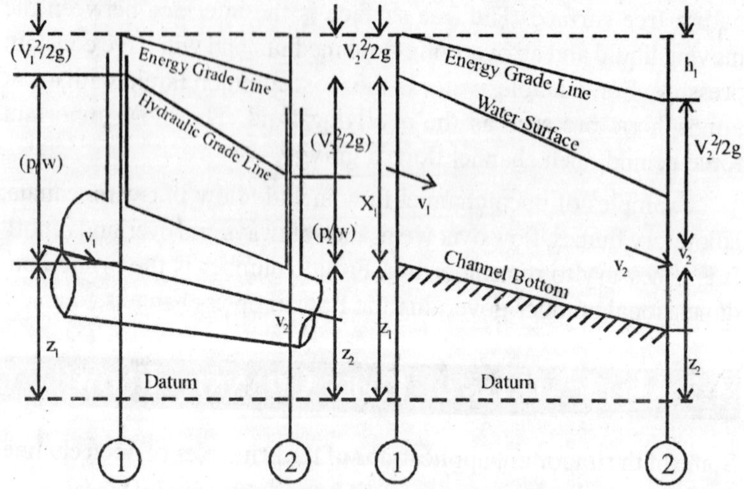

Fig. 1.1: Comparison between pipe flow and channel flow

S.No.	Points	Open Channel Flow	Pipe Flow
1.	Cause of Flow	Gravity force (provided by slope)	Pipe flow take place at the expense of hydraulic pressure.
2.	Geometry of cross-section	Open channels may have any shape Rectangular, Triangular, Trapezoidal, Circular, etc.	Pipes are generally round in cross-section.
3.	Surface roughness	The hydraulic roughness varies with depth of flow.	Roughness coefficient depends upon the material of pipe.
4.	Piezometric head surface.	(Z + y), where y is the depth of flow. Hydraulic gradient line concides with water water surface.	(Z+*p*/*w*), where p is the pressure in pipe. Hydraulic gradient line (HGL) does not concide with
5.	Velocity distribution	The maximum velocity occurs at a little distance below the water surface. The shape of velocity profile is dependent on channel roughness.	The velocity distribution is symmetrical about the pipe axis, maximum velocity occurs at pipe centre and the velocity at the wall is zero.

1.4 Basic Concepts of Free Surface Flows

The flow in open Channel or in a closed conduit having a free surface is referred as free surface flow or open Channel flow if there is no free surface and conduit is flowing full, then the flow is called pipe flow or pressurized flow. In a closed conduit it is possible to have both free surface flow and pipe flow at different times. It is also possible to have these flows at a given instant of time in different position of conduit. For example, the flow in storm sewer may be free surface flow at certain time. Then due to large inflows produced by sudden storm, the sewer may start to flow full, thereby flow changes to pipe flow. Similarly, a closed conduit may have free surface flow in part of the length and pipe flow in the remaining length. This condition usually occurs in a closed conduit when its down stream end is submerged.

For analysing free surface flow problems, it is important to know that what are the governing forces which causes free surface flow or open Channel flow. It should be kept in mind that open channel flow is governed by the following forces :

(i) Inertia force

(ii) Component of gravity $\omega \sin \theta$ due to bed Slope

(iii) Surface tension : It is negligible, but it is taken into account when the flow depth is very small, preferably below 7 cm as seen from experiment.

(iv) Viscous force : It is small for water when flow is turbulent. It is importat at low velocity and for liquids with viscosity.

(v) Force of resistance due to friction (due to surface roughness)

For free surface flow, the following concept should be kept in mind.

(i) The height of Total energy line from datum is $\left(Z + y + \frac{v^2}{2g}\right)$, where Z is datum head, y, potential head and $\frac{V^2}{2g}$ is velocity head.

(ii) Flow takes place due to component of gravity force in the flow direction

(iii) Hydraulic gradient line (HGL) concides with free surface and is at a height at (Z + Y).

1.5 TYPES OF CHANNELS

The various types of channels are as follows:

1. **Prismatic and Non-prismatic Channels:** A channel in which cross-sectional shape and size and also bottom slope are constant is termed as a prismatic channel. Most of man made channels are prismatic channels. The rectangular, circular, trapezoidal, triangular are the examples of man made channel. All natural channels generally have varying cross-section are non-prismatic channels.

2. **Rigid and Mobile Boundry Channels:** Rigid channels are those in which boundry is not deformable, therefore the shape and roughness are not the functions of flow parameters e.g., lined canal, sewer etc. No scour and erosion take place in such a channel.

   All unlined channels and natural rivers in which boundary undergo deformation due to continuous process of erosion, in such a case flows carries sediments. Such channels are classified as mobile boundary channel.

3. **Natural and Artificial Channels:** Natural Channels are those channels which have irregular sections of varying shapes, developed in a natural way. Examples: Rivers, Streams, etc.

   Artificial channels are those channels which are build artificially for carrying water. They have regular geometrical shapes. Examples: Rectangular, Trapezoidal, Parabolic Channels, etc.

1.6 Classification of Open Channel Flows

Open channel flows are classified for purposes of identification and analysis as follows:

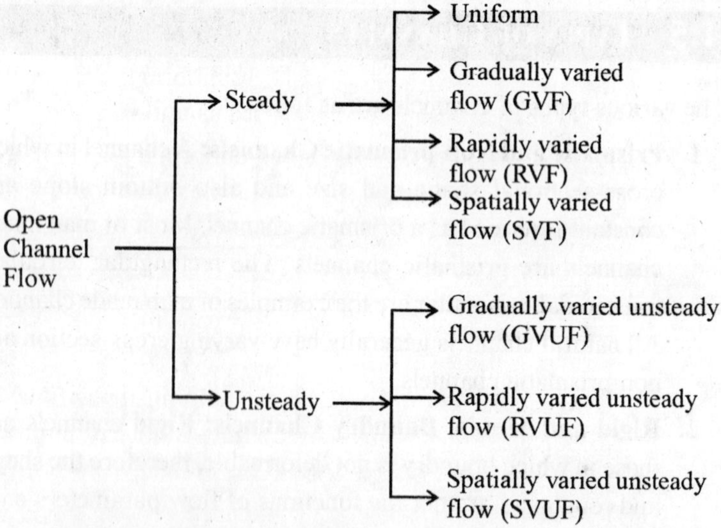

1. **Steady and Unsteady Flow:** A steady flow occurs when the flow properties, such as the depth, velocity and discharge at a section donot change with time.

   Mathematically, $$\frac{\partial y}{\partial t} = 0, \frac{\partial v}{\partial t} = 0, \frac{\partial Q}{\partial t} = 0$$

   where y, v and Q are depth, velocity and discharge respectively. In general, it is almost impossible to have a strictly steady flow. No matter how good a control exists on the flow conditions, there is bound to be some change with times.

   Unsteady Flow occurs when the flow properties, such as the depth, velocity and discharge at a section changes with time.

   Mathematically, $$\frac{\partial y}{\partial t} \neq 0, \frac{\partial v}{\partial t} \neq 0, \frac{\partial Q}{\partial t} \neq 0$$

   Flood flows in rivers and rapidly varying surges in canals are some examples of unsteady flows.

2. **Uniform and Non-uniform Flow:** Flow in channel is said to be uniform if the depth of flow, velocity and discharge donot change in longitudinal direction.

Mathematically, $\frac{\partial y}{\partial l} = 0, \frac{\partial v}{\partial l} = 0, \frac{\partial Q}{\partial l} = 0$

It would be rare to achieve a strictly uniform flow but a number of flow situations may be idealised as uniform flow.

Flow in channel is said to be non-uniform if the depth of flow, velocity and discharge change in longitudinal directions.

Mathematically, $\frac{\partial y}{\partial l} \neq 0, \frac{\partial v}{\partial l} \neq 0, \frac{\partial Q}{\partial l} \neq 0$

Flow over a wier and flow upstream of a dam are some examples of non-uniform flow.

It should be kept in mind that the term unsteady flow is used to indicate unsteady-nonuniform flow. Similarly, uniform flow implies steady uniform flow. Also unsteady uniform flow is non-existent and non-uniform implies steady non-uniform flow.

Non-uniform flow in further classified as:

1. Gradually varied flow
2. Rapidilly varied flow
3. Spatially varied flow

1. **Gradually varied flow:** In this type of flows, the change depth take place gradually along the length of channel.

   Examples: The backing up of water in a stream due to dam or droping of water surface due to sudden drop in channel bed are an example of steady GVF.

   The passage of a flood wave in a river is a case of unsteady GVF. (or GVUF).

2. **Rapidly varied flow:** If the curvature in varied flow is large and depth changes appreciably over short length. Such flow is called rapidly varied flow.

Example: Hydraulic jump occuring below a spillway or sluice gate is example of steady RVF. A surge moving up a canal and bore travelling up a river are example of unsteady RVF.

Remember frictional resistance plays important role in GVF while it is insignificant in case of RVF.

3. **Spatially varied flow:** If some flow is added or subtracted from the system the resultant flow is known as SVF.

   Examples: The flow over side weir is an example of steady SVF. The production of surface runoff due to rainfall is an example of unsteady SVF.

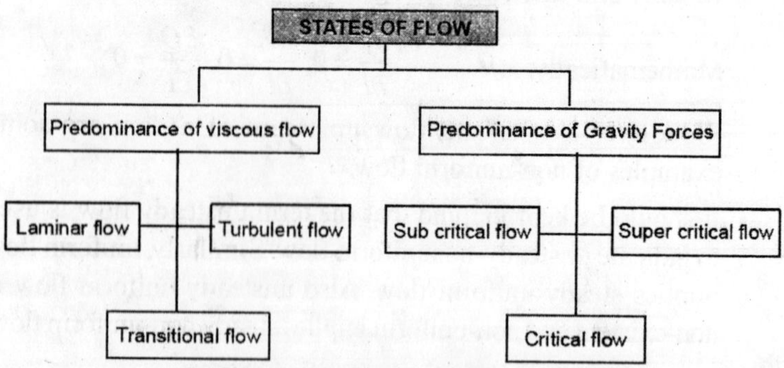

Laminar Flow and Turbulent Flow

It depends upon Reynolds number defined as

$$R_e = \frac{\rho VR}{\mu}$$

V = average velocity of flow in channel

R = Hydraulic radius, R_e = Reynold number

When $R_e < 500$ flow is laminar

$R_e > 2000$ flow is turbulent

$500 < R_e < 2000$ flow is transitional.

Subcritical Flow, Critical Flow and Supercritical Flow

As we know that gravitational force is the predominant force in open channel flow, therefore, Froude number $F = \frac{V}{\sqrt{gD}}$ is an important parameter for analysing open channel flow.

Depending upon Froude number the channel may be characterised as:

(*i*) When $F < 1$ or $V < \sqrt{gD}$: The flow is described as subcritical or tranquil or streaming.

(*ii*) When $F = 1$: The flow is said to be in critical state.

(*iii*) When $F > 1$: The flow is said to be supercritical or rapid or shooting or torrential.

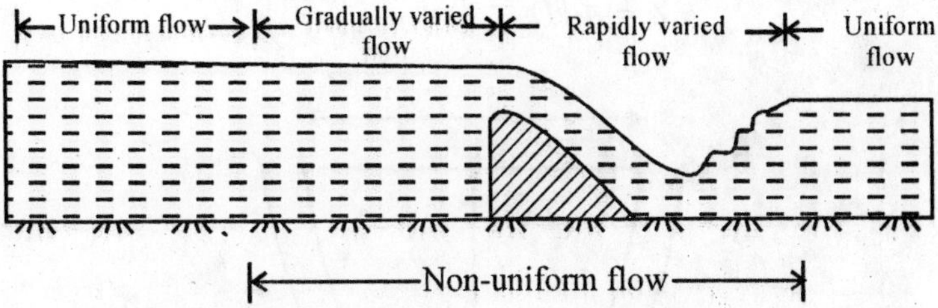

Fig. 1.2a: Uniform and non-uniform flow

1.7 Geometrical Parameters of a Channel

The Geometrical properties of a channel section can be defined by the Geometry of the channel section and depth of flow. Some of the Geometrical parameters of basic importance are given below:

1. **Depth of flow (y):** It is the vertical distance of the lowest point of channel section from free surface.
2. **Top Width (T):** It is the width of channel section at free surface.
3. **Wetted Area (A):** It is the cross-sectional area of flow normal to the direction of flow.
4. **Wetted Perimeter (P):** It is the length of channel boundry in contact with flowing water at any section.
5. **Hydraulic Radius (R) or Hydraulic Mean depth:** It is the ratio of the wetted area to its wetted perimeter.

$$R = \frac{A}{P}$$

6. **Hydraulic depth(D):** It is the ratio of wetted area to top width *T*.

$$D = \frac{A}{T}$$

7. **Section factor (Z):** It is product of wetted area and square root of hydraulic depth.

$$Z = A\sqrt{D} = A\sqrt{\frac{A}{T}} = \left(\frac{A^3}{T}\right)^{1/2}$$

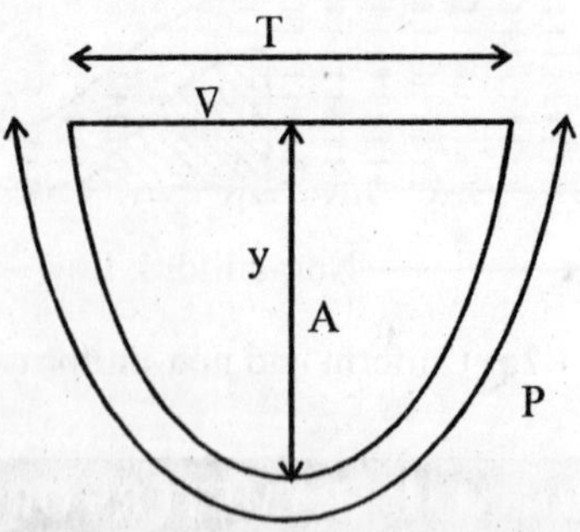

Fig. 2b Channel Cross-section

1.8 Mass, Energy and Momentum Principle for Prismatic and Non Prismatic Channel

The three basic equations of fluid mechanics are the continuity, energy and momentum equations. These three basic equation that governs flow in different condition are:

1. Equation of continuity based on principle of conservation of mass.
2. Energy equation based on conservation of energy.
3. Momentum equation based on Newton second law of motion, i.e. change in momentum per unit time in the body of water in a flowing channel which is equal to the resultant of all external forces that are acting on the mass of liquid. We will discuss one by one all of the three:

1) *Equation of continuity:* The law of conservationof mass between different channel sections gives that the volumetric flow rates Q, i.e., discharges in those section, remains equal provided there is no lateral inflow or outflow. For sake of simplicity only one dimensional flows are considered.

   a) ***Steady continuity equaion in one dimension:*** Consider the flow of incompressible water in channel.

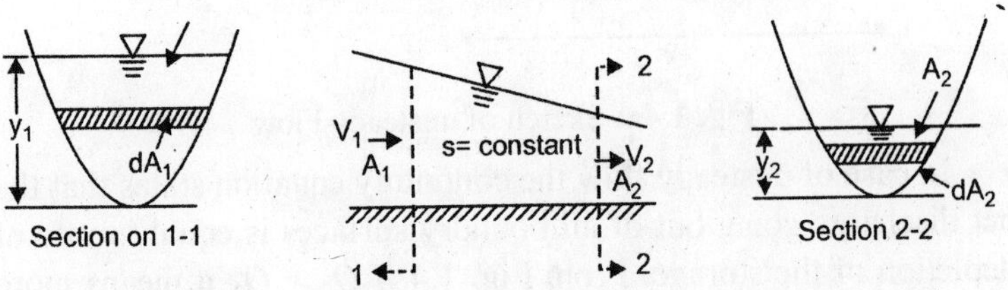

Fig. 1.3: Views of open channel flow to derive continuity equation

It is asumed that there is no inflow or outflow in the channel under consideration for study and flow is steady. Let us suppose A_1, V_1, dA_1 and A_2 V_2 and dA_2 are the total area, mean velocity and infinitesimal area at section 1-1 and 2-2 respectively.

Rate of mass flow through section 1–1 = $\rho V_1 dA_1$

Rate of mass flow through section 2–2 = $\rho V_2 dA_2$

Applying conservation of mass law at section 1-1 and 2-2 we have

$$\int_{A_1} \rho V_1 dA_1 = \int \rho V_2 dA_2$$

or
$$\int_{A_1} V_1 dA_1 = \int V_2 dA_2$$ (since liquid is in compressible)

$$V_1 A_1 = V_2 A_2 = Q$$

or
$$Q_1 = Q_2$$

b) Unsteady continuity equation in one dimension:

2. For unsteady flow

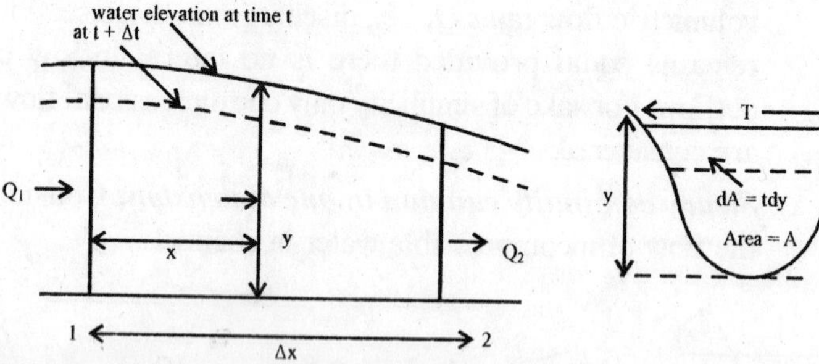

Fig. 1.4a: Sketch of unsteady Flow

In case of unsteady flow the continuity equation states that the net discharge going out of all boundry surfaces is equal to rate of depletion of the storage. From Fig. 1.4 if $Q_2 > Q_1$ it means more flow goes out. The excess volume of outflow causes depletion in storage within bourdary 1 and 2 in a time Δt. As a result, water surface will start falling.

We can say $Q_2 - Q_1 = \dfrac{\partial Q}{\partial x}\Delta x$

Excess volume rate in time Δt

$$= \left(\frac{\partial Q}{\partial x}\right)\Delta x.\Delta t$$

Storage volume at depth $y = A.\Delta x$

The rate of decrease of storage $= \left(-\Delta x \dfrac{\partial A}{\partial y}.\dfrac{\partial y}{\partial t}\right)$

We know $\dfrac{\partial A}{\partial y} = T$

The rate of decrease of storage $= -T\Delta x \dfrac{\partial y}{\partial t}$

The rate of decrease in Δt time $= \left(-T\Delta x \dfrac{\partial y}{\partial t}\right)\Delta t$

By continuity condition

$$\frac{\partial Q}{\partial x}\Delta x\Delta t = -T\frac{\partial y}{\partial t}\Delta x.\Delta t$$

or
$$\frac{\partial Q}{\partial x}+T\frac{\partial y}{\partial t} = 0$$

2. Energy Equation for Prismatic Channel

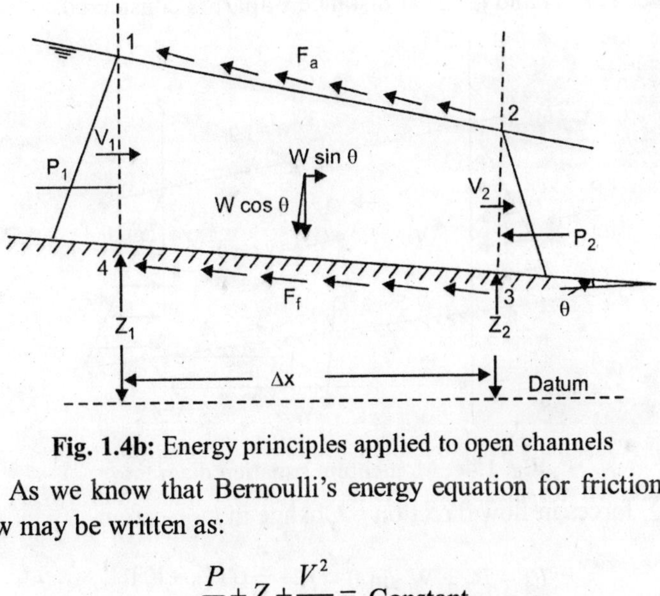

Fig. 1.4b: Energy principles applied to open channels

As we know that Bernoulli's energy equation for frictionless flow may be written as:

$$\frac{P}{\gamma}+Z+\frac{V^2}{2g} = \text{Constant}$$

Consider non uniform flow in channel. Let us consider section 1–1 and 2-2. Applying Bernoullis energy equations

$$y_1+\frac{V_1^2}{2g}+Z_1 = y_2+\frac{V_2^2}{2g}+Z_2+E_L \qquad \text{(a)}$$

E_L is the head loss between two section. The loss of head is due to boundry frction, excessive turbulence in hydraulic jump

As we know $y_1+\frac{V_1^2}{2g} = E_1$

and $$y_2 + \frac{V_2^2}{2g} = E_2$$

Put in (a) we get

$$E_1 - E_2 = Z_2 - Z_1 + E_L$$

3. Momentum equation for prismatic channel: This is based on Newton Second Law of Motion. i.e. summation of forces under consideration in direction of flow is equal to net change in momentum in same direction. To derive the relation control volume of liquid between (1–1) and (2–2) at distance x apart is considered.

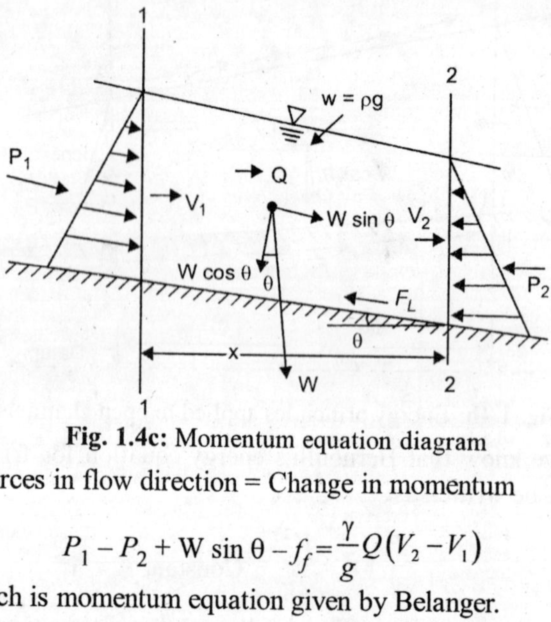

Fig. 1.4c: Momentum equation diagram

Σ forces in flow direction = Change in momentum

$$P_1 - P_2 + W \sin\theta - f_f = \frac{\gamma}{g} Q(V_2 - V_1)$$

Which is momentum equation given by Belanger.

2a. Energy equation in non prismatic channel : For non prismatic channel section, the channel section varies along the length of channel, hence specific energy curve differ from section to section. For example a non prismatic channel with variable slope is taken. For variable slope channel it is more convenient to plot totalenergy head $H = Z + y + \frac{V^2}{2g}$ against the depth of flow. Four channel section are then selected and four energy curves for these sections are plotted. The initial section 0 is an upstream section in subcritical region. The two depth corresponding to given energy H_0 can be obtained from energy curve. Since this section is in sub critical flow region. The higher value of y_0 should be actual depth of flow, whereas lower value is alternate depth. Similarly the alternate depth in other section can be obtained. In the downstream section 1 and 2, the lower value of y_1 and y_2 are the actual depth of flow since they are in supercritical flow region.

The critical depth at each section can also be obtained from energy curve at point of minimum energy. At section C the critical flow occurs, and the depth y_c is the critical depth.

- In this nonprismatic channel a gradually varied flow is carried out from subcritical state to supercritical flow state.

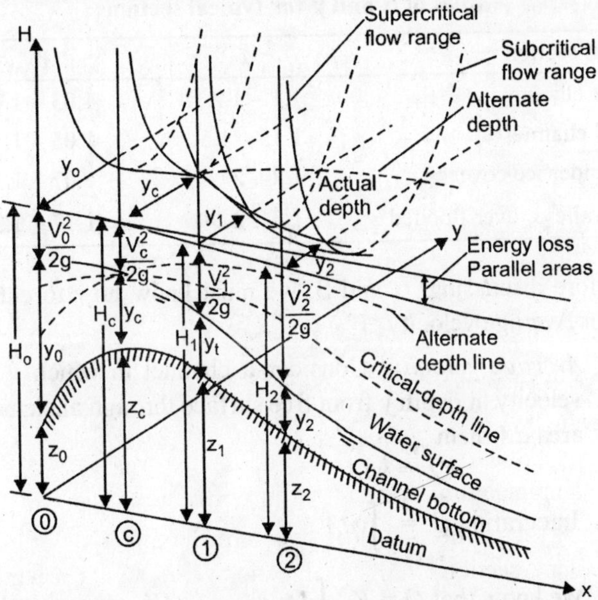

Fig. 1.5: Energy in a nonprismatic channel of variable-slope carrying gradually varied flow from subcritical to supercritical state.

3a. Momentum equation in nonprismatic channel

The specific force, like specific energy varies with the shape of the channel section. In applying the momentum equation to nonprismatic channels, a three dimensional plot similar to energy equation can be constructed. For practical purposes, however, this is rarely necessary.

1.9 Velocity Distribution Coefficients (α & β)

The velocity distribution coefficient α (kinetic energy coefficient) and β (Momentum coefficient) is slightly greater than unity. For channels of regular cross-section and fairly straight alignment the effect of non-uniform velocity distribution is usually negligible. Therefore, α and β are assumed to be unity. In the complex cross section, the value of both coefficient may be higher.

Values of α and β for typical sections

Channel section	α	β
Regular channel	1.1 – 1.2	1.03 – 1.07
Natural channel	1.15 – 1.5	1.05 – 1.17
River under ice cover	1.2 – 2	1.07 – 1.33
River, valleys, over flooded	1.5 – 2	1.17 – 1.33

Before discussing, α and β, we must know how to calculate Mean or Average velocity '*V*'.

(a) *Average velocity:* Consider a channel in which v is the velocity at depth y from free surface through anelementary area dA them

$$dQ = vdA$$

Integrating $Q = \int vdA$

We know that $Q = V_m \int dA$ $\qquad$ (V_m = mean velocity)

$$V_m \int dA = \int vdA$$

or

$$V_m = \frac{\int vdA}{\int dA}$$

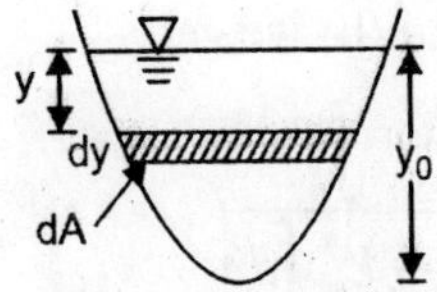

Let us suppose channel is rectangular, $A = By_0$, $dA = Bdy$

$$V = \frac{\int_0^{y_0} v dy}{\int_0^{y_0} dy}$$

(b) *Kinetic Energy Coefficient* α: Consider an elementary area ΔA of whole cross-section A of channel. Let us suppose γ is the specific weight of liquid flowing in the channel. The weight of water pasing through ΔA with velocity v is $\gamma \Delta A v$.
Now mass of liquid = ρ ($\Delta A v$)
Kinetic energy of liquid passing through ΔA

$$\text{Kinetic energy} = \frac{1}{2} m v^2$$

$$\text{Kinetic energy} = \frac{1}{2} (\rho \Delta A v) v^2$$

$$\text{Kinetic energy of whole area } A = \sum \frac{1}{2} (\rho \Delta A v^3) \qquad \ldots(a)$$

Now let us consider whole area, mean velocity V and corrected velocity head $\alpha \dfrac{V^2}{2g}$

$$\text{Total K.E.} = \alpha\rho \cdot V_m \cdot \frac{V_m^2}{2} \int dA$$

$$= \frac{\alpha \rho A V_m^3}{2} \qquad \ldots(b)$$

Equating (a) and (b) we get

$$\alpha = \frac{\int v^3 dA}{V_m^3 \int dA}$$

If channel is rectangular, then $A = By_0$ $dA = bdy$

Now

$$\alpha = \frac{\int_0^{y_0} v^3 dy}{V^3 \int_0^{y_0} dy}$$

(c) *Momentum correction (β)*: The momentum passing though elemental area ΔA with velocity v at depth y

$$M = \rho\,(\Delta A v)\,v$$

Total momentum = $\Sigma(\rho \Delta A v^2)$ ...(a)

Now taking whole area and mean velocity V_m and momentum correction β, we have

Momentum transfer rate through area $A = \beta \rho V_m^2 \int dA$ ...(b)

Now equating (a) and (b) we get

$$\beta = \frac{\int v^2 dA}{V_m^2 \int dA}$$

If we take rectangula channel $A = bdy$

and $dA = Bdy$

$$\beta = \frac{\int_0^{y_0} v^2 dy}{V_m^2 \int_0^{y_0} dy}$$

1.10 Specific Energy and Specific Energy Curve

As we know the total energy of flow per unit weight of liquid is given by

$$\text{Total Energy} = z + y + \frac{v^2}{2g}$$

where z = Elevation of channel bottom above horizontal bottom

y = Depth of flow

v = Average velocity of flow

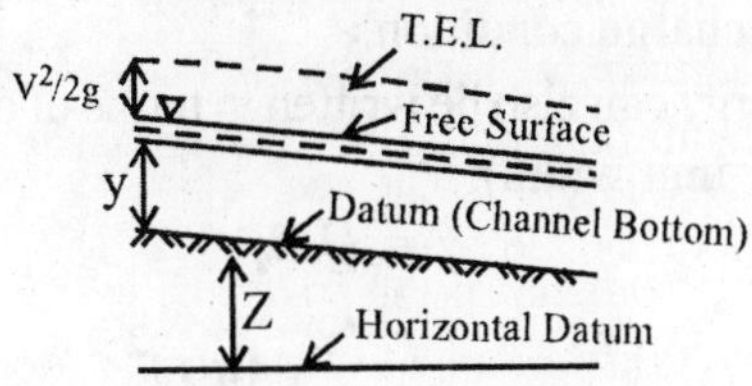

Fig. 1.6: Notation for specific energy

If the channel bottom concides with datum, the resulting expression is known as specific energy.

$$E = y + \frac{v^2}{2g}$$

Thus, specific energy can be defined as energy per unit weight of flowing liquid above the channel bottom.

It should be remember that total energy is a real fluid flow always decrease in downstream direction but specific energy is constant for uniform flow. In case of non-uniform flow the total energy will be constant at all section while specific energy is constant only in horizontal bed channel and in all other cases the specific energy will vary.

Specific energy comprises of:

(*i*) Potential energy of flow $E_p = y$

(*ii*) Kinetic energy of flow $E_K = \frac{v^2}{2g}$

$$E = y + \frac{v^2}{2g} = E_P + E_K$$

Specific Energy Curve

We know that $E = y + \frac{v^2}{2g} = y + \frac{Q^2}{2gA^2}$ $(\because Q = AV)$

For a given channel geometry:

E is the function of depth and discharge, $E = f(y, Q)$

The two condition arises therefore constant discharge condition and variable discharge condition.

Specific energy can also be written in terms of discharge intensity q (discharge per unit width):

Where $q = Q/B$

$$E = y + \frac{(q/y)^2}{2g} = y + \frac{q^2}{2gy^2}$$

where $V = \frac{\text{Discharge}}{\text{Area}} = \frac{Q}{by} = \frac{q}{y}$

(*a*) **Constant Discharge Condition:** For a given Geometry (width constant), keeping the discharge as constant $= Q_1$. The variation of E and y is represented by cubic parabola. Specific energy curve contain following information:

(*i*) The curve of potential energy ($E_p = y$) is straight line passing through origin making an angle of 45° with each of the two axes.

(*ii*) There are two positive roots of E, which means at any particular discharge Q_1 in a channel, there are two depth which is having same specific energy called as alternative depth or congugate depth.

(*iii*) Specific energy is asymptotic to horizontal axes for small values of y and asymptotic to 45° line for high value of y.

(*iv*) At certain depth y_c called critical depth the specific energy curve has a point of minimum specific energy, the corresponding flow velocity is called critical velocity V_C.

Consider the Fig. 1.6.

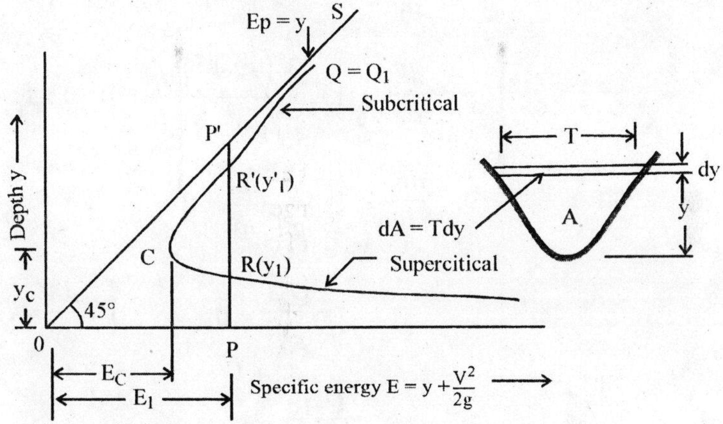

Fig. 1.7: Definition sketch of specific energy

The ordinate PP' represent the condition of specific energy E_1. The depths of flow $PR = y_1$ or $PR' = y_1'$ having same specific energy known as alternative depths. The Intercept $P'R'$ or $P'R$ represents the velocity head. Remember smaller alternative depth have larger velocity head and bigger alternative depth have smaller velocity head. From the figure, It is clear that if specific energy decreases the difference between $(y_1' - y_1)$ will decrease and at a certain value $E = E_C$, the two depth will merge with each other. (Point C). When $E < E_C$, the flow under given conditions is not possible in this region.

The condition of minimum specific energy is known as the critical flow condition and corresponding depth y_C is known as critical depth.

(*a*) **Condition for minimum specific energy**

We know that, $E = y + \dfrac{Q^2}{2gA^2}$

For a given discharge (constant discharge, Q) the condition for minimum specific energy is obtained by differentiating specific energy equation with respect to y and equating to zero.

$$\frac{dE}{dy} = 0$$

$$\frac{dE}{dy} = 1 + \frac{Q^2\left(-2A^{-1}\right)}{gA^2}\frac{dA}{dy} = 0$$

$$\frac{dE}{dy} = 1 - \frac{Q^2}{gA^3}\frac{dA}{dy} = 0$$

$$\Rightarrow \qquad 1 - \frac{Q^2 T}{gA^3} = 0 \qquad \left(\because \frac{dA}{dy} = T\right)$$

or

$$\frac{Q^2 T_C}{gA_C^3} = 1$$

which is condition for critical flow (suffix c represent critical flow condition)

or

$$\frac{Q^2}{g} = \frac{A_C^3}{T_C}$$

It should be kept in mind at critical flow $y = y_C$, the Froude number is unity and specific energy is minimum. At the lower limb, CR of the specific energy curves, the depth $y_1 < y_C$ and $V_1' > V_C$ (critical velocity), therefore $F_1 > 1$. This region is called supercritical flow region. In the upper limb CR', $y_1' > y_C$. As such $V_1' > V_C$ and $F_1' < 1$. This region is known as subcritical flow region.

(*b*) **Variable discharge condition**

Consider the figure below, $Q_1 < Q_2 < Q_3 < ....$ and is constant along respectivity E verses y. let us suppose section PP' in the figure. Where $E = E_1 =$ constant. It is seen from the figure that different Q curve give different intercepts. The difference between alternative depths decreases as the Q value increases.

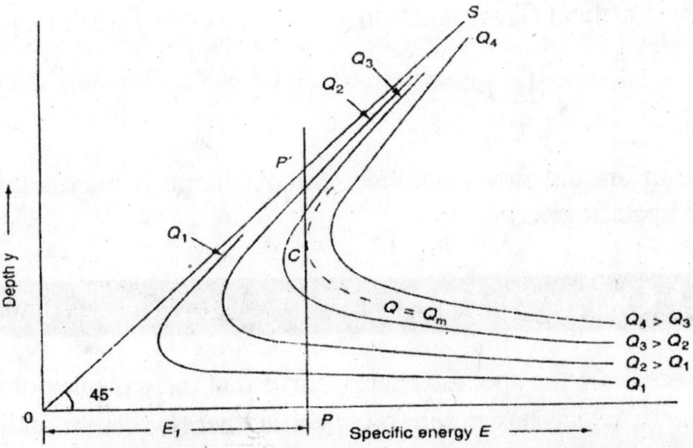

Fig. 1.8: Specific energy for varying discharges

and it will reach maximum at $Q = Q_m$ at point C at which the specific energy curve is just tangential to the ordinate PP'. (having one depth called critical depth). If $Q > Q_m$ the specific energy curve has no intercept with ordinate PP' and hence there will be no depth at which discharge can be passed with given specific energy (flow is not possible)

Condition for maximum discharge

The condition for maximum discharge can be obtained by differentiating equation (*i*) with respect to *y* and equating it to zero while keeping E = constant

$$E = y + \frac{Q^2}{2gA^2}$$

or $$Q = A\sqrt{2g(E-y)} \qquad (i)$$

$$\frac{dQ}{dy} = \sqrt{2g(E-y)}\frac{dA}{dy} - \frac{gA}{\sqrt{2g(E-y)}} = 0$$

$$\left(\because \frac{dA}{dy} = T \text{ and } \frac{Q}{A} = \sqrt{2g(E-y)}\right)$$

we get $$\frac{Q}{A}T - \frac{gA^2}{Q} = 0 \quad \text{or} \quad \frac{Q^2 T_C}{gA_C^3} = 1$$

which is critical flow condition

or $$\frac{Q^2}{g} = \frac{A_C^3}{T_C}$$

hence at critical flow condition, the discharge is maximum for a given specific energy.

1.11 Critical Depth y_C

It is seen from the specific energy curve that there is one point C on the curve which has minimum specific energy, thereby indicating that below this value of specific energy , the given discharge cannot occur. The depth of flow at which specific energy is minimum is called critical depth y_C (Constant discharge) critical depth y_C can also be defined as the depth of flow at which maximum discharge occurs for a fixed specific energy (specific energy constant) For example, the critical depth of rectangular channel can be obtained as

$$E = y + \frac{q^2}{2gy^2} \qquad \left(q = \frac{Q}{B}\right)$$

$$\frac{dE}{dy} = \frac{d}{dy}\left[y + \frac{q^2}{2gy^2}\right] = 0$$

$$1 + \frac{q^2}{2g}\left(\frac{-2}{y^3}\right) = 0 \text{ or } \frac{q^2}{gy^3} = 1$$

$$y_C = \left(\frac{q^2}{g}\right)^{1/3}$$

(specific energy is minimum)

1.12 Critical Velocity

The velocity of flow at critical depth is known as critical velocity denoted by V_C. Its value is obtained as

$$\text{Velocity} = \frac{\text{Discharge}}{\text{area}} = \frac{Q}{By} = \frac{q}{y} \text{ (rectangular channel)}$$

$$V_C = \frac{q}{y_C} = \frac{q}{\left(\frac{q^2}{g}\right)^{1/3}} = q^{\frac{1}{3}} g^{\frac{1}{3}}$$

or $$V_C^3 = qg$$

or $$V_C^3 = V_C y_C g \qquad (\because q = V_C y_C)$$

$$V_C = \sqrt{g y_C}$$

1.13 Critical Flow

The critical flow is one in which specific energy is minimum. (At a constant discharge). A flow corresponding to critical depth is also known as critical flow.

we know $$V_C = \sqrt{g y_C}$$

or $$\frac{V_C}{\sqrt{g y_C}} = 1$$

But $\frac{V_C}{\sqrt{g y_C}}$ = Froude number for rectangular channel

$$\left(\because F = \frac{V}{g\sqrt{\frac{A}{T}}} = \frac{V}{\sqrt{gy}}\right)$$

$$F = \frac{V_C}{\sqrt{g y_C}}$$

here $F = 1$ for critical flow

1.14 Calculation of Critical Depth

The critical depth in channel of various geometric shapes can be obtained as follows:

1. **Rectangular Channel**

Here $$A = By \text{ and } T = B$$

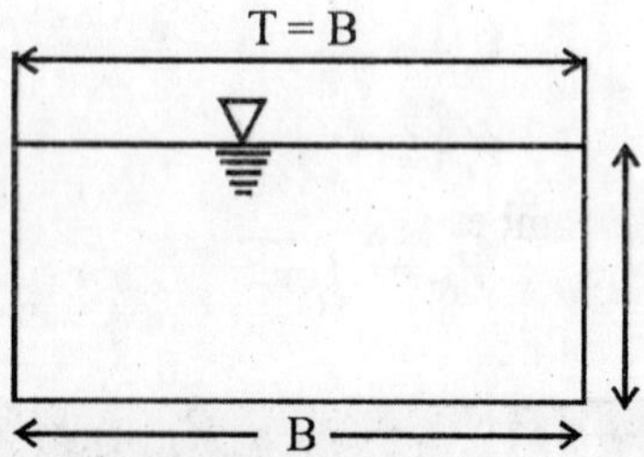

Fig. 1.9: Rectangular Channel

using critical flow condition

$$\frac{Q^2 T_C}{gA_C^3} = \frac{V_C^2}{gy_C} = 1 \qquad (\because Q = A_C V_C \text{ and } T_C = B)$$

Specific energy at critical depth

$$E_C = y_C + \frac{V_C^2}{2g}$$

From equation (*i*), $$\frac{V_C^2}{gy_C} = 1$$

or $$\frac{V_C^2}{2g} = \frac{y_C}{2}$$

Now, $$E_C = y_C + \frac{y_C}{2} = \frac{3}{2} y_C$$

$$E_C = \frac{3}{2} y_C$$

Also $$\frac{Q^2}{g} = \frac{A^3}{T}$$

or
$$\frac{Q^2}{g} = \frac{y_C^3 B^3}{B} \quad \text{or} \quad \frac{Q^2}{gB^2} = y_C^3$$

Discharge intensity $\left(q = \frac{Q}{B}\right)$ put in equation

$$y_C^3 = \frac{q^2}{g}$$

or
$$y_C = \left(\frac{q^2}{g}\right)^{\frac{1}{3}}$$

We know Froude number is given by

$$F = \frac{V}{\sqrt{g\left(\frac{A}{T}\right)}}.$$

For rectangular channel

$$F = \frac{V}{\sqrt{gy}} \qquad (\because A = By, T = B)$$

2. **Triangular Channel**

From figure having side slope *m* horizontal and 1 vertical. We get

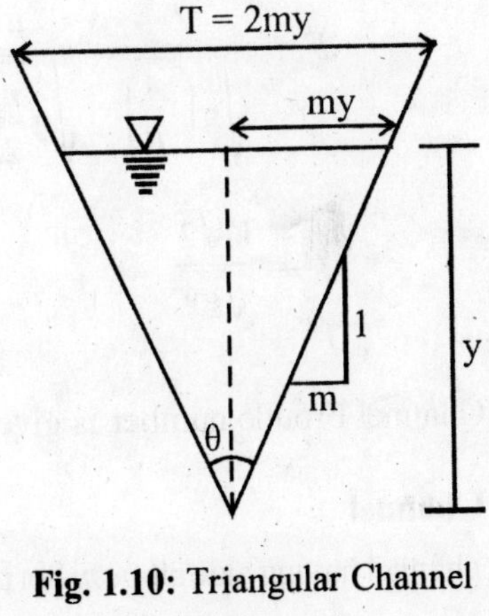

Fig. 1.10: Triangular Channel

$$A = my^2,\ T = 2my$$

Since $$A = \frac{1}{2} \times 2my \times y$$

Applying critical flow condition

$$\frac{Q^2}{g} = \frac{A^3}{T} = \frac{m^3 y_C^6}{2my_C} = \frac{m^2 y_C^5}{2}$$

Hence $$y_C = \left(\frac{2Q^2}{gm^2}\right)^{\frac{1}{5}}$$

The specific energy at critical depth

$$E_C = y_C = \frac{V_C^2}{2gy_C} = y_C + \frac{Q^2}{2gA_C^2}$$

or $$E_C = y_C + \frac{m^2 y_C^5}{4m^2 y_C^4}$$

$$\left(\because \frac{Q^2}{g} = \frac{m^2 y_C^5}{2},\ A_C = my_C^2\right)$$

$$E_C = 1.25\, y_C$$

Now, $$F = \frac{V}{\sqrt{g\left(\frac{A}{T}\right)}} = \frac{V}{\sqrt{g\frac{my^2}{2my}}}$$

$\Rightarrow$ $$F = \frac{V\sqrt{2}}{\sqrt{gy}}$$

For triangular Channel Froude number is given by $F = \frac{V\sqrt{2}}{\sqrt{gy}}$

3. Trapezoidal Channel

For trapezoidal channel having side slope *m* horizontal : 1 vertical

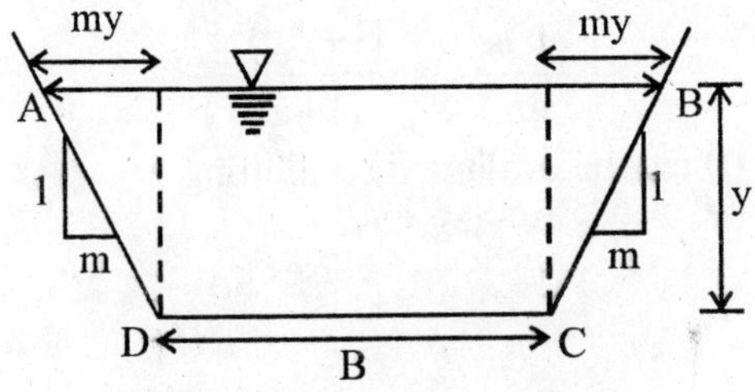

Fig. 1.11: Trapezoidal Channel

$$A = (B + my)y$$

$$\therefore \qquad A = \frac{1}{2}\left(AB + CD\right)y$$

$$T = B + 2my$$

Applying Critical flow condition

$$\frac{Q^2}{g} = \frac{A_C^3}{T_C} = \frac{\left(B + my_C\right)^3 y_C^3}{\left(B + 2my_C\right)}$$

or
$$\frac{\left(B + my_C\right)^3 y_C^3}{\left(B + 2my_C\right)} = \frac{B^3\left(1 + \frac{my_C}{B}\right)^3 y_C^3}{B\left(1 + \frac{2my_C}{B}\right)}$$

$$= \frac{B^6\left(1 + \frac{my_C}{B}\right)^3\left(\frac{my_C}{B}\right)^3}{Bm^3\left(1 + \frac{2my_C}{B}\right)}$$

or
$$= \frac{B^5}{m^3}\frac{\left(1 + \xi_C\right)^3 \xi_C^3}{\left(1 + 2\xi_C\right)}$$

where $\left(\xi_C = \frac{my_C}{B}\right)$

which on equating with $\frac{Q^2}{g}$ gives

$$\frac{Q^2 m^3}{gB^5} = \frac{(1+\xi_C)^3 \xi_C^3}{(1+2\xi_C)} \qquad (i)$$

Equation (*i*) can be evaluated by plotting

$$\frac{Q^2 m^3}{gB^5} \text{ verses } \xi_C$$

The specific energy at critical depth, E_C is a function of $\left(\frac{my_C}{B}\right)$ and is given by

$$\frac{E_C}{y_C} = \frac{(3+5\xi_C)}{2(1+2\xi_C)} \quad \text{where} \left(\xi_C = \frac{my_C}{B}\right)$$

Froude number for Trapezoidal Channel

$$F = \frac{V}{\sqrt{g\,{}^{A}\!/_{T}}} \qquad \left(\because \frac{A}{T} = \frac{(B+my)\,y}{(B+2my)} = \frac{\left(1+\frac{my}{B}\right)y}{\left(1+2\frac{my}{B}\right)}\right)$$

$$F = \frac{{}^{Q}\!/_{A}}{\sqrt{g\,{}^{A}\!/_{T}}} = f\left(\frac{my}{B}\right),$$

for given discharge Q.

4. Circular Channel

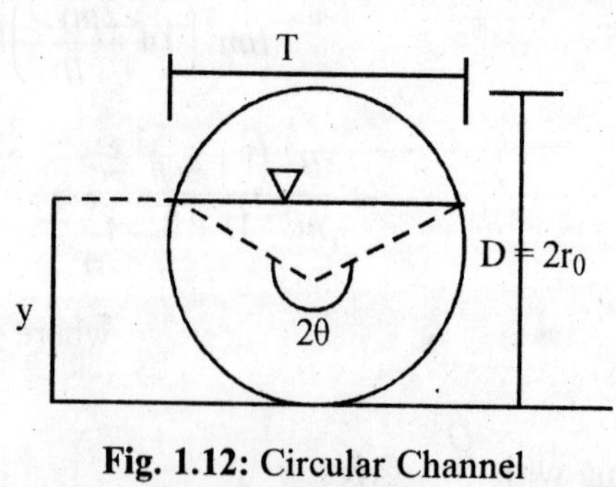

Fig. 1.12: Circular Channel

A = are of sector + area of triangular portion

$$= \left[\frac{1}{2}r_0^2 2\theta + \frac{1}{2}2r_0 \sin(\pi-\theta) r_0 \times \cos(\pi-\theta)\right]$$

$$A = \frac{1}{2}\left[r_0^2 2\theta - r_0^2 \sin 2\theta\right]$$

$$= \frac{D^2}{8}[2\theta - \sin 2\theta] \qquad \left(\because r_0 = \frac{D}{2}\right)$$

Top width = $T = D \sin\theta = 2r_0 \sin\theta$

Applying critical flow condition

$$\frac{Q^2}{g} = \frac{A^3}{T} = \frac{\left[\frac{D^2}{8}(2\theta_C - \sin 2\theta_C)\right]^3}{D \sin\theta_C}$$

1.15 FIRST HYDRAULIC EXPONENT 'M'

In GVF computation it is convenient to express variation of section factor (Z) with y in an exponential form.

It is given by $$Z^2 = C_1 y^M \qquad (i)$$

in equation (*i*) C_1 = coefficient,

M = first hydraulic exponent.

The value of M for a given channel can be determined by plot of Z verses y on log—log scale. If M is constant between two points (Z_1, y_1) and (Z_2, y_2) in the graph, the value of M is calculated as

$$M = 2\frac{\log\left(\frac{Z_2}{Z_1}\right)}{\log\left(\frac{y_2}{y_1}\right)}$$

An estimate of M can also be obtained by relation

$$M = \frac{y}{A}\left[3T - \frac{A}{T}\frac{dT}{dy}\right]$$

1.16 Concept of Specific Force

Consider a channel of large slope as shown in figure and apply the momentum principle to the water contained between section 1 and 2. (The Newton's second law when expressed in terms of momentum, states that change in momentum per unit time in a given direction is equal to resultant of all external forces acting the body in the same direction).

The momentum of flow entering section (1) per second is $\left(\frac{\gamma Q}{g}\right)V_1$, where γ is the specific weight of flowing liquid and momentum per second leaving section 2 is $\left(\frac{\gamma Q}{g}\right)V_2$. The external force acting on the body enclosed between section 1 and 2 are

1. P_1 and P_2 (Hydrostatic force at section 1 and 2
2. w sinθ (the component of weight of water enclosed)
3. F_f, the frictional force.

Applying momentum principle we get

$$\frac{\gamma Q}{g}(V_2 - V_1) = P_1 - P_2 = \text{w}\sin\theta - F_f$$

where V_1 and V_2 are mean velocity

When momentum principle is applied in horizontal channel, the force of friction F_F and weight component w sin θ can be ignored because θ = 0 for horizontal channel.

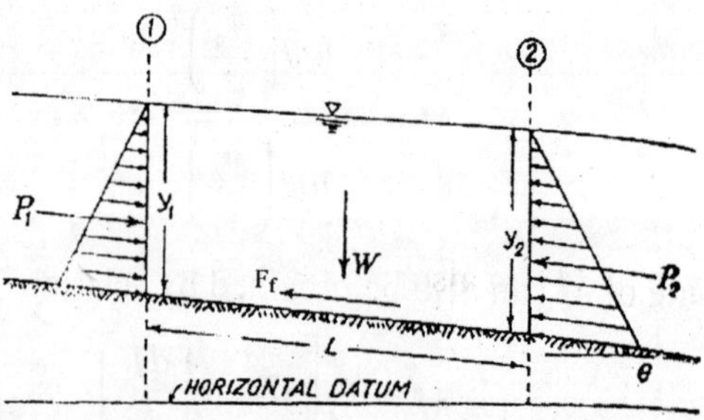

Fig. 1.13: Application of the momentum principle

Now equation become

$$\frac{\gamma Q}{g}(V_2 - V_1) = P_1 - P_2$$

Assuming hydrostatic distribution at section 1 and 2.

$$P_1 = \gamma A_1 \overline{Z_1}, \qquad P_2 = \gamma A_2 \overline{Z_2}$$

where A_1 and A_2 are cross section area and $\overline{Z_1}$ and $\overline{Z_2}$ are vertical depth of centroid of the area below Surface of flow. Also $V_1 = \frac{Q_1}{A_1}$ and $V_2 = \frac{Q_2}{A_2}$ thus equation becomes.

$$\gamma \frac{Q^2}{g}\left(\frac{1}{A_2} - \frac{1}{A_1}\right) = \gamma \overline{Z}_1 A_1 - \gamma \overline{Z}_2 A_2$$

Rearranging the terms we get

$$\frac{Q^2}{gA_1} + A_1 \overline{Z}_1 = \frac{Q^2}{gA_2} + A_2 \overline{Z}_2$$

The two sides equation are Identical and are represented by F. and it may be written as general equation

$$F = \frac{Q^2}{gA} + A\overline{Z}$$

The first term $\frac{Q^2}{gA}$ is the momentum of flow passing the channel per unit time per specific weight of the liquid. The second term $A\overline{Z}$ is the hydrostatic force per specific weight of liquid. Their sum denoted by F is known as specific force. For horizontal Channel specific force at section 1 is equal to 2.

therefore $\qquad F_1 = F_2$

From specific force equation it is evident that for a given channel section and discharge Q the specific force depends only on the depth of flow. Thus by plotting the depth of flow against specific force for a given channel section, a specific force curve is obtained.

It can be seen from the curve that for a given value of specific force there are two possible depths y_1 and y_2. At point C on the curve the two depths become one and corresponding to this depth the specific force is minimum and the depth is known as critical depth. (y_C).

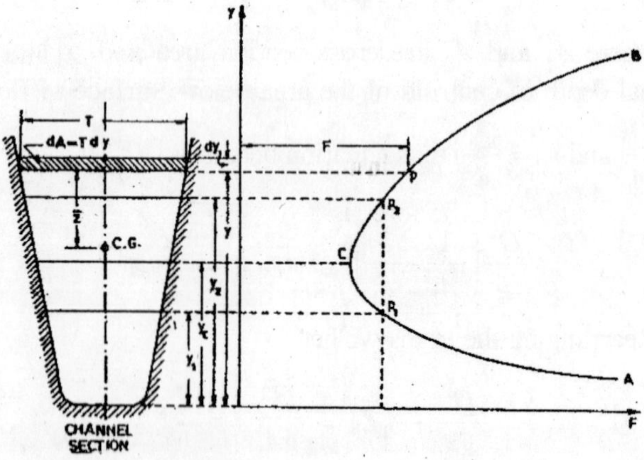

Fig. 1.14: Specific force curve

(*i*) Condition for Minimum Specific Force

For a given discharge the condition for minimum specific force can be obtained by differentiating equation of specific force with respect to y and then considering it as $\frac{dF}{dy}=0$

$$\left(F=\frac{Q^2}{gA}+A\overline{Z}\right),\quad \frac{dF}{dy} = -\frac{Q^2}{gA^2}\frac{dA}{dy}+\frac{d\left(A\overline{Z}\right)}{dy}=0$$

$$d\left(A\overline{Z}\right) = \left(A\left(\overline{Z}+dy\right)+Tdy.\frac{dy}{2}\right)-A\overline{Z}$$

Neglecting smaller term, $\frac{T\left(dy\right)^2}{2}$ we get

$$d\left(A\overline{Z}\right) = Ady$$

Now we can write

$$\frac{dF}{dy} = -\frac{Q^2}{gA^2}\frac{dA}{dy} + \frac{Ady}{dy} = 0$$

or $$\frac{Q^2T}{gA^3} = 1 \qquad \left(\because \frac{dA}{dy} = T\right)$$

The above condition is critical flow condition. Thus it can be said at critical depth, like specific energy, specific force also attain minimum value.

(*ii*) Condition for maximum discharge

We know specific force is given by

$$F = \frac{Q^2}{gA} + A\overline{Z},$$

or $$Q = \sqrt{\left(F - A\overline{Z}\right)gA} \qquad (i)$$

In a given channel section for a given specific force, the condition for maximum discharge can be obtained by putting $\left(\frac{dQ}{dy} = 0\right)$

$$\frac{dQ}{dy} = \frac{g\left[F\left(\frac{dA}{dy}\right) - A\frac{d\left(A\overline{Z}\right)}{dy} - \left(A\overline{Z}\right)\frac{dA}{dy}\right]}{2\sqrt{\left(F - A\overline{Z}\right)gA}} = 0$$

or $$\text{F} = \frac{A^2}{T} + A\overline{Z}$$

Substituting value of F in equation (i), we get,

$$Q = \sqrt{\left(\frac{A^2}{T} + A\overline{Z} - A\overline{Z}\right)gA}$$

Squaring and rearranging we get

$$\frac{Q^2T}{gA^3} = 1$$

Which is again criteria for critical flow condition. Therefore it may be stated that for given specific force the discharge in a given channel is maximum. When flow is in critical state.

1.17 Channel Transitions

The concept of specific energy find its application for solving many open channel problem. The most important among them are analysis of flow through channel transition, flow over raised channel floor-broad crested weir, flow through sluice gate openings, etc.

Various channel transitions may be broadly classified as sudden transitions are those in which the change in cross-section takes place gradually in a relatively long length of channel. The main functions of channel transition are metering of flow, dissipation of energy, reduction or increase of velocity, change in channel section or alignment with minimum energy dissipation, etc.

Some of the devices which are used for measuring discharge through channel are ventriflume, standing wave flume and parshall flume are based on specific energy concept.

We will discuss in detail the two important transactions:

1. Vertical Contraction or Channel with a hump
2. Horizontal Contraction or width constriction

1. Vertical Contraction (Channel with hump): We will discuss here two conditions:

(*a*) Subcritical flow

(*b*) Supercritical flow

(*a*) Subcritical flow: Let us suppose horizontal frictionless rectangular channel of width B and carrying discharge Q at a depth y_1. Suppose the flow to be subcritical. At section (2), a smooth hump of height Δz is built. Since there are no energy loss between section 1 and 2 and also the hump causes the specific energy at 1 and 2 and also the hump causes the specific energy at section to decrease by Δz. Thus specific energy are given by

At sections 1, $$E_1 = y_1 + \frac{V_1^2}{2g}$$

At section 2, $$E_2 = E_1 - \Delta z$$

Also $$E_2 = y_2 + \frac{V_2^2}{2g}$$

From the specific energy curve in Figure 1.16. As the value of Δz increased, the depth y_2 at section 2 will decrease. The minimum depth will reach when point R concides with critical depth at C. At this point C the value of hump height will be maximum (ΔZ_m). Also at $y_2 = y_C$ and $E_2 = E_C$.

$$E_2 = E_2 - \Delta Z_m = E_C = y_C + \frac{Q^2}{2gB^2 y_C^2}$$

It also clear from Figure 1.17 that if $\Delta Z > \Delta Z_m$ then flow is not possible with given specific energy. The upstream depth has to increase to cause in crease in specific energy at section (1). If now depth is given by y_1', then at section 2, the flow will continue at minimum specific energy level E_C.

Now we can conclude that when $0 < \Delta Z > \Delta Z_m$ the upstream water level will be constant at y_1 while y_2 decrease with Δz and goes on decreasing till y_C at ΔZ_m (when ΔZ is maximum). When $\Delta Z > \Delta Z_m$, y_1 will change to y_1' while y_2 will continue to remain y_C.

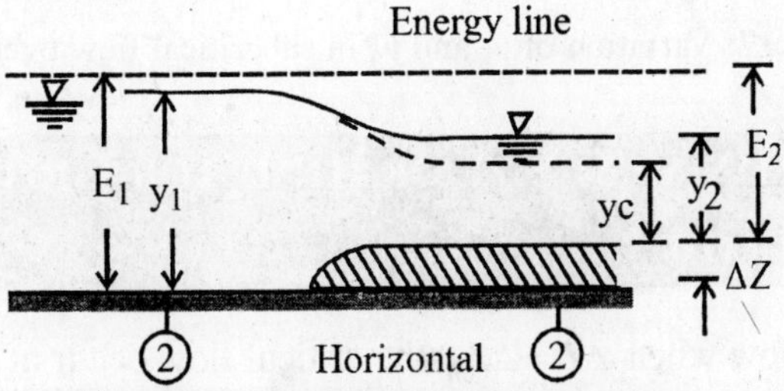

Fig. 1.15: Channel transition with hump

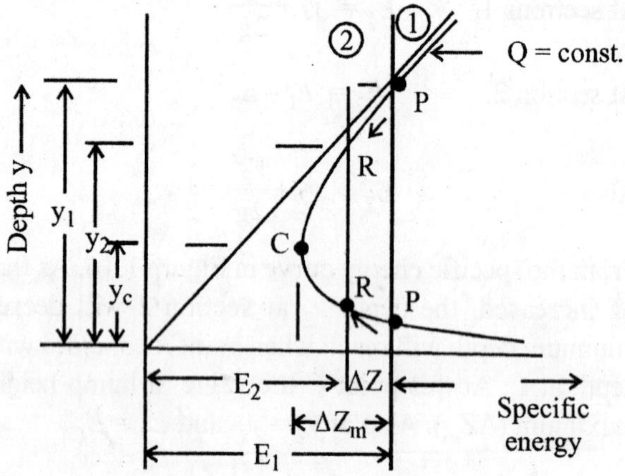

Fig. 1.16: Specific-energy diagram for Fig. 1.6

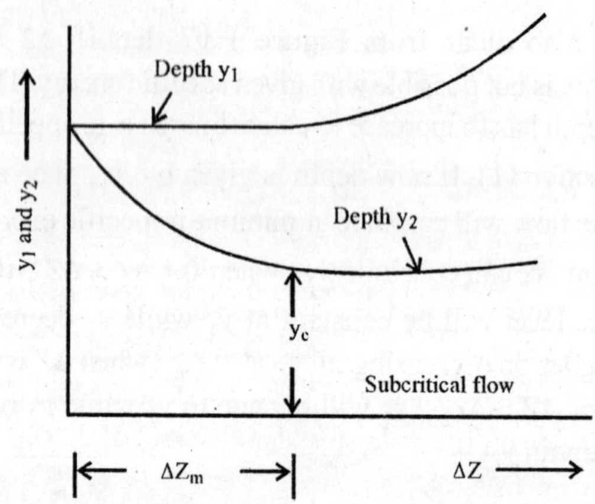

Fig. 1.17: Variation of y_1 and y_2 in subcritical flow over a hump

1.18 Minimum Size of Hump for Critical Flow

As we know when $\Delta Z = \Delta Z_m$ the critical flow occur at section (2), but as $\Delta Z > \Delta Z_m$ the specific energy at section (1) will be changing thus ΔZ_m is called as minimum size of hump for critical flow to occur.

At $\Delta Z = \Delta Z_m$, $E_2 = E_C = \frac{3}{2} y_C$

$$E_1 = E_C + \Delta Z_m$$

$$\Delta Z_m = E_1 - \frac{3}{2} y_C$$

$$\frac{\Delta Z_m}{y_1} = \frac{E_1}{y_1} - \frac{3}{2}\frac{y_C}{y_1}$$

We know $\frac{E_1}{y_1} = \left(1 + \frac{f_1^2}{2}\right)$ and $y_C = \left(\frac{q^2}{g}\right)^{1/3}$ put in equation

$$\frac{\Delta Z_m}{y_1} = \left(1 + \frac{f_1^2}{2} - \frac{3}{2}\left(\frac{q^2}{g}\right)^{1/3} \cdot \frac{1}{y_1}\right)$$

$$\frac{\Delta Z_m}{y_1} = \left(1 + \frac{f_1^2}{2} - \frac{3}{2} f_1^{2/3}\right)$$

(*ii*) When there is energy loss due to hump

Let h_L be energy loss. Then

$$E_1 = E_2 + \Delta Z + H_L$$

then $\frac{\Delta Z_m}{y_1} + \frac{h_L}{y_1} = \left(1 + \frac{f_1}{2} - \frac{3}{2}\frac{f}{2} f_1^{3/2}\right)$

Now it is clear that effect of energy loss due to shape and friction is equivalent to that of hump placed in the downstream section.

(*b*) **Supercrtical flow condition:** If y_1 is in supercriticial flow condition then depth of flow will increase due to reduction in specific energy in figure. P' correspond to y_1 and point T' to depth at section 2 up critical depth y_2 increases to reach y_C at $\Delta Z = \Delta Z_m$ for $\Delta Z > \Delta Z_m$, the depth over hump $y_2 = y_C$ will remain constant and upstream depth y_1 will change. It will decreases to have higher specific energy E_1'. The variation y_1 and y_2 with ΔZ is given by figure 1.18.

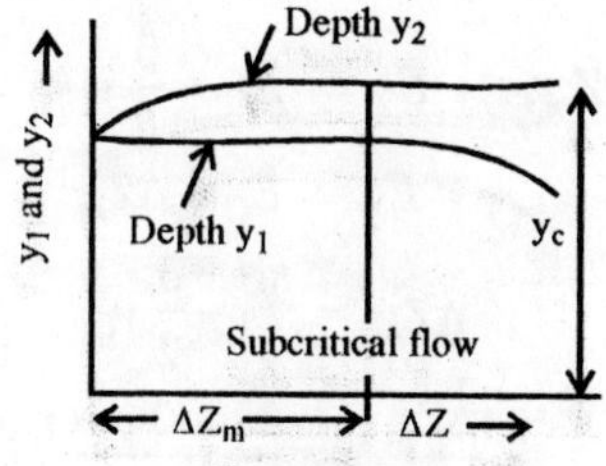

Fig. 1.18

Horizontal contraction or width contraction: We will discuss here two conditions.

(*a*) **Subcritical flow in width constriction:** Let us consider frictionless horizontal channel of width B_1 carrying discharge Q at depth y_1. At section 2 the channel width is constricted to B_2 by smooth transition. Since there is no energy loss the specific energy at section 1 is equal to specific energy at 2.

$$E_1 = y_1 + \frac{V_1^2}{2g} = y_1 + \frac{Q^2}{2gB_1^2 y_1^2}$$

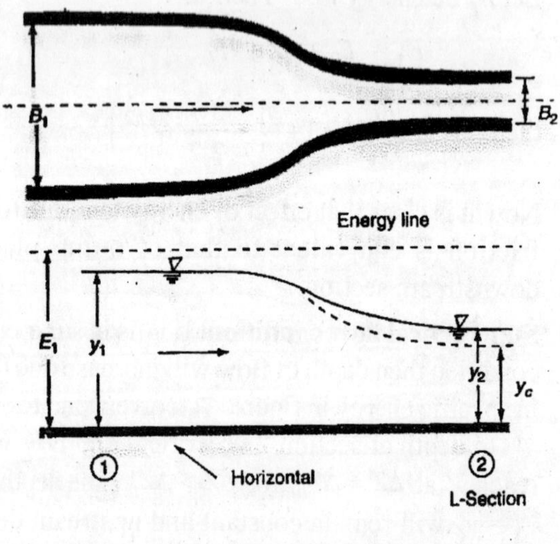

Fig. 1.19: Transition with width constriction

$$E_2 = y_2 + \frac{V_2^2}{2g} = y_2 + \frac{Q^2}{2gB_2^2 y_2^2}$$

At section 1 $$q_1 = \frac{Q}{B_1}$$

At section 2 $$q_2 = \frac{Q}{B_2}$$

Since $B_2 < B_1$, $q_2 > q_1$. In the specific energy diagram the discharge intensity is taken in place of discharge as width changes in the channel point P on the curve q_1 correspond to depth y_1 and specific energy E_1. At section 2, $E_2 = E_1$ and $q = q_2$, the point p will move vertically downward to point R on the curve q_2 to reach the depth y_2.

If B_2 is made smaller, q_2 will increase and y_2 will decrease. When B_2 is contracted till it last limit $B_2 = B_{2min}$. Then discharge intensity is maximum $q_2 = q_m$. Therefore, critical flow condition prevail. At $B_2 = B_{2min}$, y_2 = critical depth at section 2.

$$E_1 = E_{cm} = y_C + \frac{Q^2}{2g(B_{2\,min})^2 \, y_C^2}$$

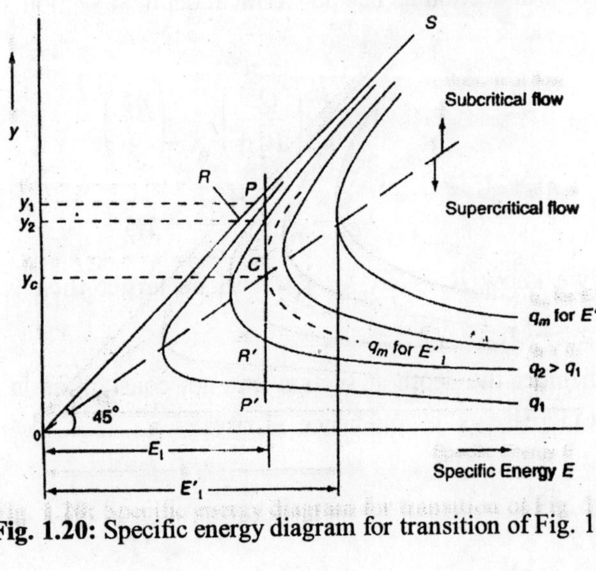

Fig. 1.20: Specific energy diagram for transition of Fig. 1.19

For rectangular channel at critical flow

$$y_C = \frac{2}{3}E_C$$

Since $\qquad E_1 = E_C$

$$y = y_C = \frac{2}{3}E_C = \frac{2}{3}E_1$$

$$y_C = \left(\frac{Q^2}{B_{2\min}\, g}\right)^{1/3} \quad \text{or} \quad B_{2\min} = \sqrt{\frac{Q^2}{g y_C^3}}$$

or $\qquad B_{2\min} = \sqrt{\frac{27Q^2}{8gE_1^3}} \qquad \text{Put } y_c = \frac{2}{3}E_1$

If $B_2 < B_{2\min}$, the discharge intensity will be larger than q_m. The flow will not be possible for given upstream condition. The upstream depth will have to increase to y_1' so that new specific energy

$E_1' = y_1'^2 + \frac{Q^2}{\left(2gB_1^2 y_1'^2\right)}$ is formed which will be sufficient to cause critical flow at section 2. The new critical depth at section 2 given by

$$y_{C_2} = \left(\frac{Q^2}{B_2^2 g}\right)^{1/3} = \left(\frac{q_2^2}{g}\right)^{1/3}$$

$$E_{C_2} = y_{C_2} + \frac{V_{C2}^2}{2g} = 1.5$$

As we know $B_2 < B_{2\min}$, y_{C2} will be larger then y_C also $E_1' = E_{C2} = 1.5 y_{C_2}$.

Remember the depth at section 2 is not constant as in hump case but increase as y_1' and hence E_1' rises.

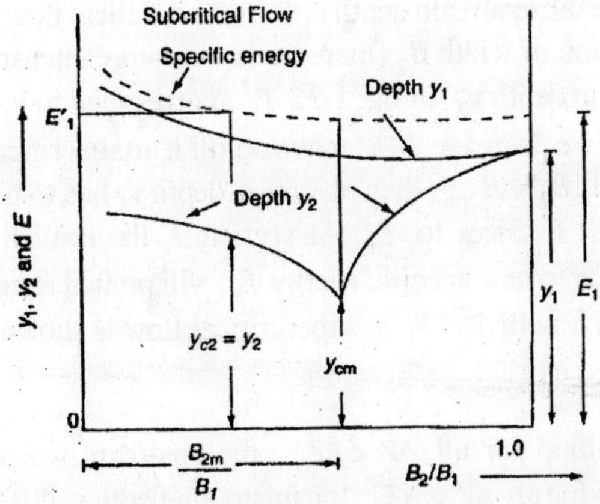

Fig. 1.21: Variation of y_1 and y_2 in subcritical flow in a width constriction

Supercritical Flow in Width Constriction

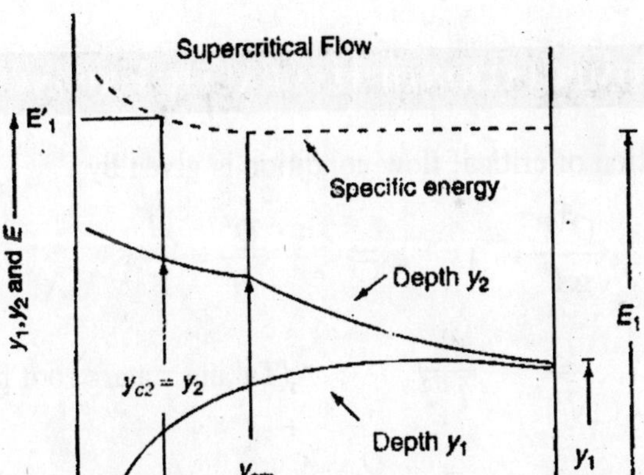

Fig. 1.22: Variation of y_1 and y_2 in subcritical flow in a width constriction

Suppose the upstream depth y_1 is in supercritical flow condition. Then reduction of width B_2 (increase in discharge intensity q_2) will cause a rise in depth y_2. In fig. 1.22 P_1' . correspond to y_1 and point R' to y_2. As we decrease $B_2' R'$ move up till it attain critical depth at $B_2 = B_{2min}$. If $B_2 < B_{2min}$ then upstream depth y_1 has to be decrease to y_1' to that E_1 rises to E_1'. At section 2, the critical depth y_C corresponding to new specific energy E_1' will prencil. The variation of y_1, y_2 and E with B_2 / B_1 in supercritical flow is shown in figure.

Choking

In case of hump for all $\Delta Z < \Delta Z_m$, the upstream water depth is constant and for all $\Delta Z > \Delta Z_m$, the upstream depth is different from y_1 similarly in case of width constriction. If $B_2 > B_{2\,min}$, the upstream depth will change to cause flow.

The onset of critical condition at section 2 is prerequisite to choking. Thus all cases with $\Delta Z > \Delta Z_m$ or $B_2 < B_{min}$ are known as choked condition.

1.19 Computation of Hydraulic Exponent (M)

The condition of critical flow condition is given by

$$\frac{Q^2 T}{g A^3} = 1 \quad \Rightarrow \quad \frac{Q^2}{g} = \frac{A^3}{T}$$

or
$$\frac{Q}{\sqrt{g}} = \frac{A^{3/2}}{T^{1/2}} \qquad \text{(Taking square roof both side)}$$

Also we know that $Z = \dfrac{Q}{\sqrt{g}} = A^{3/2}\, T^{-1/2}$ ...(i)

Z = Section factor and is a function of depth, hence it is written as

$$Z^2 = CyM \qquad \text{...}(ii)$$

where C is consant and M is hydraulic exponent for computation of critical flow. Taking log of above eqn. (*ii*)

$$2 \log_e Z = \log_e C + M \log_e y$$

Differentiating with respect to y

$$\frac{2(d\log_e Z)}{dy} = 0+\frac{M}{y}$$

$$\frac{d(\log_e Z)}{dy} = \frac{M}{2y} \quad \text{...(iii)}$$

Now taking log of eqn. (i)

$$\log_e Z = \frac{3}{2}\log_e A - \frac{1}{2}\log_e T$$

Differentiating with respect to y

$$\frac{d(\log_e Z)}{dy} = \frac{3}{2}\frac{d}{A}\frac{(\log_e A)}{dy} - \frac{1}{2T}\frac{dT}{dy}$$

or

$$\frac{d(\log_e Z)}{dy} = \frac{3T}{2A} - \frac{1}{2T}\frac{dT}{dy} \quad \text{(iv)}$$

$\therefore$

$$\frac{dA}{dy} = T$$

Equating equations (iii) and (iv)

$$\frac{M}{2y} = \frac{3T}{2A} - \frac{1}{2T}\frac{dT}{dy}$$

$$M = \frac{3yT}{A} - \frac{y}{T}\frac{dT}{dy} \quad \text{...(v)}$$

This is general equation of hydraulic exponent 'M'.

(i) When channel is rectangular

$$T = B, A = B_{y'} \frac{dT}{dy} = 0$$

$$M = \frac{3By}{By} - 0$$

$$M = 3$$

(ii) When channel is triangular

$$T = 2\,my, \quad \frac{dT}{dy} = 2m, \quad A = my^2$$

$$M = \frac{3y \times 2my}{my^2} - \frac{y}{2my} \times 2m$$

$$M = 6 - 1$$

$$M = 5$$

(iii) When channel is trapezoidal

$$T = B + 2\,my, \quad \frac{dT}{dy} = 2m\,, A = (B + my)\,y$$

$$M = \frac{3(B + 2my)y}{(B + my)y} - \frac{y}{(B + 2my)} \cdot 2m$$

$$M = \frac{3(B + 2my)y}{(B + my)y} - \frac{2my}{(B + 2my)}$$

Putting $\gamma = \dfrac{y}{B} \quad \therefore \quad y = B\gamma$

$$M = \frac{3(B + 2m\,B\gamma)B\gamma}{(B + m\,B\gamma)\,B\gamma} - \frac{2m\,B\gamma}{B + 2m\,B\gamma}$$

$$= \frac{3B^2(1 + 2m\,\gamma)\gamma}{B^2(1 + m\,\gamma)\,\gamma} - \frac{2m\,B\gamma}{B\,(1 + 2m\,\gamma)}$$

$$\therefore \quad M = \frac{3(1 + 2m\,\gamma)}{(1 + m\,\gamma)} - \frac{2m\,\gamma}{1 + 2m\,\gamma} \qquad (vi)$$

Which shows the value of M varies with y/B varies. Direct solution of critical depth does not exist.

Solved Questions

Problem 1: ***While measuring the discharge in a small stream it was found that the depth of flow increases at the ratio of 0.10 m/h. If the discharge at that section was 20 m³/s and the surface width of the stream was 20 m, estimate the discharge at a section 1 km upstream.***

Solution: This is a case of unsteady flow and the continuity equation will be used

$$T\frac{\partial y}{\partial t} = \frac{20 \times 0.10}{60 \times 60} = 0.000556$$

By continuity equation

$$\frac{Q_2 - Q_1}{\Delta x} = \frac{\partial Q}{\partial x} = -T\frac{\partial y}{\partial t}$$

$$Q_1 = \text{discharge at upstream section}$$

$$= Q_2 + T\frac{\partial y}{\partial t}\Delta x$$

$$= 20 + 1000 \times 0.000556$$

$$= 20.556 \text{ m}^3/\text{s}.$$

Problem 2: ***Figure shows a hydraulic jump in a horizontal apron aided by two-dimensional block on a apron. Obtain an expression for the drag force per unit length of the block.***

Solution: Consider a control volume surrounding the block as shown in Fig. 1.23. A unit width of apron is considered. The drag force on the block would have a reaction force = F_D on a control surface, acting in upstream direction. Assume a frictionless, horizontal channel and hydrostatic pressure distribution at sections 1 and 2.

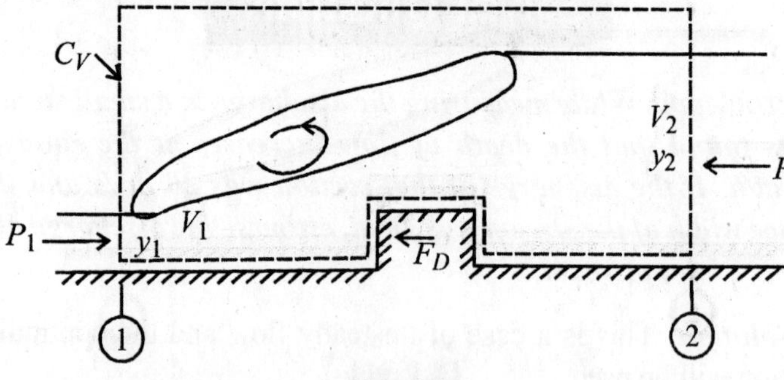

Fig. 1.23

By momentum equation, in direction of the flow

$$P_1 - F_D - P_2 = M_2 - M_1$$

$$\frac{1}{2}\gamma y_1^2 - F_D - \frac{1}{2}\gamma y_2^2 = \rho q(\beta_2 V_2 - \beta_1 V_1)$$

Where $q =$ discharge per unit width of apron

$$= y_1 V_1 = y_2 V_2$$

Assuming $\beta_1 = \beta_2 = 1.0$

$$F_D = \frac{1}{2}\gamma y_1^2 - \frac{1}{2}\gamma y_2^2 = \rho q^2\left(\frac{1}{y_2} - \frac{1}{y_1}\right)$$

$$= \frac{\gamma}{2}\left[y_1^2 - y_2^2 + \frac{q^2}{g}\left(\frac{y_2 - y_1}{y_1\, y_2}\right)\right]$$

Problem 3: ***The velocity distribution in a channel section may be approximated by the equation*** $V = V_o\left(\dfrac{y}{y_o}\right)^n$***, in which V is the flow velocity at depth y,*** V_0 ***is the flow velocity at depth*** y_0 ***and n is constant.Derive expression for the energy and momentum coefficient?***

Solution: Consider unit width of channel, then Area A, can be replace by y in momentum and kinetic energy coefficient equation:

$$V_{\text{mean}} = \frac{\int V dA}{\int dA}$$

As width is unit, so

$$V_{\text{mean}} = \frac{\int V dy}{\int dy}$$

Now substituting the value of V in this equation

$$V_{\text{mean}} = \int_0^{y_0} \frac{V_0 \left(\dfrac{y}{y_0}\right)^n dy}{\int_0^{y_0} dy}$$

$$V_{\text{mean}} = \frac{\left[\dfrac{V_o}{y_0^n} \cdot \dfrac{y^{n+1}}{n+1}\right]_0^{y_0}}{y_0}$$

$$V_{\text{mean}} = \frac{V_o \cdot y_0^{n+1}}{y_0^{n+1} \cdot n+1}$$

$$V_{\text{mean}} = \frac{V_o}{n+1}$$

Now kinetic energy correction factor α is equal

$$\alpha = \frac{\int V^3 dA}{V_m^3 \int dA}$$

or

$$\alpha = \frac{\int_0^{y_0} V_0^3 \left(\dfrac{y}{y_0}\right)^{3n} dy}{\left(\dfrac{V_0}{n+1}\right)^3 \int_0^{y_0} dy}$$

$$\alpha = \frac{\left(\frac{V_0^3}{y_0^{3n}}\right)\left(\frac{y_0^{3n+1}}{3n+1}\right)}{\left(\frac{V_0}{n+1}\right)^3 \cdot y_0}$$

$$\alpha = \frac{(n+1)^3}{3n+1}$$

Now momentum factor β is given by

$$\beta = \frac{\int V^2 dA}{V_m^2 \int dA}$$

$$\beta = \frac{\int_0^{y_0} V_0^2 \left(\frac{y}{y_0}\right)^{2n} dy}{\left(\frac{V_0}{n+1}\right)^2 \int_0^{y_0} dy}$$

$$\beta = \frac{\left(\frac{V_0^2}{y_0^{2n}}\right)\left(\frac{y_0^{2n+1}}{2n+1}\right)}{\left(\frac{V_0}{n+1}\right)^2 \cdot y_0}$$

$$\beta = \frac{(n+1)^2}{(2n+1)}$$

Problem 4: ***Velocity distribution in a rectangular channel of width B and depth y_0 was approximated as $V = \sqrt{y}$ where k is*** constant. Compute average or mean velocity V_m, ***α and β of the channel?***

Solution: We know that

$$V_{\text{mean}} = \frac{\int V dA}{\int dA}$$

As the channel is rectangular, so Area $A = By_0$. Therefore V_{mean} can be written as

$$V_{mean} = \frac{\int_0^{y_0} V dy}{\int_0^{y_0} dy}$$

$$V_{mean} = \frac{\int_0^{y_0} K\sqrt{y} dy}{\int_0^{y_0} dy}$$

$$V_{mean} = \frac{k}{y_0} \cdot \frac{2}{3} \left[y^{3/2} \right]_0^{y_0}$$

$$V_{mean} = \frac{2}{3} k y_0^{1/2}$$

Now $$\alpha = \int \frac{V^3 dA}{V_{mean}^3 dA}$$

or $$\alpha = \frac{\int_0^{y_0} V^3 dy}{V_{mean}^3 \int_0^{y_0} dy}$$

$$\alpha = \frac{\int_0^{y_0} \left(k y^{1/2} \right)^3 dy}{y_0 \left(\frac{2}{3} k y_0^{1/2} \right)^3}$$

$$\alpha = \frac{\frac{2}{5}k^3 \cdot y_0^{5/2}}{y_0\left(\frac{2}{3}ky_0^{1/2}\right)^3}$$

$$\alpha = \frac{2}{5} \times \frac{27}{8} = \mathbf{1.35\ Ans.}$$

Now $$\beta = \frac{\int V^2 dA}{V_m^2 \cdot \int dA}$$

As channel is rectangular $A = By_0$, so

$$\beta = \frac{\int_0^{y_0} V^2 dA}{V_m^2 \cdot \int_0^{y_0} dy}$$

$$\beta = \frac{\int_0^{y_0} \left(ky^{1/2}\right)^2 dy}{\left(2/3\, ky_0^{1/2}\right)^2 \cdot y_0}$$

$$\beta = \frac{k^2 \cdot y_0^2/2}{4/9\, k^2 y_0^2}$$

$$\beta = \frac{9}{4 \times 2} = \mathbf{1.125\ Ans.}$$

Problem 5: ***The velocity distribution in a rectangular channel of 1.5 m depth is given by*** $v = \left(\frac{y}{y_0}\right)^{1/2}$. ***Find*** V_m, α ***and*** β***?***

Solution: $V_{mean} = \dfrac{\int v\,dA}{\int dA}$

As the channel is rectangular $A = By_0$, where y_0 is equal to 6.5 m.

So V_{mean} can be written as

$$V_{mean} = \frac{\int_0^{y_0} v\,dy}{\int_0^{y_0} dy}$$

Now $$V_{mean} = \frac{\int_0^{y_0} \left(\frac{y}{y_0}\right)^{1/2} dy}{y_0}$$

$$V_{mean} = \frac{\frac{1}{y_0^{1/2}} \int_0^{y_0} (y)^{1/2}\, dy}{y_0}$$

$$V_{mean} = \frac{\frac{1}{y_0^{1/2}} \cdot \left(y_0^{3/2}\right) \cdot \frac{2}{3}}{y_0}$$

$$V_{mean} = \frac{y_0^{3/2}}{y_0^{3/2}} \cdot \frac{2}{3}$$

$$V_{mean} = \frac{2}{3} \text{ m/sec}$$

Now $$\alpha = \frac{\int v^3 dA}{V_m^3 \int dA}$$

As channel is rectangular

$$\alpha = \frac{\int_0^{y_0} v^3 dy}{V_m^3 \int_0^{y_0} dy}$$

$$\alpha = \frac{\int_0^{1.5} \left[\left(\frac{y}{1.5}\right)^{1/2}\right]^3 dy}{\left(\frac{2}{3}\right)^3 \cdot 1.5}$$

$$\alpha = \frac{\frac{1}{(1.5)^{3/2}} \left[y_0^{5/2}\right]_0^{1.5} \cdot \frac{2}{5}}{\left(\frac{2}{3}\right)^3 \cdot 1.5}$$

$$\alpha = \frac{\frac{1}{(1.5)^{3/2}} \cdot (1.5)^{5/2} \cdot \frac{2}{5}}{\left(\frac{2}{3}\right)^3 \cdot 1.5}$$

$$\alpha = \frac{2}{5} \times \frac{27}{8} = \textbf{1.35 Ans.}$$

Now $$\beta = \frac{\int v^2 dA}{V_m^2 \int dA}$$

As channel is rectangular, therefore

$$\beta = \frac{\int_0^{y_0} v^2 dy}{V_m^2 \int_0^{y_0} dy}$$

$$\beta = \frac{\int_0^{1.5} \left(\frac{y}{1.5}\right)^2 dy}{\left(\frac{2}{3}\right)^2 \cdot 1.5}$$

$$\beta = \mathbf{1.125}$$ **Ans.**

Problem 6: ***The velocity distribution in a rectangular channel may be approximated as $v = 0.4 + 0.6\, y/y_0$. Find V_{mean}, α and β take $y_0 = 1$ m.***

Solution: As we know

$$V_{\text{mean}} = \frac{\int v dA}{\int dA}$$

As the channel is rectangular $A = By_0$, therefore

$$V_{\text{mean}} = \frac{\int_0^{y_0} v dy}{\int_0^{y_0} dy}$$

Now $$V_{\text{mean}} = \frac{\int_0^{1} \left(0.4 + 0.6\frac{y}{1}\right) dy}{\int_0^{1} dy}$$

$$V_{\text{mean}} = \frac{\left[0.4y + 0.6\frac{y^2}{2}\right]_0^1}{1}$$

$$V_{\text{mean}} = \mathbf{0.7\ m/s\ Ans.}$$

Now $$\alpha = \frac{\int_0^{y_0} v^3 dy}{v_m^3 \int_0^{y_0} dy}$$

$$\alpha = \frac{1}{0.343}\int_0^1 (0.4 + 0.6y)^3\, dy$$

$$\alpha = \frac{1}{0.343}\int_0^1 (0.4^3 + 0.6^2 y^3 +$$

$$3 \times 0.4 \times 0.6^2 y^2 + 3 \times 0.4^2 \times 0.6y)dy$$

$$\alpha = 1.18$$

Now $$\beta = \frac{\int_0^{y_0} v^2 dy}{V_{\text{mean}}^2 \int_0^{y_0} dy}$$

$$\beta = \frac{1}{0.7^2}\int_0^1 (0.4 + 0.6y)^2\, dy$$

$$\beta = 1.06\ \textbf{Ans.}$$

Problem 7: ***The velocity distribution in an open channel could be approximated as shown in Fig. Determine the kinetic energy correction factor α and momentum correction factor β for the velocity profile.***

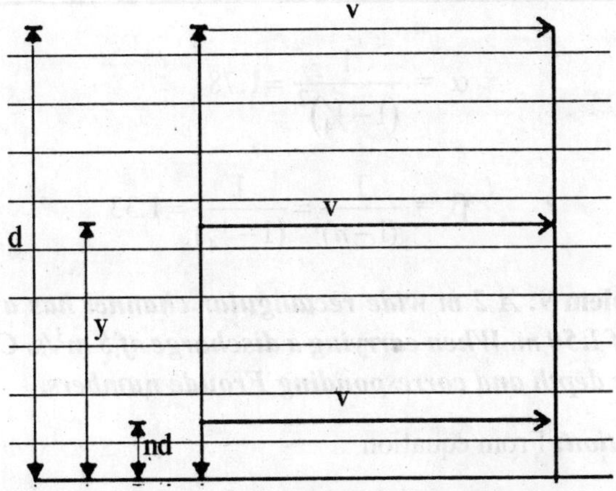

Fig.

Solution: From figure we have, $V \cdot d = v \cdot (d - nd)$

$$V = (1 - n)v$$

$$\alpha = \frac{\int_{nd}^{d} v^3 dy}{V^3 d} = \frac{v^3(d - nd)}{v^3(1 - n^3)d} = \frac{1}{(1-n)^2}$$

$$\beta = \frac{\int_{nd}^{d} v^2 dy}{V^2 d} = \frac{u^2(d - nd)}{v^2(1-n)^2 d} = \frac{1}{(1-n)}$$

Problem 8: ***At a section in a channel expansion, the velociity over a quarter of cross-section is zero and is uniform our remaining three-fourth of the area find α and β.***

Solution: This problem is similar to previous one so for α, we have $\alpha = \dfrac{1}{(1-n)^2}$

here $n = \dfrac{1}{4}$ as in question.

$$\alpha = \frac{1}{\left(1-\frac{1}{4}\right)^2} = 1.78$$

$$\beta = \frac{1}{(1-n)} = \frac{1}{\left(1-\frac{1}{4}\right)} = 1.33$$

Problem 9: ***A 2 m wide rectangular channel has a specific energy of 1.50 m. When carrying a discharge of 5 m³/s. Calculate alternate depth and corresponding Froude numbers.***

***Solution*:** From equation

$$E = y + \frac{V^2}{2g} = y + \frac{Q^2}{2gB^2y^2}$$

$$1.5 = y + \frac{(5)^2}{2\times 9.81\times (2)^2 y^2}$$

$$1.5 = y + \frac{0.31855}{y^2}$$

Solving this equation by trial and error, the alternate depth y_1 and y_2 are obtained as $y_1 = 1.3161$ m and $y_2 = 0.592$ m

Froude number $$F = \frac{V}{\sqrt{gy}} = \frac{5}{2y\sqrt{9.81\times y}}$$

$$= \frac{0.798188571}{y^{3/2}}$$

At $y_1 = 1.3161$ m, $F_1 = 0.529$; and
at $y_2 = 0.592$ m, $F_2 = 1.752$

The depth $y_1 = 1.3161$ m is in subcritical flow region and the depth $y_2 = 0.592$ is in supercritical flow region.

Problem 10: ***Show that at the critical rate of flow, the specific energy in a rectangular channel is equal to 1.5 times the depth of***

flow. Will the depth of flow be greater or less that $\frac{2}{3}\times$ specific energy for the trapezoidal channel and why?

Solution: For Rectangular channel

At critical flow,
$$\frac{Q^2}{g} = \frac{A^3}{T}$$

$$\frac{V^2}{g} = \frac{A}{T} = \frac{by}{b} = y$$

The specific energy

$$E = y + \frac{V^2}{2g} = y + \frac{y}{2} = \frac{3}{2}y$$

Hence, specific energy = 1.5 × Depth of flow

Trapezoidal channel

$$A = y\,(b + my)$$
$$T = b + 2my$$

At critical flow,

$$\frac{V^2}{g} = \frac{A}{T} = \frac{y(b + ym)}{(b + 2my)} < y$$

$$\frac{V^2}{2g} < \frac{y}{2} \text{ and } E = y + \frac{V^2}{2g} < \frac{3}{2}y$$

Therefore, depth of flow

$$y < \frac{2}{3} \times \text{ specific energy}$$

Problem 11: ***Using the concept of the specific energy; obtain for flow at a constant specific energy in a rectangular channel.***

$$\frac{q}{q_c} = \sqrt{3\left(\frac{y}{y_c}\right)^2 - 2\left(\frac{y}{y_c}\right)^3}$$

Where q_c is the critical discharge.

Solution:

$$E = y + \frac{V^2}{2g} = y + \frac{Q^2}{2gA^2} = y + \frac{q^2}{2gy^2}$$

At constant E, for maximum q, $\frac{dq}{dy} = 0$.

$$q^2 = 2gy^2(E - y)$$

differentiating $\quad 2q\frac{dq}{dy} = 2g\left(2Ey - 3y^2\right)$

setting, $\quad \frac{dq}{dy} = 0, \; E = \frac{3}{2}y$

(which is the condition for critical flow)

But we know that when $y = y_c$, $q = q_c$

$$E = \frac{3}{2}y_c = y + \frac{q^2}{2gy^2}$$

$$V_C = \sqrt{gy_c}$$

$$q_c = V_c y_c = \sqrt{gy_c} . y_c = y_c^{3/2}\sqrt{g}$$

or $q_c^2 = gy_c^3$

$$\frac{3}{2}y_c = y + \frac{q^2 \times gy_c^3}{q_c^2 \times 2gy^2} = y + \frac{1}{2}\frac{y_c^3}{y^2}.\frac{q^2}{q_c^2}$$

$$\frac{3}{2} = \frac{y}{y_c} + \frac{1}{2}\left(\frac{y_c}{y}\right)^2\frac{q^2}{q_c^2}$$

Multiplying throughout by $2\left(\frac{y}{y_c}\right)^2$

$$3\left(\frac{y}{y_c}\right)^2 = 2\left(\frac{y}{y_c}\right)^3 + \left(\frac{q}{q_c}\right)^2$$

from which

$$\frac{q}{q_c} = \sqrt{3\left(\frac{y}{y_c}\right)^2 - 2\left(\frac{y}{y_c}\right)^3}$$

Problem 12: ***Find the specific energy of flowing water through a rectangular channel of width 5 m when the discharge 10 m³/s and depth of water 3 m.***

Solution: Given

Width of rectangular channel,

$$B = 5 \text{ m}$$

Discharge $\quad Q = 10 \text{ m}^3/\text{s}$

Depth of water $\quad y = 3 \text{ m}$

Specific energy is given by equation

$$E = y + \frac{V^2}{2g},$$

where $\quad V = \dfrac{Q}{Area} = \dfrac{10}{5 \times 3} = \dfrac{2}{3}$

$$E = 3 + \frac{\left(\frac{2}{3}\right)^2}{2 \times 9.81} = 3 + 0.0226$$

$$= 3.0226 \text{ m}$$

Problem 13: ***The discharge of water through a rectangular channel of width 8 m, is 10 m³/s when depth of flow of water is 1.2 m. Calculate.***

(i) critical depth
(ii) critical velocity
(iii) value of minimum specific energy

Solution: Given:

Discharge, $\quad Q = 10 \text{ m}^3/\text{s}$

Width, $B = 8$ m

Depth, $y = 1.2$ m

∴ Discharge per unit width

$$= \frac{Q}{B} = \frac{10}{8} = 1.25\,\text{m}^2/\text{s}$$

Velocity of flow

$$= \frac{Q}{Area} = \frac{10}{8 \times 1.2}$$

$$= 1.041667 \text{ m/s}$$

(*i*) Critical depth (y_c) is given by equation

$$y_c = \left(\frac{q^2}{g}\right)^{1/3} = \left(\frac{(1.25)^2}{9.81}\right)^{1/3}$$

$$\doteq 0.542 \text{ m}$$

(*ii*) Critical velocity (V_c) is given by equation

$$V_c = \sqrt{g \times y_c} = \sqrt{9.81 \times 0.542}$$

$$= 2.306 \text{ m/s}$$

(*iii*) Minimum specific energy $(E_{\min})$ given by equation

$$E_{\min} = \frac{3y_c}{2} = \frac{3 \times 0.542}{2} = 0.813\,\text{m}$$

Problem 14: ***The specific energy for a 3 m wide channel is to be 2 kg-m/kg. What would be the maximum possible discharge?***

Solution**:** Given

Width of channel, $B = 3$ m

Specific energy, $E = 2$ m

For the given value of specific energy, the discharge will be maximum, when depth of flow is critical. Hence from equation for maximum discharge

$$y_c = y = \frac{2}{3}E = \frac{2}{3} \times 2 = 1.3333\,\text{m}$$

Critical velocity (V_c) is given by

$$V_c = \sqrt{g \times y_c} = \sqrt{9.81 \times 1.3333}$$

$$= 3.6166 \text{ m/s}$$

$$Q_{max} = \text{Area} \times \text{velocity}$$

$$= (B \times y_c) \times V_c$$

$$= (3 \times 1.3333) \times 3.6166$$

$$= 14.466 \text{ m}^3/\text{s}.$$

Problem 15: ***A flow of 5 m³/s is passing at a depth of 1 m through a rectangular channel of 2.5 m width. The kinetic energy correction factor α is found to be 1.20. What is the specific energy of the flow? What is the value of the depth alternate to the existing depth if α = 1.0 is assumed for the alternate flow?***

Solution:

$$V_1 = \frac{Q}{A_1} = \frac{5}{2.5 \times 1} = 2 \text{ m/s}$$

$$\alpha_1 \frac{V_1^2}{2g} = 1.20 \times \frac{(2)^2}{2 \times 9.81} = 0.2446 \text{ m}$$

Specific energy

$$E_1 = y_1 + \alpha_1 \frac{V_1^2}{2g} = 1 + 0.24465$$

$$= 1.24465 \text{ m}$$

For the alternate depth y_2,

$$y_2 + \frac{(5.0)^2}{2 \times 9.81 \times (2.5 y_2)^2} = 1.24465 \qquad (\because \alpha_2 = 1.0)$$

$$y_2 + \frac{0.2039}{y_2^2} = 1.24465$$

By trial and error,

$$y_2 = 0.54 \text{ m}$$

Problem 16: ***Obtain the value of the first hydraulic exponent M for (a) rectangular channel and (b) an exponential channel where the area A is given as $A = k_1 y^a$.***

Solution:

(a) for a rectangular channel $A = By$ and $T = B$

By equation,
$$z^2 = \frac{A^3}{T} = B^2 y^3 = C_1 y^M$$

By equating the exponent on both sides, $M = 3.0$

(b)
$$A = K_1 y^a$$

$$T = \frac{dA}{dy} = K_1 a y^{(a-1)}$$

By equation

$$M = \frac{y}{A}\left(3T - \frac{A}{T}\frac{dT}{dy}\right)$$

$$= \frac{y}{K_1 y^a}\left[3K_1 a y^{(a-1)} - \frac{K_1 y^a}{K_1 a y^{(a-1)}}\left\{K_1 a(a-1)y^{a-2}\right\}\right]$$

Equating the coefficient of $y = 3a - a + 1 = 2a + 1$

Problem 17: ***Calculate the critical depth and the corresponding specific energy for a discharge of 10 m³/s in the following channel.***

(a) *Rectangular channel, B = 5.0 m*

(b) *Triangular channel, m = 0.5*

(c) *Circular channel, D = 2.0 m*

Solution: (a) Rectangular channel

$$q = {}^{Q}\!/_{B} = {}^{10}\!/_{5} = 2 \text{ m}^2\text{/s}$$

$$y_c = \left({}^{q^2}\!/_{g}\right)^{1/3} = \left[\frac{(2)^2}{9.81}\right]^{1/3}$$

$$= 0.7415 \text{ m}$$

Since for a rectangular channel

$E_c = 1.5\, y_c,\ E_c = 1.1123$ m

(b) Triangular Channel

From equation

$$y_c = \left(\frac{2Q^2}{gm^2}\right)^{1/5} = \left[\frac{2\times(10)^2}{9.81\times(0.5)^2}\right]^{1/5}$$

$$= 2.4115 \text{ m}$$

Since for a triangular channel

$$E_c = 1.25\, y_c,\ E_c = 3.01435 \text{ m}$$

(*c*) By Swamme's equation

$$F_D = \frac{10.0}{(2.0)^2\sqrt{9.81\times 2.0}} = 0.5644$$

$$\frac{y_c}{D} = \left[0.77\, F_D^{-6} + 1.0\right]^{-0.085}$$

$$\frac{y_c}{2.0} = \left[0.77(.5644)^{-6} + 1.0\right]^{-0.085}$$

$$\frac{y_c}{2.0} = 0.7611$$

$$y_c = 1.522 \text{ m}$$

Problem 18: ***Find the critical depth for a specific energy heat 2 m in the following channels.***

(a) Rectangular channel, B = 2 m

(b) Triangular channel, m = 1.5

(c) Trapezoidal channel, B = 2.0 m and m = 1.0

Solution: (*a*) Rectangular channel

$$E_c = \frac{3}{2}y_c = 2$$

$$y_c = \frac{2\times 2}{3} = 1.333 \text{ m}$$

(*b*) Triangular channel

By equation $E_c = 1.25\ y_c = 2$

$$y_c = \frac{2}{1.25} = 1.6 \text{ m}$$

(*c*) Trapezoidal channel

$$E_c = y_c + \frac{V_c^2}{2g} = y_c + \frac{Q^2}{2gAc^2}$$

$$2 = y_c + \frac{A_c}{2T_c} \qquad \left(\because \frac{Q^2}{g} = \frac{A_c^3}{T_c}\right)$$

$$2 = y_c + \frac{(2.0 + y_c)\,y_c}{2\,(2.0 + 2y_c)}$$

Solving by trial-and-error, $y_c = 1.481$ m

Problem 19: ***Calculate the bottom width of a trapezoidal*** channel section required to carry a discharge of 15.0 m^3/s ***as a critical flow at a depth of 1 m, if side slope 1.5 horizontal: 1 vertical.***

***Solution*:** The solution in this case is by trial-and-error

$$A_c = (B + 1.5 \times 1) \times 1 = (B + 1.5)$$

$$T_c = (B + 2 \times 1.5 \times 1) = (B + 3)$$

$$\frac{Q^2}{g} = \frac{A_c^3}{T_c}$$

$$\frac{(B+1.5)^3 \times (1)^3}{(B+3)} = \frac{(15)^2}{9.81}$$

By trial-and-error

$$B = 3.91225 \text{ m}$$

Problem 20: ***Water is flowing a critical depth in a Δ shaped channel, with side slope of 0.5 H: 1 V (fig. 1.24). If the critical***

depth is 1.5 m, estimate the discharge in the channel and the specific energy at the critical depth section.

Solution: (*i*) Here $m = -0.5$

$$T_c = 3.0 - (2 \times 0.5 \times 1.5) = 1.5 \text{ m}$$

$$A_c = \frac{(3.0+1.5)}{2} \times 1.50 = 3.375 \text{ m}^2$$

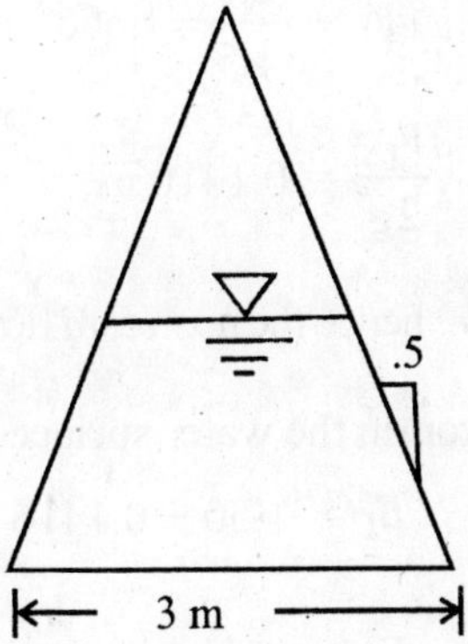

Fig. 1.24

$$\frac{Q^2}{g} = \frac{A_c^3}{T_c} = \frac{(3.375)^3}{1.50} = 25.629$$

Discharge, $$Q = 15.856 \text{ m}^3/\text{s}$$

(*ii*) $$V_c = \frac{Q}{A_c} = \frac{15.856}{3.375} = 4.6981 \text{ m/s}$$

$$\frac{V_c^2}{2g} = \frac{(4.6981)^2}{2 \times 9.81} = 1.125 \text{ m}$$

$$E_c = y_c + \frac{V_c^2}{2g}$$

$$1.5 + 1.125 = 2.625 \text{ m}$$

Problem 21: ***A rectangular channel has a width of 2.0 m and carries a discharge of 5 m³/s with a depth of 1.50 m at a certain section a small, smooth hump with a flat top and of height***

0.10 m is proposed to be built. Calculate the likely change in the water surface. Neglect the energy loss.

Solution: Let the suffixes 1 and 2 rebel to the upstream and downstream sections respectively as in Fig. 1.23 .

$$q = \frac{5}{2} = 2.5 \text{ m}^3/\text{s/m}$$

$$V_1 = \frac{2.5}{1.5} = 1.6667 \text{ m/s},$$

$$\frac{V_1^2}{2g} = 0.1416 \text{ m}$$

$F_1 = \dfrac{V_1}{\sqrt{gy_1}} = 0.4345$, hence the upstream flow is subcritical and the hump will cause a drop in the water surface elevation.

$$E_1 = 1.50 + 0.1416 = 1.6416 \text{ m}$$

At section 2,

$$E_2 = E_1 - \Delta z = 1.6416 - 0.10$$
$$= 1.5416 \text{ m}$$

$$y_c = \left(\frac{(2.5)^2}{9.81}\right)^{1/3} = 0.861 \text{ m}$$

$$E_c = 1.5\, y_c = 1.291 \text{ m}$$

The minimum specific energy at section 2, E_{c_2} is less than E_2, the available specific energy at that section. Hence $y_2 > y_c$ and the upstream depth y_1 will remain unchanged. The depth y_2 is calculated by solving the specific energy relation.

$$y_2 + \frac{V_2^2}{2g} = E_2$$

i.e., $$y_2 + \frac{(2.5)^2}{2 \times 9.81 \times y_2^2} = 1.5416$$

Solving by trial-and-error, $y_2 = 1.37$ m.

Problem 22: *(a) In Problem 15, if the weight of a hump is 0.5 m, estimate the water surface elevation on the hump and at a section upstream of the hump.*

(b) Estimate the minimum size of the hump to cause critical flow over the hump.

Solution: (*a*) From example

$$F_1 = 0.4345, \; E_1 = 1.6416 \text{ m}$$

and $$y_c = y_{c_2} = 0.861 \text{ m}$$

Available specific energy at the section 2

$$= E_2 = E_1 - \Delta z$$

$$E_2 = 1.6416 - 0.5 = 1.1416 \text{ m}$$

$$E_2' = 1.5\, y_{c_2} = 1.291 \text{ m}$$

The minimum specific energy at section 2 is greater than E_2, the available specific energy at that section. Hence, the depth at section 2 will be at the critical depth. Thus $E_2 = E_{c_2} = 1.291$ m.

The upstream depth y_1 will increase to a depth y_1'. Such that the new specific energy at upstream section 1 is

$$E_1' = E_{c_2} + \Delta z$$

$$= y_1' + \frac{V_1'^2}{2g} = E_{c_2} + \Delta z$$

$$y_1' + \frac{q^2}{2gy_1'^2} = 1.291 + 0.500 = 1.791$$

$$y_1' + \frac{(2.5)^2}{2 \times 9.81 \times y_1'^2} = 1.791$$

Solving by trial-and-error and $y_1' > y_2$, $y_1' = 1.678$ m

(*b*) $$\Delta Z_m = E_1 - E_c$$

$$= 1.6416 - 1.291$$

$$= 0.3506 \text{ m}$$

Problem 23: ***A 2.5 m wide rectangular channel carries 6.0 m³/s of flow at a depth of 0.60 m. Calculate the minimum height of a streamlined, flat-topped hump required to be placed at a section to cause critical flow over the hump. The energy loss over the hump can be taken as 0.05% of the upstream velocity head.***

***Solution*:** Discharge intensity

$$q = 6/2.5 = 2.40 \text{ m}^2/\text{s}$$

$$V_1 = 2.40/0.6 = 4 \text{ m/s}$$

$$\frac{V_1^2}{2g} = 0.8155 \text{ m}$$

Energy loss, $$h_L = .05 \times \frac{V_1^2}{2g} = 0.0408 \text{ m}$$

$$\Delta Z_m = E_1 - E_C - h_L$$

$$E_C = 1.5\, y_C = 1.5 \times \left(\frac{q^2}{g}\right)^{1/3}$$

$$= 1.5 \times \left[\frac{(2.4)^2}{9.81}\right]^{1/3} = 1.256 \text{ m}$$

$$E_1 = y_1 + \frac{V_1^2}{2g} = 0.6 + 0.8155$$

$$= 1.4155 \text{ m}$$

$$\Delta Z_m = 1.4155 - 1.256 - 0.0408$$

$$= .1187 \text{ m}$$

Problem 24: ***A rectangular channel is 3.5 m wide and conveys a discharge of 15.0 m³/s at a depth of 2.0 m. It is proposed to reduce the width of the channel at a hydraulic structure. Assuming the transition to be horizontal and the flow to be frictionless, determine the water surface elevations upstream and downstream of the construction when the constricted width is 2.5 m.***

Solution: Let suffixes 1 and 2 denote sections upstream and downstream of the transition respectively, Discharge

$$Q = B_1 y_1 V_1$$

$$V_1 = \frac{15.0}{3.5 \times 2.0} = 2.143 \text{ m/s}$$

$$F_1 = \text{Froude number}$$

$$= \frac{V_1}{\sqrt{g y_1}} = \frac{2.143}{\sqrt{9.81 \times 2.0}}$$

$$= 0.484$$

The upstream flow is subcritical and the transition will cause a drop in the water surface

$$E_1 = y_1 + \frac{V_1^2}{2g} = 2.0 + \frac{(2.143)^2}{2 \times 9.81}$$

$$= 2.234 \text{ m}$$

Let B_{2m} = minimum width at section 2 which does not cause choking

then $$E_{cm} = E_1 = 2.234 \text{ m}$$

$$y_{cm} = \frac{2}{3} E_{cm} = \frac{2}{3} \times 2.234 = 1.489 \text{ m}$$

Since $$y_{cm}^3 = \left[\frac{Q^2}{g B_2^2 \text{ m}}\right]$$

$$B_{2m} = \left[\frac{Q^2}{g y_{C_2}^3}\right]^{1/2}$$

$$= \left[\frac{(15.0)^2}{9.81 \times (1.989)^3}\right]^{1/2}$$

$$= 2.636 \text{ m}$$

when $$B_1 = 2.4 \text{ m}$$

$B_2 < B_{2m}$ and hence choking conditions prevail. The depth at the section 2 = $y_2 = y_{c_2}$

The upstream depth y_1 will increase to y_1'

Actually $$q_2 = \frac{15.0}{2.4} = 6.25 \text{ m}^2/\text{s}$$

$$y_{c_2} = \left(\frac{q_2^2}{g}\right)^{1/2}$$

$$= \left[\frac{(6.25)^2}{9.81}\right]^{1/3} = 1.585 \text{ m}$$

$$E_{c_2} = 1.5 y_{c_2}$$

$$1.5 \times 1.585 = 2.3775 \text{ m}$$

At the upstream section 1:

$$E_1' = E_{c_2} = 2.3775 \text{ m}$$

Now,

$$y_1' + \frac{V_1'^2}{2g} = 2.3775$$

New upstream depth y_1' such that

$$q_1 = y_1' V_1' = 14/3.5$$
$$= 4.2857 \text{ m}^2/\text{s}$$

Hence

$$y_1' + \frac{V_1'^2}{2g} = 2.3775$$

$$y_1' + \frac{(4.2857)^2}{2 \times 9.81 \times y_1'^2} = 2.3775$$

Solving by the trial-and-error and selecting a root that gives subcritical flow such that $y_1' > y_1$

Hence, $$y_1' = 2.1806 \text{ m}$$

Problem 25: ***In a rectangular channel, F_1 and F_2 are the Froude numbers corresponding to the alternate depths of a certain discharge show that***

$$\left(\frac{F_2}{F_1}\right)^{2/3} = \frac{2+F_2^2}{2+F_1^2}$$

***Solution*:** Let us consider y_1 and y_2 be the alternate depths and F_1 and F_2 be the Froude numbers respectively.

Let B = bottom width of rectangular channel

Q = discharge through channel

A = Area of cross-section

T = Top width

Now $$F_1 = \frac{Q}{By_1\sqrt{gy_1}} \qquad \left(\because \frac{A_1}{T} = y_1\right)$$

Squaring on both side, we get

$$F_1^2 = \frac{Q^2}{B^2 g\, y_1^3} \qquad ...(i)$$

Similarly $$F_2 = \frac{Q}{By_2\sqrt{gy_2}}$$

Squaring on both side, we get

$$F_2^2 = \frac{Q^2}{B^2 g\, y_2^3} \qquad ...(ii)$$

Dividing equation (*i*) and (*ii*), we get

$$\frac{y_1}{y_2} = \left(\frac{F_2}{F_1}\right)^{2/3} \qquad ...(iii)$$

If E be the specific energy, since specific energy remain constant

Hence, $$E = y_1 + \frac{V_1^2}{2g} = y_2 + \frac{V_2^2}{2g}$$

$$E = y_1 + \frac{Q^2}{\left(B^2 y_1^2\right) 2g}$$

$$= y_2 + \frac{Q^2}{\left(B^2 y_2^2\right) 2g}$$

$$= y_1 + \left(\frac{F_1^2}{2}\right) y_1 = y_2 + \left(\frac{F_2^2}{2}\right) y_2$$

$$\Rightarrow \qquad \frac{y_1}{y_2} = \frac{1 + \frac{F_2^2}{2}}{1 + \frac{F_1^2}{2}}$$

or

$$\frac{y_1}{y_2} = \frac{2 + F_2^2}{2 + F_1^2} \qquad \text{...(iv)}$$

Using equations (*iii*) and (*iv*), we get

$$\left(\frac{F_2}{F_1}\right)^{2/3} = \frac{2 + F_2^2}{2 + F_1^2}$$

Problem 26: ***Show that in triangular channel, the Froude numbers corresponding to alternate depth are given by.***

$$(F_2/F_1) = (4 + F_2^2)^{5/2} / (4 + F_1^2)^{5/2}$$

***Solution*:** Let F_1 and F_2 be the Froude numbers corresponding to alternate depth y_1 and y_2 respectively.

Let $m =$ Side slope of triangular channel with 1 vertical : m horizontal

$Q =$ discharge through channel.

$$T = \frac{dA}{dy} = 2my = \text{Top width}$$

Now, $$F_1 = \frac{V_1}{\sqrt{g\frac{A_1}{T_1}}} = \frac{Q}{A_1\sqrt{g\left(\frac{A_1}{T_1}\right)}}$$

$$F_1 = \frac{Q}{my_1^2\sqrt{\left(g\, y_{1/2}\right)}}$$

Squaring on both side, we get

$$F_1^2 = \frac{2Q^2}{gm^2\, y_1^5} \qquad ...(i)$$

Similarly $$F_2 = \frac{V_2}{\sqrt{g\frac{A_2}{T_2}}} = \frac{Q}{A_2\sqrt{g\left(\frac{A_2}{T_2}\right)}}$$

$$= \frac{Q}{my_2^2\sqrt{\left(g\, y_{2/2}\right)}}$$

Squaring on both side, we get

$$F_2^2 = \frac{2Q^2}{gm^2\, y_2^5} \qquad ...(ii)$$

Dividing equation (*i*) and (*ii*) we get,

$$y_1/y_2 = (F_2/F_1)^{2/5} \qquad ...(iii)$$

if E be the specific energy, since specific energy remain constant

hence, $$E_1 = E_2 \text{ or } y_1 + v_1^2/2g = y_2 + v_2^2/2g$$

or $$y_1 + Q^2/2g(m\, y_1^2)^2 = y_2 + Q^2/2g(m\, y_2^2)^2$$

or $$y_1 + 2Q^2 y_1/4gm^2\, y_1^5 = y_2 + 2Q^2 y_2/4gm^2\, y_2^5$$

$$y_1 + y_1 F_1^2/4 = y_2 + y_2 F_2^2/4$$

$$y_1/y_2 = 4 + F_2^2 / 4 + F_1^2 \qquad ...(iv)$$

from (iii) and (iv) we get,

$$(F_2/F_1) = (4 + F_2^2)^{5/2} / (4 + F_1^2)^{5/2}$$

Problem 27: ***If y_1 and y_2 are alternative depth in a rectangular channel show that,***

$$\frac{2y_1^2 y_2^2}{y_1 + y_2} = y_c^3$$

Hence the specific energy, $E = \frac{y_1^2 + y_1 y_2 + y_2^2}{(y_1 + y_2)}$

Solution:

Let B = Bottom width of rectangular channel

and y_1, y_2 = Alternate depth

Q = Discharge through rectangular channel

We know that for rectangular channel,

$$y_C^3 = \frac{Q^2}{B^2 g} \quad ...(i)$$

Where y_C = critical depth.

Since specific energy remain constant

Hence,
$$E = y_1 + \frac{V_1^2}{2g} = y_2 + \frac{V_2^2}{2g}$$

$$= y_1 + \frac{Q^2}{B^2 2g\ y_1^2} = y_2 + \frac{Q^2}{B^2 2g\ y_2^2}$$

$$\Rightarrow \quad y_1 + \left(\frac{Q^2}{B^2 g}\right)\frac{1}{2y_1^2} = y_2 + \left(\frac{Q^2}{B^2 g}\right)\frac{1}{2y_2^2} \quad ...(ii)$$

Now,
$$E = y_1 + \frac{y_c^3}{2y_1^2} = y_2 + \frac{y_c^3}{2y_2^2} \quad ...(iii)$$

$$\frac{y_c^3}{2y_1^2} - \frac{y_c^3}{2y_2^2} = y_2 - y_1$$

$$y_c^3\left(\frac{y_2^2 - y_1^2}{2y_1^2\ y_2^2}\right) = (y_2 - y_1)$$

$$y_c^3 = \frac{2y_1^2\ y_2^2}{y_1 + y_2}$$

Putting value of y_c^3 in equation (*iii*), we get

$$E = y_1 + \frac{2y_1^2\ y_2^2}{(y_1 + y_2)2y_1^2}$$

$$= y_2 + \frac{2y_1^2\ y_2^2}{(y_1 + y_2)2y_1^2}$$

$$E = y_1 + \frac{y_2^2}{y_1 + y_2}$$

$$E = \frac{y_1^2 + y_2^2 + y_1 y_2}{y_1 + y_2}$$

Hence, specific energy $= \dfrac{y_1^2 + y_2^2 + y_1 y_2}{y_1 + y_2}$

Problem 28: ***If y_1 and y_2 are the alternate depth in a triangular channel show that,***

$$\frac{4y_1^4 y_2^4}{\left(y_1^2 + y_2^2\right)(y_1 + y_2)} = y_C^5$$

Where y_C = critical depth show further

$$E = \frac{n^4 + n^3 + n^2 + n + 1}{\left(1 + n^2\right)(1 + n)}$$

where $\qquad n = y_2/y_1$

Solution: Let y_1, y_2 be the alternate depth respectively

Let Q = Discharge through triangular channel

Side slope with m horizontal : 1 vertical of triangular channel

For triangular channel, critical depth is given by equation

$$y_C^5 = \frac{2Q^2}{gm^2} \quad \ldots(i)$$

Since specific energy remain constant

Hence, $$E = y_1 + \frac{V_1^2}{2g} = y_2 + \frac{V_2^2}{2g}$$

$$= y_1 + \frac{Q^2}{A_1^2(2g)} = y_2 + \frac{Q^2}{2g\left(A_2^2\right)}$$

$$E = y_1 + \frac{Q^2}{2gm^2 y_1^4} = y_2 + \frac{Q^2}{2gm^2 y_2^4}, \left(\frac{Q^2}{g} = \frac{m^2 y_C^5}{2}\right)$$

$$E = y_1 + \frac{y_C^5}{4\,y_1^4} = y_2 + \frac{y_C^5}{4\,y_2^4} \quad \ldots(ii)$$

$$\frac{y_C^5}{4\,y_1^4} - \frac{y_C^5}{4\,y_2^4} = y_2 - y_1$$

$$y_C^5\left(\frac{y_2^4 - y_1^4}{4\,y_1^4\,y_2^4}\right) = y_2 - y_1$$

$$y_C^5 = \frac{4y_1^4\,y_2^4}{\left(y_1^2 + y_2^2\right)\left(y_1 + y_2\right)}$$

Putting the value of y_C^5 in equation (*ii*), we get

$$E = y_1 + \frac{4y_1^4\,y_2^4 \Big/ \left(y_1^2 + y_2^2\right)\left(y_1 + y_2\right)}{4y_1^4}$$

$$= \frac{y_1\left(y_1^2 + y_2^2\right)\left(y_1 + y_2\right) + y_2^4}{\left(y_1^2 + y_2^2\right)\left(y_1 + y_2\right)}$$

$$= \frac{\left(y_1^3 + y_2^2 y_1\right)\left(y_1 + y_2\right) + y_2^4}{\left(y_2^2 + y_1^2\right)\left(y_1 + y_2\right)}$$

$$= \frac{y_1^4 + y_2^2 y_1^2 + y_1^3 y_2 + y_2^3 y_1 + y_2^4}{\left(y_2^2 + y_1^2\right)\left(y_1 + y_2\right)}$$

$$= \frac{\left(y_2/y_1\right)^4 + \left(y_2/y_1\right)^3 + \left(y_2/y_1\right)^2 + \left(y_2/y_1\right)^1 + 1}{\left(1 + \left(y_2/y_1\right)^2\right)\left(1 + y_2/y_1\right)}$$

$$E = \frac{n^4 + n^3 + n^2 + n + 1}{\left(1 + n^2\right)\left(1 + n\right)}$$

where $n = y_2/y_1$

Problem 29: *Show that for a trapezoidal channel, the minimum specific energy E_C is related to critical depth as:.*

$$E_C = \frac{y_C}{2}\left[\frac{3 + 5\tau_C}{1 + 2\tau_C}\right] \qquad \tau_C = \frac{m y_C}{B}$$

Solution:

Let B = Bottom width of trapezoidal channel

y_C = Critical depth with 1 vertical : m horizontal side slope.

The equation for critical depth is given by

$$\frac{Q^2 T_C}{g A_C^3} = 1$$

$$\frac{Q^2\left(B+2my_C\right)}{g\left(B+my_C\right)^3 y_C^3} = 1$$

$$\Rightarrow \qquad \frac{Q^2}{g} = \frac{\left(B+my_C\right)^3 y_C^3}{\left(B+2my_C\right)}$$

$$\frac{Q^2}{g}\left(\frac{m^3}{B^5}\right) = \frac{\left(1+\frac{my_C}{B}\right)^3}{\left(1+\frac{2my_C}{B}\right)} \times \left(\frac{my_C}{B}\right)^3$$

$$\frac{Q^2 m^3}{gB^5} = \frac{\left(1+\tau_C\right)^3}{\left(1+2\tau_C\right)} \times \left(\tau_C\right)^3 \qquad \text{...}(i)$$

Now, Specific energy $= y + \dfrac{V^2}{2g}$

Minimum specific energy $= y_C + \dfrac{V_C^2}{2g}$

$$E_C = y_C + \frac{Q^2}{2gA_C^2}$$

$$E_C = y_C + \frac{Q^2}{2g\left(B+my_C\right)^2 y_C^2}$$

Put the value of $\dfrac{Q^2}{g}$

$$E_C = y_C + \frac{\left(B+my_C\right)^3 y_C^3}{2\left(B+2my_C\right)\left(B+my_C\right)^2 y_C^2}$$

$$E_C = y_C + \frac{y_C\left(B+my_C\right)}{2\left(B+2my_C\right)}$$

$$= \frac{y_C}{2}\left[\frac{2B+4my_C+B+my_C}{B+2my_C}\right]$$

$$= \frac{y_C}{2}\left[\frac{3B + 5my_C}{B + 2my_C}\right]$$

$$E_C = \frac{y_C}{2}\left[\frac{3 + 5\frac{my_C}{B}}{1 + 2\frac{my_C}{B}}\right]$$

$$E_C = \frac{y_C}{2}\left[\frac{3 + 5\tau_C}{1 + 2\tau_C}\right]$$

Problem 30: ***Prove that in an exponential channel*** $\left(A = k_1 y^a\right)$ ***are given by***

$$\frac{2ay_1^{2a}\ y_2^{2a}\left(y_1 - y_2\right)}{\left(y_1^{2a} - y_2^{2a}\right)} = y_C^{2a+1}$$

Solution:

Area, $A = k_1 y^a$

Differentiating both side w.r.t. y, we get

$$T = \frac{dA}{dy} = ak_1 y^{a-1}$$

We know that

$$\frac{Q^2 T_C}{g\ A_C^3} = 1$$

$$\Rightarrow \qquad \frac{Q^2}{g} = \frac{A_C^3}{T_C} = \frac{k_1^3\ y_C^{3a}}{a\ k_1 y_C^{a-1}}$$

$$\frac{Q^2 a}{g\,k_1^2} = y_C^{2a+1} \qquad \ldots(i)$$

Since specific energy

$$E = y_1 + \frac{Q^2}{A_1^2(2g)} = y_2 + \frac{Q^2}{A_2^2(2g)}$$

$$E = y_1 + \frac{Q^2}{2g\left(k_1 y_1^a\right)^2} = y_2 + \frac{Q^2}{2g\left(k_1 y_1^a\right)^2}$$

$$\Rightarrow \quad y_1 + \frac{k_1^2\, y_C^{2a+1}}{2ak_1^2\, y_1^{2a}} = y_2 + \frac{k_1^2\, y_C^{2a+1}}{2ak_1^2\, y_2^{2a}}$$

$$\Rightarrow \quad \frac{y_C^{2a+1}}{2a}\left[\frac{1}{y_1^{2a}} - \frac{1}{y_2^{2a}}\right] = y_2 - y_1$$

$$\frac{y_C^{2a+1}}{2a}\left[\frac{y_2^{2a} - y_1^{2a}}{y_1^{2a}\, y_2^{2a}}\right] = (y_2 - y_1)$$

$$y_C^{2a+1} = \frac{2a\, y_1^{2a}\, y_2^{2a}\,(y_1 - y_2)}{y_1^{2a} - y_2^{2a}}$$

Problem 31: ***Show that in an exponential channel*** $\left(A = k_1 y^a\right)$ ***the minimum specific energy*** E_C ***and the critical depth are related as***

$$\frac{E_C}{y_C} = 1 + \frac{1}{2a}$$

Solution: Specific energy is given by equation

$$E = y + \frac{V^2}{2g}$$

Now minimum specific energy is given by equation

$$E_C = y_C + \frac{V_C^2}{2g}$$

$$E_C = y_C + \frac{Q^2}{A_C^2\, 2g} \qquad \text{...}(i)$$

Critical depth is given by

$$\frac{Q^2 T_C}{g A_C^3} = 1$$

$$\frac{Q^2}{g} = \frac{A_C^3}{T_C} \qquad \text{...(ii)}$$

Putting these value in equation (*i*)

$$E_C = y_C + \frac{A_C^3 / T_C}{A_C^2 \times 2} = y_C + \frac{A_C}{2T_C}$$

$$E_C = y_C + \frac{A_C}{2T_C}$$

$$E_C = y_C + \frac{k_1 y_C^a}{2ak_1 y_C^{a-1}} \qquad \left(T_C = \frac{dA_C}{dy_c} = ak_1 y_C{}^{a-1}\right)$$

$$E_C = y_C + \frac{y_C}{2a}$$

$$\frac{E_C}{y_C} = 1 + \frac{1}{2a}$$

Problem 32: ***Show that in parabolic channel*** $\left(x = C\sqrt{y}\right)$ ***area can be expressed in term of top width T.***

$$A = \frac{2}{3} Ty$$

***Solution*:** Let us consider area is divided in no. of strip. Let us consider small strip of thicknses *dy*

Hence total Area of channel

$$A = \int x dy$$

$$A = \int C\sqrt{y} dy$$

$$A = \frac{Cy^{3/2}}{3/2} + \text{Constant}$$

at $y = 0, \Rightarrow A = 0$

Hence constant $= 0$

$$A = \frac{2}{3}Cy^{3/2} \qquad ...(i)$$

differentiating both side, we get

$$T = \frac{dA}{dy} = Cy^{1/2}$$

Now, $$T.y = Cy^{3/2} \qquad ...(ii)$$

Putting these value in equation (*i*), we get

$$A = \frac{2}{3}T.y$$

Problem 33: ***Show that the approximate expression for the first hydraulic exponent M is given by:***

$$M = \frac{y}{A}\left[3T - \frac{A}{T}\frac{dA}{dy}\right]$$

Solution: We know that first hydraulic exponent is given by equation.

$$Z^2 = C_1 y^M$$

$$\frac{A^3}{T} = C_1 y^M$$

Taking log on both side, we get

$$3\log A - \log T = \log C_1 + M \log y$$

Differentiating on both side, we get

$$\frac{3}{A}\left(\frac{dA}{dy}\right) - \frac{1}{T}\frac{dT}{dy} = \frac{M}{y}$$

$$\frac{3T}{A} - \frac{1}{T}\frac{dT}{dy} = \frac{M}{y}$$

$$\Rightarrow \qquad M = \frac{y}{A}\left[3T - \frac{A}{T}\frac{dT}{dy}\right]$$

Problem 34: ***The average velocity of flow of water through a rectangular channel 6m wide is 1.5 m/s. If the discharge through the channel is 18 m³/s. State the type of flow.***

Solution: Given $B = 6$ m $V = 1.5$ m/s, $Q = 18$ m³/s

To find : Type of flow

$$q = \frac{Q}{B} = \frac{18}{6} = 3 \text{ m}^3\text{/s/m}$$

$$\text{Depth of flow} \quad y = \frac{q}{v} = \frac{3}{1.5} = 2 \text{ m}$$

$$y_c = \text{Critical depth} = \left(\frac{q^2}{g}\right)^{1/3} = \left(\frac{3^2}{9.81}\right)^{1/3} = 0.97 \text{ m}$$

Since $y > y_c$, the flow is subcritical.

Problem 35: ***A trapezoidal channel with base of 6 m and side slope of 2 horizontal to 1:vertical convey water of 20 m³/s with depth of 1.6 m. Is the flow condition is sub or super critical flow?***

Solution: Given $B = 6$ m, $Q = 20$ m³/s

$y = 1.6$ m, Side slope = 2:1

To find: Type of flow

Area = $(B + my)\,y = (6 + 2 \times 1.6) \times 1.6 = 14.72$ m²

Top width $T = B + 2\,my = 6 + 2 \times 2 \times 1.6 = 12.4$ m

$$F = \frac{Q^2 T}{gA^3} = \frac{20^2 \times 12.4}{9.81 \times (14.72)^3} = 0.158$$

Since $F < 1$, the flow is sub-critical.

Problem 36: ***A Triangular channel is to carry a discharge of 6 m³/s if the side slope are IV to 2H find the critical depth, critical velocity and critical slope?***

Solution: Given $Q = 6$ m³/s Side slope = $H{:}V = 2{:}1$

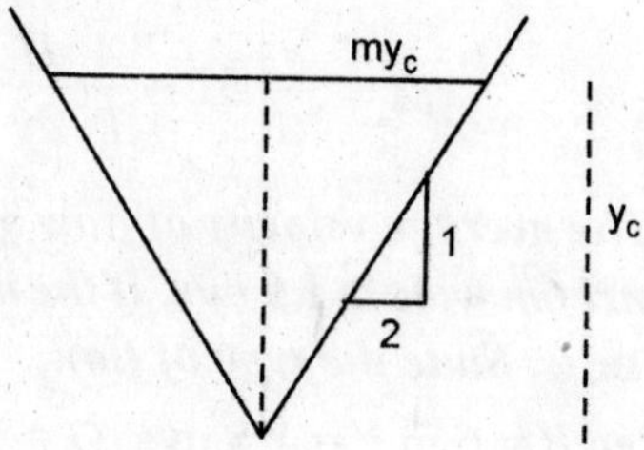

Top width $= 2\,my_c = 2 \times 2 \times y_c = 4\,y_c$

$$\text{Area } A_c = \frac{1}{2} B \times y_c = \frac{1}{2} \times 4y_c \times y_c = 2y_c^2$$

Since froude number for critical depth $F = 1$

$$\text{Therefore} \qquad F = \frac{Q^2 T}{gA^3} = 1$$

$$\text{or} \qquad \frac{Q^2}{g} = \frac{A_c^3}{T_c} \qquad \Rightarrow \frac{6^2}{9.81} = \frac{\left(2y_c^2\right)^3}{4y_c}$$

$$\Rightarrow \qquad 2y_c^5 = 3.67$$

$$\therefore \qquad y_c = 1.13 \text{ m}$$

$$\text{Critical velocity } V_c = \frac{Q}{A_c} = \frac{6}{2 \times 1.13^2} = 2.35 \text{ m/s}$$

Critical slope: Assume Chezy's $C = 55$

$$A_c = 2 \times 1.13^2 = 2.55 \text{ m}^2$$

$$P_c = 2y_c \sqrt{m^2 + 1} = 2 \times 1.13\sqrt{2^2 + 1}$$

$$Pc = 5.05 \text{ m}$$

$$R = \frac{A_c}{P_c} = \frac{2.55}{5.05} = 0.505 \text{ m}$$

$$V = C\sqrt{RS_0}$$

$$2.35 = 55\sqrt{0.505\, S_0}$$

$$S_0 = 1/315.6$$

Problem 37: ***The specific energy of flowing water through rectangular channel of 6 m wide is 6 m. Determine the maximum possible discharge through the channel?***

Solution: Given $B = 6$ m, $E = 6$ m

To find Q_{max}.

For rectangular $y_c = \frac{2}{3}E_c = \frac{2}{3}\times 6 = 4$ m

$$A_c = By_c = 6 \times 4 = 24 \text{ m}^2$$

Critical velocity $V_c = \sqrt{gy_c} = \sqrt{9.81\times 4} = 6.26$ m/s

Maximum discharge $= A_c \times V_c = 24 \times 6.26$

$$Q_{max} = 150.24 \text{ m}^3/\text{s}$$

Problem 38: ***Calculate the critical depth, critical velocity and critical specific energy for a flow of 12 m³/s in a rectangular channel of width 3.5 m and energy coefficienty α = 1.1. What is the type of flow when the depth is 0.9 m? What is the froude number at this depth?***

Solution: Given $Q = 12$ m³/s, $B = 3.5$ m

$\alpha = 1.1$

For rectangular channel

$$y_c = \left(\frac{\alpha q^2}{g}\right)^{1/3}$$

$$q = \frac{Q}{B} = \frac{12}{3.5} = 3.45 \text{ m}^3/\text{s/m}$$

$$y_c = \left(\frac{1.1\times 3.43^2}{9.81}\right)^{1/3} = 1.096$$

$$V_c = \frac{q}{y_c} = \frac{3.43}{1.096} = 3.129 \text{ m/s}$$

$$E_c = \frac{3}{2}y_c = 1.5 \times 1.096 = 1.644 \text{ m}$$

when $y = 0.9$ m, $y < y_c$

Hence flow is supercritical

When $$y = 0.9,\ V = \frac{Q}{A} = \frac{12}{3.5 \times 0.9} = 3.81 \text{ m/s}$$

$$F = \frac{V}{\sqrt{gy/\alpha}} = \frac{3.81}{\left(9.81 \times 0.9/1.1\right)} = 1.345$$

Problem 39: ***A rectangular channel of width 4m carries a discharge of 10 m³/s. Plot specific energy diagram and use it to find critical depth and alternative depth of 0.7 m. What is specific energy for this depth? Assume α = 1.09 use the following depth of flow for computations and for drawing the graph:***

0.6, 0.7, 0.8, 1.1 and 1.4 m?

Solution: Given $Q = 10$ m³/s, $B = 4$ m, $\alpha = 1.09$

$$E = y + \frac{\alpha Q^2}{2gA^2}$$

$$E = y + \frac{1.09 \times 10^2}{2 \times 9.81 \times 4^2 \times y^2}$$

$$E = y + \frac{0.347}{y^2}$$

$y(m)$	$\frac{0.347^2}{y^2}$	$E = y + \frac{0.347^2}{y^2}$
0.6	0.9644	1.564
0.7	0.786	1.409
0.8	0.5425	1.343
0.9	0.4286	1.329
1.1	0.2869	1.387
1.4	0.1771	1.577

A graph of E verses y is plotted as shown in figure.

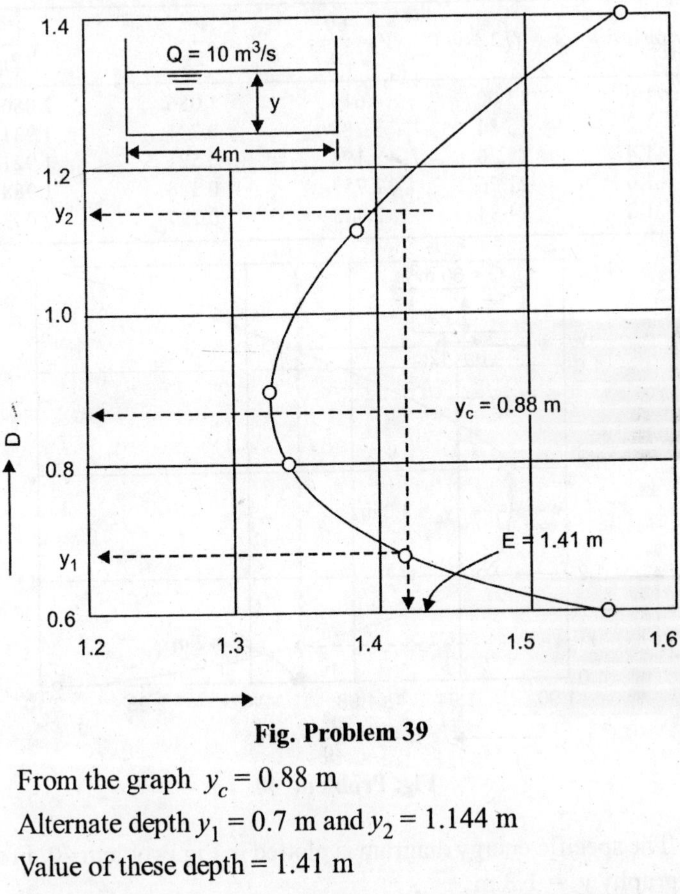

Fig. Problem 39

From the graph $y_c = 0.88$ m

Alternate depth $y_1 = 0.7$ m and $y_2 = 1.144$ m

Value of these depth = 1.41 m

Problem 40: ***Draw the specific energy diagram for a flow of 60 cumecs in a trapezoidal channel of bottom width 12 m and side slopes 1:1. From the graph, determine the critical depth and alternate depths for a specific energy of 2 m. Use the following depths for calculations: 1.0, 1.2, 1.4, 1.6 and 1.8 m.***

Solution: $A = (12 + y)y$; $V = \dfrac{Q}{A} = \dfrac{60}{(12+y)y}$. The calculations are shown in Table.

Table

y (metres)	$A = (12 + y)y$	$V = \frac{60}{A}$	$\frac{V^2}{2g}$	$y + \frac{V^2}{2g}$
1.0	13.00	4.615	1.086	2.086
1.2	15.84	3.788	0.731	1.931
1.4	18.76	3.198	0.521	1.921
1.6	21.76	2.757	0.388	1.988
1.8	24.84	2.415	0.297	2.097

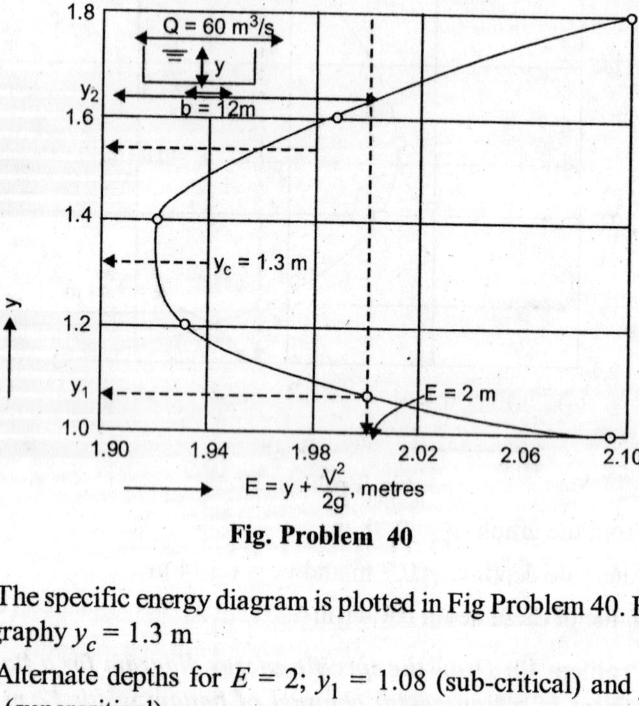

Fig. Problem 40

The specific energy diagram is plotted in Fig Problem 40. From the graphy $y_c = 1.3$ m

Alternate depths for $E = 2$; $y_1 = 1.08$ (sub-critical) and $y_2 =$ 1.63 (supercritical)

Problem 41: ***Construct the specific force diagram for a discharge of 2.5 m³/s through a rectangular channel of width 2m. Using the diagram, estimate the critical depth minimum speciic force and conjugate depths for a specific force of 1.1 m³. The following depths of flow may be used in computations and in drawing the graph: 0.3, 0.4, 0.5, 0.6, 0.7 and 0.9 m.***

Solution:

$$F = A\bar{z} + \frac{Q^2}{gA} = \text{by}\frac{y}{2} + \frac{Q^2}{\text{gby}} = 2\frac{y}{2} + \frac{6.25}{9.81 \times 2y} = y^2 + \frac{0.31855}{y}$$

Table

y (metres)	y^2	$\frac{0.31855}{y}$	$F = y^2 + \frac{0.31855}{y}$
0.3	0.09	1.062	1.152
0.4	0.16	0.796	0.956
0.5	0.25	0.627	0.887
0.6	0.36	0.531	0.891
0.7	0.79	0.455	0.945
0.9	0.81	0.354	1.164

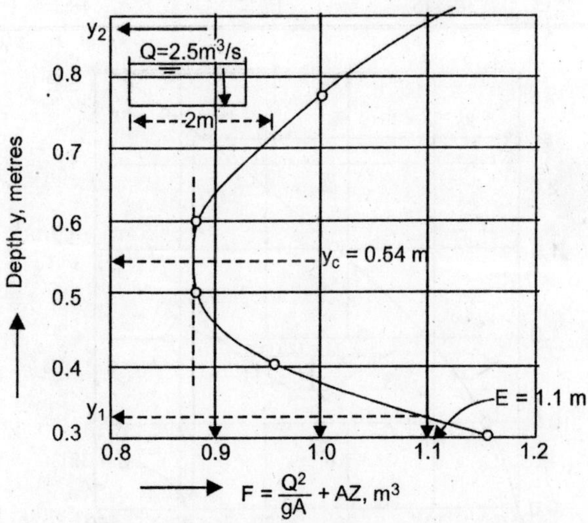

Fig. Problem 41

From the graph : $y_c = 0.85$ m

Conjugate Depths or $F = 1.1$ m $y_1 = 0.32$·m (super critical depth) and $y_2 = 0.36$ m (sub-critical depth).

Problem 42: ***A trapezoidal channel has a bed width of 10 m***

and side slopes 1 upon 1.5. Determine graphically the depth of critical state for a discharge of 120 m³/s. Use the following depths of flow for calculating the section factor: 0.5, 1.5, 2.0 and 2.5.

Solution: $A = (10 + 1/5y)y$ and $T = 10 + 3y$

$$Z = A\sqrt{\frac{A}{T}} \text{ and } Z_c = \frac{Q}{\sqrt{g}} = \frac{120}{\sqrt{9.81}} = 38.31$$

Table

y (metres)	$A = (10 + 1.5y)y$	$T = 10 + 3y$	$Z = A\sqrt{\frac{A}{T}}$
0.5	5.375	11.5	3.675
1.0	22.5	13	10.82
1.5	18.375	14.5	20.685
2.0	26	16	33.14
2.5	34.375	17.5	48.18

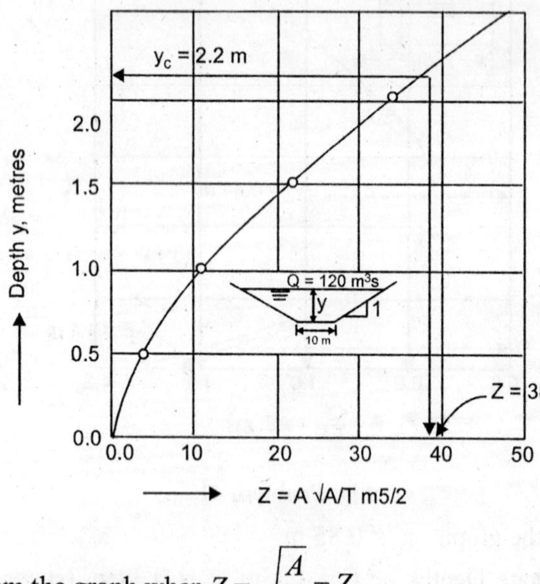

From the graph when $Z = \sqrt{\frac{A}{T}} = Z_c$

$$y = y_c = 2.2 \text{ m}$$

Problem 43: ***It is desired to have channel in which flow is critical at all stages, show that the cross section of such a channel is***

$$T^2 h^2 = \frac{Q^3}{8g}$$

in which T = top width and h = depth of water surface below the energy line.

Solution: At critical condition

$$\frac{Q^2}{g} = \frac{A^3}{T}$$

Squasing both side

$$\frac{Q^4}{g^2} = \frac{A^6}{T^2}$$

From the question $h = \frac{V^2}{2g}$

on cubing $$h^3 = \frac{V^6}{8g^3} = \frac{Q^6}{8g^3 A^6}$$

$$h^3 = \frac{Q^4 \cdot Q^2}{8g^2 \cdot gA^6} \quad \text{...(a)}$$

Put the value of $\frac{Q^4}{g^2}$

We get

$$h^3 = \frac{A^6 \cdot Q^2}{8T^2 gA^6} = \frac{Q^2}{8T^2 g}$$

or $$h^3 = \frac{Q^2}{8T^2 g}$$

or $$T^2 h^3 = \frac{Q^2}{8g}$$. **Proved**

Problem 44: ***It is required to have channel in which froude number 'F' remain constant at all depth. If the specific energy is E is also kept constant. Show that for such a channel.***

$$\frac{T}{B} = \left(\frac{E}{E-y}\right)^{1+\frac{F^2}{2}}$$

Solution: We know that $E = y + \dfrac{V^2}{2g}$

or
$$E = y + \frac{Q^2}{2gA^2} \times \frac{A}{T} \times \frac{T}{A} \quad \text{(Multiplying and dividing by } A/T\text{)}$$

$$E = y + \frac{Q^2 T}{2gA^3} \times \frac{A}{T}$$

We know that $\dfrac{Q^2 T}{gA^3} = F^2$

$$E = y + \frac{F^2 A}{2T}$$

or
$$E - y = \frac{AF^2}{2T}$$

Differentiating wrt y we get

$$\frac{dE}{dy} - 1 = \frac{\frac{F^2}{2}\left[\frac{TdA}{dy} - \frac{AdT}{dy}\right]}{T^2}$$

$\because$
$$\frac{dA}{dy} = T$$

So
$$\frac{dE}{dy} - 1 = \frac{F^2}{2}\left[1 - \frac{A}{T^2}\frac{dT}{dy}\right]$$

Put
$$\frac{dE}{dy} = 0$$

$$\frac{AF^2}{2T} \times \frac{1}{T}\frac{dT}{dy} = 1 + \frac{F^2}{2}$$

As we know $\frac{AF^2}{2T} = E - y$

Put in equation (a) we get

$$(E-y)\frac{1}{T}\frac{dT}{dy} = 1+\frac{F^2}{2}$$

$$\frac{dT}{T} = \left(1+\frac{F^2}{2}\right)\frac{dy}{(E-y)}$$

Taking log

$$\ln T = 1+\frac{F^2}{2}[-\ln(E-y)]+C$$

- At $y = 0$, $T = B$ Put in (b) we get

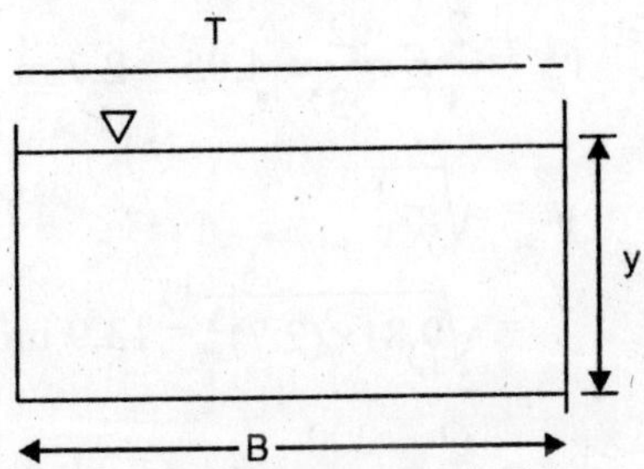

$$C = \ln B + \left(1+\frac{F^2}{2}\right)\ln E \quad \text{...(c)}$$

Put value of C in equation (b)

$$\ln\frac{T}{B} = \left(1+\frac{F^2}{2}\right)\ln\frac{E}{(E-y)}$$

or we can say

$$\frac{T}{B} = \left(\frac{E}{E-y}\right)^{1+\frac{F^2}{2}} \quad \text{Proved.}$$

Problem 45: ***A bridge is planned on a 50 m wide rectangular channel carrying a flow of 200 m3/s at a flow depth of 4.0 m. For reducing the length of the bridge, what is the minimum channel with such that the upstream water level is not influenced for this discharge?***

Solution: Given $Q = 200$ m^3/s

$$B = 50 \text{ m}$$

$$y = 4.0 \text{ m}$$

$$y = \frac{200}{50 \times 4} = 1.0 \text{ m/s}$$

$$E = 4 + \frac{(1)^2}{2 \times 9.81} = 4.05 \text{ m}$$

For discharge to be maximum, the flow should be critical

$$y_c = \frac{2}{3}E = \frac{2}{3} \times 4.05 = 2.7 \text{ m}$$

Now
$$q = \sqrt{gy_c^3}$$

$$= \sqrt{9.81 \times (2.7)^3} = 13.9 \text{ m}^3\text{/s/m}$$

$$B_c = \frac{Q}{q} = \frac{200}{13.9} = 14.4$$

Therefore the channel width may be reduced from 50 to 14.4 m without affecting the upstream level for flow of 200 m^3/s.

Review Questions

1. Define "open channel flow". Discuss its application.
2. What is difference between open channel flow and pipe flow?
3. Discuss various types of channel.
4. Describe different types of flow in channels.
5. Discuss various geometrical parameters of a channel.

6. Derive equation of continuity for unsteady flow.
7. Derive relation for energy and momentum correction factor.
8. Explain the term critical depth.
9. Explain the significance of specific energy and specific force.
10. What is specific energy curve?
11. Derive expressions for critical depth and critical velocity.
12. Discuss the applications of specific energy.
13. While measuring the discharge in a small stream it was found that the depth of flow increases at the ratio of 0.1 m/h. If the discharge at that section was 25 m^3/s and the surface width of the stream was 20 m, estimate the discharge at a section 1 km upstream?
14. At a section in a channel expansion, the velocity over the quarter of the cross section is zero and is uniform over the remaining three fourths of the area. Calculate the kinetic energy correction factor (α) and momentom correction factor (β)?
15. The two alternative depths in a 4 m wide rectangular channel are 3.9 m and 1.2 m respectively. Calculate the discharge of the channel?
16. Show that for a trapezoidal channel the minimum specific energy E_C is related to critical depth y_c as

$$E_C = \frac{y_C}{2}\left(\frac{3+5\xi_C}{1+2\xi_C}\right) \qquad \text{where } \xi_C = \frac{my_C}{B}$$

17. A triangular channel has an apex angle of 75° and carried a flow with the velocity of 2 m/s and depth of 1.2 m. (*a*) Is the flow subcritical or supercritical? (*b*) What is critical depth? (*c*) What is specific energy? (*d*) What is alternative depth?
18. What is critical depth corresponding to discharge of 6 m^3/s in (*a*) a trapezoidal channel of B = 0.7 m, m = 1.5 and (*b*) a circular channel of D = 1.5 m?
19. A rectangular channel is 4.5 m wide and carries a discharge of 18 m^3/s at a depth of 2.2 m. At a certain section it is proposed to built a hump. Calculate the water surface elevation at upstream of the hump and over the hump if the hump height is

(*a*) 0.33 m and (*b*) 0.2 m (Assume no loss of energy at hump)?

20. A rectangular channel is 3 m wide and carries a flow of 1.85 m^3/s at a depth of 0.5 m. A contraction of channel width is required at a certain section. Find the greatest allowable contraction in the width for the upstream flow to be possible as specified.

Unit 2

Uniform Flow

2.1 Introduction

Flow in a channel is said to be uniform if its properties remain constant with respect to space. In uniform flow ,depth, cross-section and velocity remain constant over a given length of the channel.

Mathematically, $$\frac{\partial y}{\partial l} = 0, \frac{\partial v}{\partial l} = 0$$

It should be remember that uniform flow is possible only in prismatic channel. The water surface and channel bottom slope is parallel in uniform flow.

As we know velocity is constant in uniform flow over the length of channel, therefore the energy line will be at constant elevation above the water surface.

It should be remember that slope of energy line S_f, slope of water surface S_w and bottom slope S_o will be equal to each other.

2.2 Chezy's Equation for Uniform Flow

According to the definition of uniform flow, velocity is constant hence there is no acceleration in uniform flow. Using momentum equation to control volume (Section 1 and Section 2), which is L distance apart. We can write

$$P_1 + W\sin\theta - F_f - P_2 = M_2 - M_1 \qquad \text{...(2.1)}$$

where P_1 and P_2 are the pressure forces and M_1 and M_2 are the momentum rate at sections 1 and 2 respectively. W is the weight of fluid in control volume and F_f= Shear force at the boundry.

As the flow is uniform

$$P_1 = P_2 \text{ and } M_1 = M_2$$

$$W = \gamma AL, \quad F_f = \tau_o PL$$

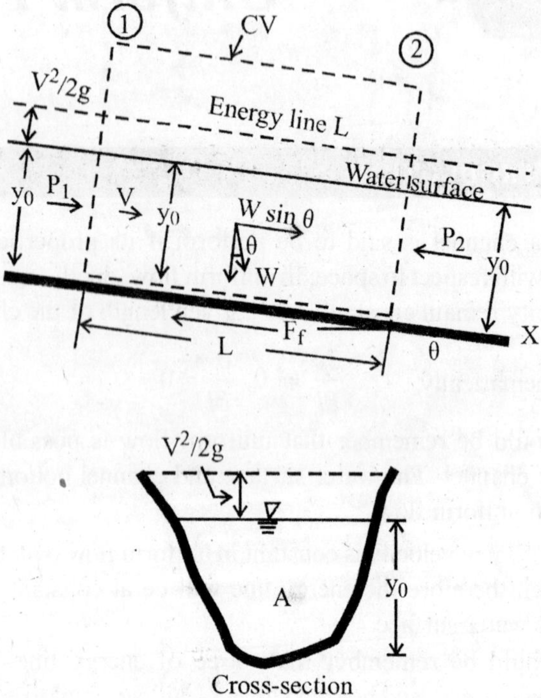

Fig. 2.1: Uniform Flow

where τ_o is average shear stress on wetted perimeter P and γ is unit weight of water.

Applying uniform flow condition to equation (2.1). It will reduce to

$$\text{W} \sin \theta = F_f, \quad (\because \sin \theta = S_o)$$

$$\gamma AL\, S_0 = \tau_0 PL$$

$$\tau_0 = \gamma\left(\frac{A}{P}\right)S_0 = \gamma R S_0 \qquad ...(2.2)$$

As we know average shear stress $= K\rho V^2$, where K = a coefficient which depends on the nature of the surface and flow parameters hence we can write

$$K\rho V^2 = \gamma RS_0$$

or $$V = C\sqrt{RS_0} \qquad ...(2.3)$$

where $$C = \sqrt{\frac{\gamma}{\rho}\frac{1}{k}}$$

where $$\gamma = \rho g,$$

C is the coefficient which depend on the nature of the surface and the flow.

Equation 2.3 is called Chezy formula. The dimensions of C are $\left[L^{1/2}T^{-1}\right]$. The coefficient C is known as Chezy coefficient.

Also we know $$Q = AV$$

$$Q = AC\sqrt{RS} \qquad ...(2.4)$$

For pipe flow, the Dacy-Weisbach equation is

$$h_f = f\frac{L}{D}\frac{V^2}{2g} \qquad ...(2.5)$$

where h_f = head loss due to friction in pipe of diameter D and length L, f = Darcy-Weisbach friction factor.

An open channel can be considered as conduit cut into two.

For circular section $$R = \frac{A}{P} = \frac{\pi D^2}{4\pi D} = \frac{D}{4}$$

$$D = 4R$$

Equation (2.5) can be written as

$$h_f = f\frac{L}{4R}\frac{V^2}{2g}$$

Which on rearranging gives

$$V = \sqrt{\frac{8g}{f}} \cdot \sqrt{R} \cdot \sqrt{\frac{h_f}{L}} \qquad \text{...(2.6)}$$

$$\frac{h_f}{L} = S_f = \text{Slope of energy line}$$

but in uniform flow $\quad S_f = S_0$

It may be seen that equation (2.6) is same as (2.3).

$$V = C\sqrt{R S_0}$$

where $\quad C = \sqrt{\frac{8g}{f}},\ S_0 = S_f = \sqrt{\frac{h_f}{L}}$

2.3 Manning's Formula for Uniform Flow

According to Robert Manning, for uniform flow the velocity is given by

$$V = \frac{1}{n} R^{2/3} S_0^{\ 1/2} \qquad \text{...(2.7)}$$

where n = *a* roughness coefficient known as Manning's n, R = hydraulic radius, S_0 = bottom slope

The dimension of n is $\left[L^{-1/3} T\right]$.

2.3.1 Comparison between Manning's Formula and Chezy's Formula

According to Chezy's equation

$$V = C\sqrt{RS_0}$$

According to Manning's equation

$$V = \frac{1}{n} R^{2/3} S_0^{\ 1/2}$$

Equating the two

$$\frac{1}{n} R^{2/3} S_0^{\ 1/2} = C\sqrt{RS_0}$$

$$C = \frac{1}{n} R^{1/6} \qquad ...(2.8)$$

Also

$$C = \sqrt{\frac{8g}{f}}$$

$$\sqrt{\frac{8g}{f}} = \frac{1}{n} R^{1/6}$$

$$f = \left(\frac{n^2}{R^{1/3}}\right) 8g \qquad ...(2.9)$$

Value of Manning's n for different materials

2.3.2 Factor Affecting n

The most important factors affecting the value of n are: (*a*) surface roughness, (*b*) vegetation, (*c*) cross-section irregularity, (*d*) irregular alignment of channel, (*e*) sediment load and obstructions, (*f*) slitting and scouring, (*g*) Depth of flow and discharge, (*h*) seasonal changes which alter fluid viscosity. The most important among these are the characteristics of surface or surface roughness the dependence of the value of n on surface roughness is given in Table 2.1.

Table 2.1

Material	n
Glass, PVC	0.010
Brass	0.011
Finesh concrete, smooth steel	0.012
Wood	0.012–0.013
Cast Iron	0.013
Brick work	0.015
Asphalt	0.016
Natural streams	0.030–0.040
Flood plains	0.040–0.150

The presence of vegetation on the channel bed and sides leads to additional resistance and increase the value of n. However, for large depth of flow and flexible vegetation, bending over of the vegetation tends to reduce the resistance and results in lower value of n.

The channel irregularities and curvature especially in natural streams, produce energy losses. As such, they are combined with the boundry resistance by suitably increasing the value of n. The sudden change in channel cross-section enhances the value of n.

The presence of obstructions, *e.g.*, bridge piers and transport of sediment load, either in suspension or near the bed, increases the value of n since it causes additional head loss.

2.3.3 Equivalent Roughness

In some of the channels different parts have different roughness coefficient (n) for such channel it is necessary to determine an equivalent roughness coefficient that can be applied to entire cross-section for use in Manning's formula. Equivalent roughness is also called composite roughness.

A simple method of calculating equivalent roughness is Horton method. Consider channel having its perimeter composed of M types of roughness. P_1, P_2,..., P_M are length of these parts $n_1, n_2, n_3, ..., n_M$ are roughness coefficient.

Let each part P_i be associated with partial area A_i so that

$$\sum_{i=1}^{N} A_i = A_1 + A_2 .... A_M = A = \text{Total area}$$

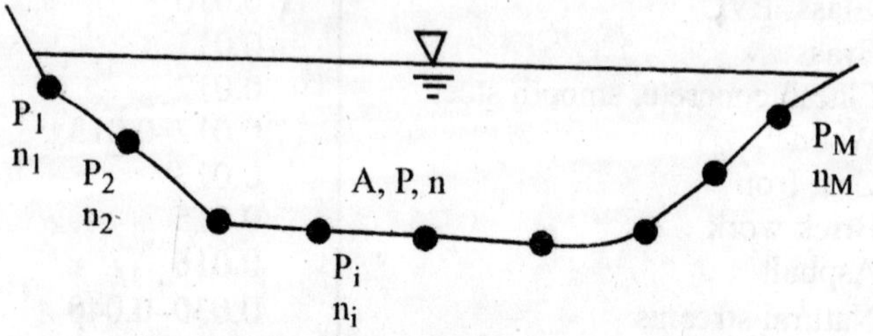

Fig. 2.2: Multiple roughness type perimeter

It is assumed that mean velocity in each area is V

$$V_1 = V_2 = ... = V_M = V$$

By Manning's formula

$$S_0^{1/2} = \frac{V_1 n_1}{R_1^{2/3}} = \frac{V_2 n_2}{R_2^{2/3}} = ... = \frac{V_M n_M}{R_M^{2/3}}$$

$$= \frac{Vn}{R^{2/3}} \qquad ...(1)$$

Where n = equivalent roughness

From equation (1) $\left(\frac{A_i}{A}\right)^{2/3} = \frac{n_i P_i^{2/3}}{nP^{2/3}}$

$$A_i = \frac{A n_i^{3/2} P_i}{n^{3/2} P}$$

$$\Sigma A_i = A = A\frac{\Sigma\left(n_i^{3/2} P_i\right)}{n^{3/2} P}$$

$$n = \frac{\left(\Sigma n_i^{3/2} P_i\right)^{2/3}}{P^{2/3}} \qquad ...(2.10)$$

2.4 Compound Channel

Compound Channel is a channel section composed of a main deep portion and one or two flood plains that carries high water flows. Figure 2.3 shows diagram of compound channel.

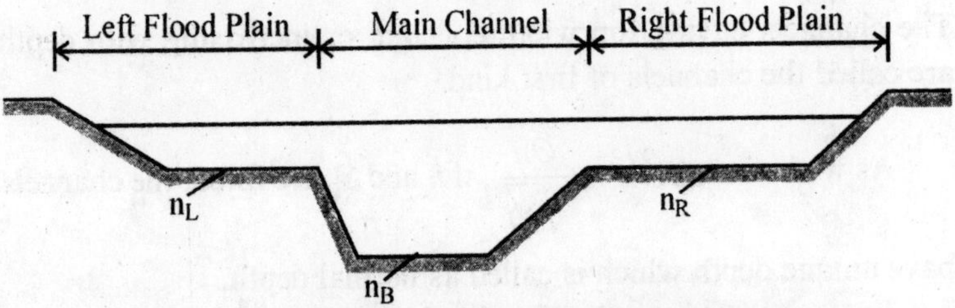

Fig. 2.3: Schematic sketch of a compound channel

The majority of natural rivers have compound sections. The flow velocity in the flood plain is lower than in the main channel due to relative smaller water depth and higher bed roughness.

2.5 Conveyance of Channel

Conveyance of channel expresses the discharge capacity of the channel per unit longitudinal slope. We know according to Manning's equation

$$Q = \frac{1}{n} A R^{2/3} S_0^{1/2}$$

$$Q = K\sqrt{S_0}$$

where $K = \frac{1}{n} A R^{2/3}$ is called conveyance of channel

For a given channel, $AR^{2/3}$ is the function of depth. Also according to Chezy's

$$Q = AC\sqrt{RS_0}$$

which can be written as $Q = K\sqrt{S_o}$

where $K = AC\sqrt{R}$ (Conveyance of channel)

For a channel of constant slope, the conveyance is directly proportional to discharge.

2.6 Channels of First Kind

The channels having top width constant or increasing with depth are called the channels of first kind.

As we know $AR^{2/3} = \frac{Qn}{\sqrt{S_0}}$, if n and S_0 are fixed, the channels have unique depth which is called as normal depth.

Examples of channel of first kind, rectangular, trapezoidal, etc.

2.7 Channels of Second Kind

Channels with a closing top width can be called as channels of second kind. For example, circular and triangular channels. The variation of $AR^{2/3}$ with depth is not single valued function of depth. They are having two values of depths for a given value of $AR^{2/3}$.

2.8 Second Hydraulic Exponent N

The conveyance of a channel is in general a function of depth of flow. For solving gradually varied flow, Bakhmeteff introduced the following assumptions.

$$K^2 = C_2 y^N$$

where C_2 = coefficient and N = exponent called second hydraulic exponent.

To determinc N for any channel, a plot of log K Versus log y is prepared. If N is constant between two points (k_1, y_1) and (k_2, y_2) it is determine 2d by

$$N = \frac{2\log\left(\frac{k_1}{k_2}\right)}{\log\left(\frac{y_1}{y_2}\right)} \qquad \text{...(2.11)}$$

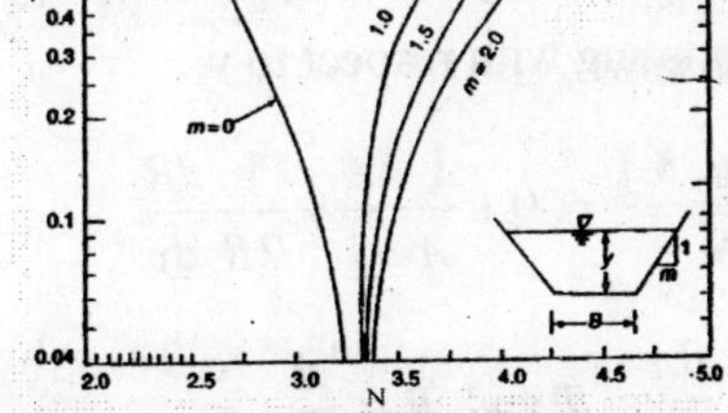

Fig. 2.4: Variation of the second hydraulic exponent N in trapezoidal channels

Computation of Hydraulic exponent (N): Hydraulic component (N) can be computed by two method:

(a) using Chezy's Equation
(b) using Manning's Equation

(a) Using Chezy's Equation

According to Chezy equation, Q is given by

$$Q = AV = AC\,R^{1/2}\,S_0^{\,1/2}$$

$$= CAR^{1/2}\sqrt{S_0} = K\sqrt{S_0}$$

where conveyance $K = CAR^{1/2}$ ...(*i*)

As A and R are functions of depth, relationship between conveyance 'K' and depth y was given by Boris Bakhmetiff as:

$$K^2 = C'y^N$$

where C' is constant and N is called hydraulic exponent for computation of uniform flow.

Taking log we have,

$$2\log_e K = \log_e C' + N \log y$$

Now differentiating with respect to y,

$$\frac{2d(\log_e K)}{dy} = 0 + \frac{N}{y}$$

$$\therefore \qquad \frac{d(\log_e K)}{dy} = \frac{N}{2y} \qquad \text{...(ii)}$$

Now taking log of Eq. (*i*)

$$\log_e K = \log_c C + \log_e A + \tfrac{1}{2}\log_e R$$

Again diferentiating with respect to y,

$$\frac{d(\log_e K)}{dy} = 0 + \frac{1}{A}\frac{dA}{dy} + \frac{1}{2R}\frac{dR}{dy}$$

$\because$ Putting $\dfrac{dA}{dy} = T$ and $R = \dfrac{A}{P}$

$$\frac{d(\log_e K)}{dy} = \frac{T}{A} + \frac{1}{2R}\frac{d\left(A/P\right)}{dy}$$

$$= \frac{T}{A} + \frac{1}{2R}\frac{1}{P}\frac{dA}{dy} - \frac{A}{P^2}\frac{dP}{dy}\frac{1}{2R}$$

$$= \frac{T}{A} + \frac{T}{2RP} - \frac{A}{P^2 \times 2A/P}\frac{dP}{dy}$$

$$= \frac{T}{A} + \frac{T}{2RP} - \frac{1}{2P}\frac{dP}{dy}$$

$$= \frac{T}{A} + \frac{T}{2A/P \, P} - \frac{1}{2P}\frac{dP}{dy}$$

$$= \frac{T}{A} + \frac{T}{2A} - \frac{1}{2A}\left(\frac{A}{P}\right)\frac{dP}{dy}$$

$$= \frac{3T}{2A} - \frac{R}{2A}\frac{dP}{dy}$$

$$\therefore \quad \frac{d(\log_e K)}{dy} = \frac{1}{2A}\left(3T - R\frac{dP}{dy}\right) \qquad \ldots(iii)$$

Equating Eqs. (*ii*) and (*iii*), we have

$$\frac{N}{2y} = \frac{1}{2A}\left(3T - R\frac{dP}{dy}\right)$$

$$N = \frac{y}{A}\left(3T - R\frac{dP}{dy}\right) \qquad \ldots(iv)$$

above equation is general equation of hydraulic exponent in uniform flow with Chezy's Equation.

Case I: When the channel is trapezoidal

$$A = (B + my)y, \; T = B + 2my, \; P = B + 2y\sqrt{1+m^2}$$

$$R = \frac{A}{P} = \frac{(B+my)y}{B+2y\sqrt{1+m^2}}, \; \frac{dp}{dy} = 2\sqrt{1+m^2}$$

Substituting R and $\frac{dp}{dy}$, in Eqn (iv)

$$N = \frac{y}{(B+my)y}\left[3(B+my) - \frac{(B+my)y}{B+2y\sqrt{1+m^2}} \cdot 2\sqrt{1+m^2}\right]$$

$$N = \frac{3(B+2my)}{(B+my)} - \frac{2y\sqrt{1+m^2}}{B+2y\sqrt{1+m^2}}$$

Dividing both numerator and denominator of R.H.S. by B,

$$N = \frac{3\left(1+\frac{2my}{B}\right)}{\left(1+\frac{my}{B}\right)} - \frac{\frac{2y}{B}\sqrt{1+m^2}}{1+\frac{2y}{B}\sqrt{1+m^2}} \qquad \ldots(v)$$

Case II: When channel is rectangular

Putting $m = 0$, in Eq (v), we get

$$N = 3 - \frac{2\left(\frac{y}{B}\right)}{1+\frac{2y}{B}} \qquad \ldots(vi)$$

Case III: When channel is wide rectangular

$$P \simeq B,\ R = \frac{A}{P} = \frac{By}{B} = y,\ T = B$$

$$\frac{dP}{dy} = 0,\ m = 0$$

Putting $\frac{dP}{dy}$ and T in Eqn (iv)

$$N = \frac{y}{By}[3B - y \times 0]$$

$$N = 3 \qquad (vii)$$

Case IV: When channel is triangular

$$A = my^2,\ T = 2my,\ P = 2y\sqrt{1+m^2}$$

$$R = \frac{my^2}{2y\sqrt{1+m^2}} = \frac{my}{2\sqrt{1+m^2}},\quad \frac{dP}{dy} = 2\sqrt{1+m^2}$$

Putting in Eqn. (*iv*)

$$N = \frac{y}{my^2}\left[3.2my - \frac{my}{2\sqrt{1+m^2}} \times 2\sqrt{1+m^2}\right]$$

$$N = \frac{1}{my}[6my - my]$$

$$N = 5 \qquad (viii)$$

(*b*) Using Manning's Equation

According to Manning's Equation, Q is given by

$$Q = AV = A\frac{1}{n}(R)^{2/3}\ (s_0)^{1/2}$$

$$Q = \left(\frac{AR^{2/3}}{n}\right)\sqrt{s_0} = K\sqrt{s_0}$$

where $$K = \frac{AR^{2/3}}{n} \qquad \ldots(ix)$$

'K' is called conveyance. If 'n' is constant, K is again unique function of depth y and is related to y as:

$$K^2 = C'yN \qquad \ldots(x)$$

C' is a constant and N is hydraulic exponent. To get a general equation of N, take log of above equation (x).

$$2\log_e K = \log_e C' + N\log_e y$$

Differentiating with respect to y,

$$\frac{2d(\log_e K)}{dy} = 0 + \frac{N}{y}$$

$$\frac{d(\log_e K)}{dy} = \frac{N}{2y} \qquad \ldots(xi)$$

Now taking the log of equation (ix)

$$\log_e K = \log_e\left(\frac{1}{n}\right) + \log_e A + \frac{2}{3}\log_e R.$$

Differentiating with respect to y,

$$\frac{d(\log_e K)}{dy} = 0 + \frac{1}{A}\frac{dA}{dy} + \frac{2}{3R}\frac{d\left(\frac{A}{P}\right)}{dy} \qquad \because R = \frac{A}{P}$$

$$= \frac{T}{A} + \frac{2}{3R}\left[\frac{1}{P}\frac{dA}{dy} - \frac{A}{P^2}\frac{dP}{dy}\right] \qquad \because T = \frac{dA}{dy}$$

$$= \frac{T}{A} + \frac{2}{3\frac{A}{P}}\frac{T}{P} - \frac{2}{3}\frac{A}{\left(\frac{A}{P}\right)P^2}\frac{dP}{dy}$$

$$= \frac{T}{A} + \frac{2T}{3A} - \frac{2}{3P}\frac{dP}{dy}$$

$$\therefore \quad \frac{d(\log_e K)}{dy} = \frac{5T}{3A} - \frac{2}{3P}\frac{dP}{dy} \qquad \ldots(xii)$$

Equating Eqn. (*xi*) and (*xii*)

$$\frac{N}{2y} = \frac{5T}{3A} - \frac{2}{3P}\frac{dP}{dy}$$

$$\therefore \quad N = \frac{10yT}{3A} - \frac{4}{3}\frac{y}{A}\left(\frac{A}{P}\right)\frac{dP}{dy}$$

$$N = \frac{10yT}{3A} - \frac{4}{3}\frac{Ry}{A}\frac{dP}{dy}$$

$$N = \frac{2y}{3A}\left[5T - 2R\frac{dP}{dy}\right] \qquad \ldots(xiii)$$

Above equation is general equation of hydraulic exponent of uniform flow when Manning's eqn. is used:

Case I: When channel is trapezolidal,

$$A = (B + my)y,\ T = B + 2my,\ P = B + 2y\sqrt{1+m^2}$$

$$R = \frac{A}{P} = \frac{(B+my)y}{B+2y\sqrt{1+m^2}},\ \frac{dp}{dy} = 2\sqrt{1+m^2}$$

Substituting above value in Equation (xiii)

$$N = \frac{2y}{3(B+my)y}\left[5(B+2my) - \frac{2(B+my)y}{B+2y\sqrt{1+m^2}} \times 2\sqrt{1+m^2}\right]$$

$$= \frac{10}{3}\frac{(B+2my)}{(B+my)} - \frac{8}{3}\frac{\left(y\sqrt{1+m^2}\right)}{B+2y\sqrt{1+m^2}}$$

Now dividing both numerator and denominator of R.H.S. of above equation by 'B',

$$N = \frac{10}{3}\frac{\left(1+\dfrac{2my}{B}\right)}{\left(1+\dfrac{my}{B}\right)} - \frac{8}{3}\frac{\left(\dfrac{y}{B}\sqrt{1+m^2}\right)}{\left(1+\dfrac{2y}{B}\sqrt{1+m^2}\right)} \quad \ldots(xiv)$$

Case II: When channel is rectangular

$$m = 0$$

$$N = \frac{10}{3} - \frac{8}{3}\frac{\dfrac{y}{B}}{\left(\dfrac{1+2y}{B}\right)} \quad \ldots(xv)$$

Case III: When the channel is wide rectangular

$$P = B,\ \frac{dP}{dy} = 0,\ A = By,\ T = B,\ R = y$$

Putting above value in Equation (xiii)

$$N = \frac{2y}{3By}[5B - 2y \times 0] = \frac{10}{3}$$

$$N = \frac{10}{3} \quad \text{...(xvi)}$$

Case IV: When the channel is triangular

$$A = my^2,\ T = 2my,\ P = 2y\sqrt{1+m^2}\ ,\ \frac{dP}{dy} = 2\sqrt{1+m^2},\ R = \frac{A}{P}$$

Putting these values in Equation (*xiii*)

$$N = \frac{2y}{3my^2}\left[5(2my) - 2\frac{my^2}{2y\sqrt{1+m^2}}2\sqrt{1+m^2}\right]$$

$$N = \frac{2y}{3my^2}\times 10my - \frac{2y}{3my^2}\times\frac{2my^2}{2y\sqrt{1+m^2}}2\sqrt{1+m^2}$$

$$N = \frac{20}{3} - \frac{4}{3}$$

$$N = \frac{16}{3} \quad (xvii)$$

Characteristics of Equation (*xiv*) and (*xvii*)

(i) When the channel is trapezium

When $y/B \rightarrow$ i.e. depth is low

$$N \rightarrow \frac{10}{3} \rightarrow 3.33$$

When $y/B \rightarrow$ very large, from eqn. (*xiv*)

$$\left(1 + 2m\frac{y}{B}\right) \simeq \frac{2y}{B}\left(1 + \frac{my}{B}\right) \simeq \frac{y}{B} \text{ etc.}$$

$$N \rightarrow \frac{10}{3}\times\frac{\frac{2y}{B}}{\frac{y}{B}} - \frac{8}{3}\frac{\frac{y}{B}}{\frac{2y}{B}}$$

$$N \rightarrow \frac{20}{3} - \frac{8}{6}$$

$$N \rightarrow \frac{16}{3} \rightarrow 5.33 \quad (xix)$$

When channel is rectangular

'*m*' will be zero, (*xv*)

Consider equation (*ix*), we have

$$N = \frac{10}{3} - \frac{8}{3}\frac{\frac{y}{B}}{\frac{1+2y}{B}}$$

Now when $\frac{y}{B}$ is very small, then

$\frac{y}{B} \to 0$, S_0

$N \to \frac{10}{3}$

$N \to 3.33$ (*xx*)

Now when $\frac{y}{B}$ is very large;

$$\left(1 + \frac{2y}{B}\right) \simeq \frac{2y}{B}$$

from equation (xv),

$N \to \frac{10}{3} - \frac{8}{6}$

$N \to 2$ (*xxi*)

Characteristics of Equation (*v*) from Chezy's Equation:

$$N = \frac{3\left(\frac{1+2my}{B}\right)}{\left(\frac{1+my}{B}\right)} - \frac{\frac{2y}{B}\sqrt{1+m^2}}{\frac{1+my}{B}\sqrt{1+m^2}}$$

(*i*) When channel is trapezium

When $\frac{y}{B}$ is very small,

Then $\frac{y}{B} \to 0$

$N \to 3$ (*xxii*)

When $\frac{y}{B}$ is very large.

$$1+\frac{2my}{B} \simeq \frac{2y}{B},\ 1+\frac{my}{B} = \frac{y}{B} \text{ and etc.}$$

$$N \to \frac{3\times\frac{2y}{B}}{\frac{y}{B}} - \frac{\frac{2y}{B}}{\frac{2y}{B}}$$

$$N \to 6-1$$

$$N \to 5 \qquad \text{(xxiii)}$$

(ii) When channel is rectangular

m will be 0, then we have equation (*xi*)

$$N = 3 - 2\,\frac{\frac{y}{B}}{1+2\frac{y}{B}}$$

When $\frac{y}{B} \to 0$

$$N \to 3-3$$

$$N \to 3$$

When $\frac{y}{B}$ is very large,

then $\frac{1+2y}{B} = \frac{2y}{B},\ \frac{1+y}{B},\ \frac{y}{B},$

We have

$$N \to 3 - \frac{\frac{2y}{B}}{\frac{2y}{B}}$$

$$N \to 3-1$$

$$N \to 2 \qquad \text{...(xxv)}$$

(iii) When channel is rectangular

$N = 5$ (According to Eqn. *viii*)

(iv) When channel is wide rectangular

$N = 3$ (According to Eqn.*vii*)

2.9 Velocity Distribution in Open Channel Flow

The velocity in open channel are retarded near solid boundries. The surface tension of the liquid-air interface (*i.e.*, free surface) produces resistance to flow and causes maximum velocity to occur at same distance below the free surface. A strong wind when blowing in upstream or downstream direction also effect the flow and modify the velocity profile.

The velocity distribution in channel section depends on various factors such as shape of section, the roughness of the channel and presence of bends in the channel.

Figure 2.5 shows typical pattern of velocity distribution in various channels.

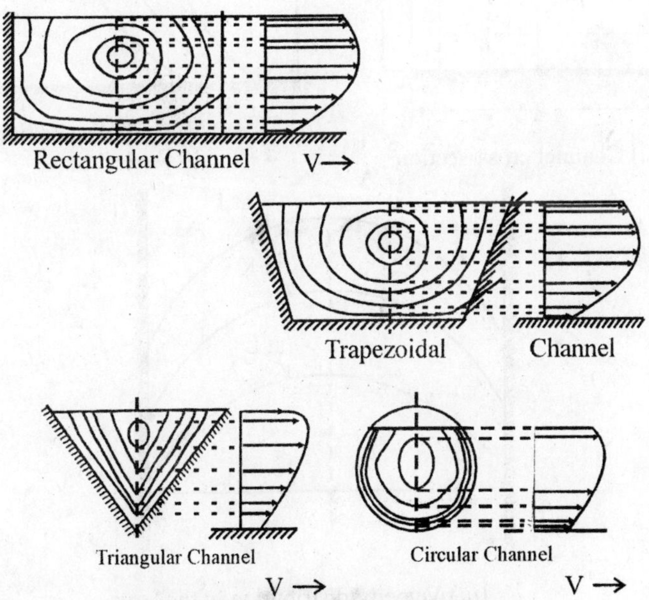

Fig. 2.5: Typical patterns of velocity distribution in various channel sections

Some of the important points of velocity distribution in open channel flow are as follows:

1. In a straight reach of channel maximum velocity usually occurs below the free surface at a distance of 0.05 to 0.15 of the depth of flow.
2. It is observed that velocity at 0.6 depth from free surface is very close to mean velocity of flow in vertical section. A still better approximation for the mean velocity of flow is obtained by taking average of velocity measured at 0.2 depth and 0.8 depth from free surface.

The sketch of Fig. 2.6 show the pattern of velocity distribution in the horizontal and vertical direction for a rectangular channel.

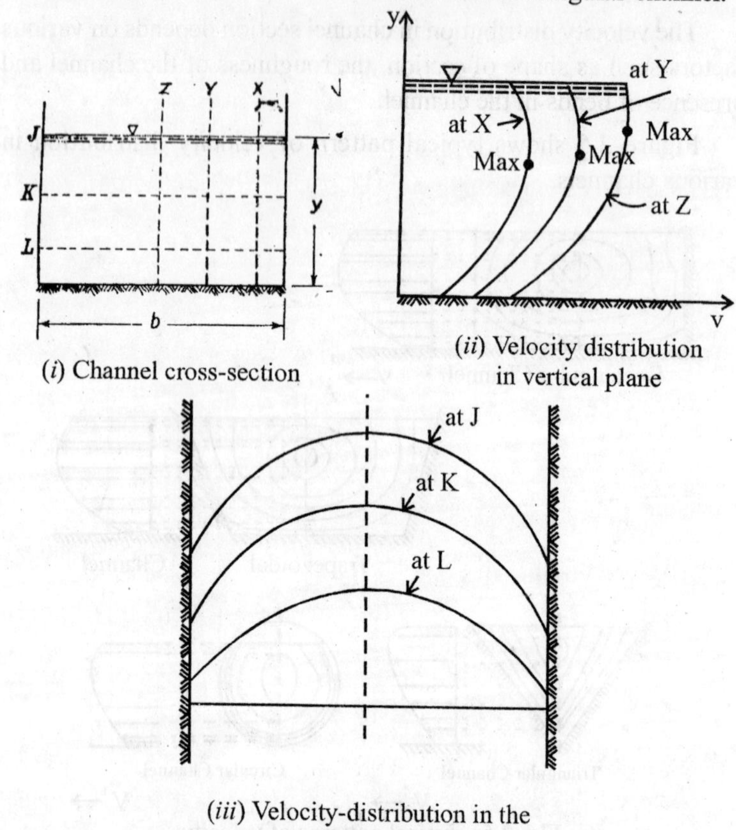

(*i*) Channel cross-section

(*ii*) Velocity distribution in vertical plane

(*iii*) Velocity-distribution in the horizontal plane

Fig. 2.6: Velocity distribution in horizontal channel.

The Fig. 2.6 reveals the fact that as we go away from the location of X, Y and Z, the point of maximum velocity shifts upwards and at the central section Z the maximum velocity occurs almost at the free surface.

The effect of roughness is to increase the curvature of the velocity distribution Fig. 2.7 shows the comparison between the velocity distribution curve for smooth and rough channels.

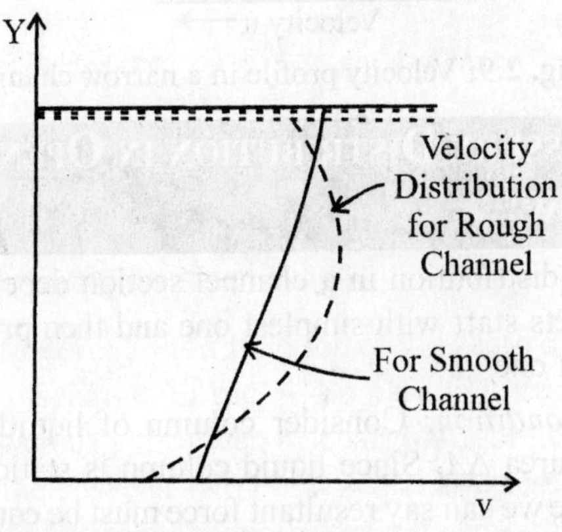

Fig. 2.7: Velocity distribution in rough and smooth channels

Figures 2.8 and 2.9 shows the velocity distribution of wide and narrow channel.

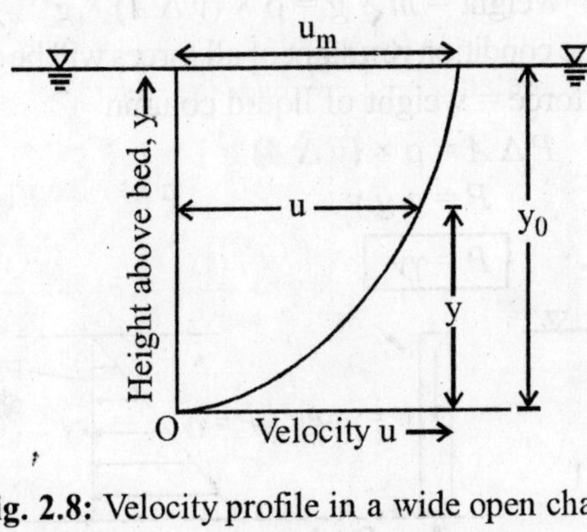

Fig. 2.8: Velocity profile in a wide open channel

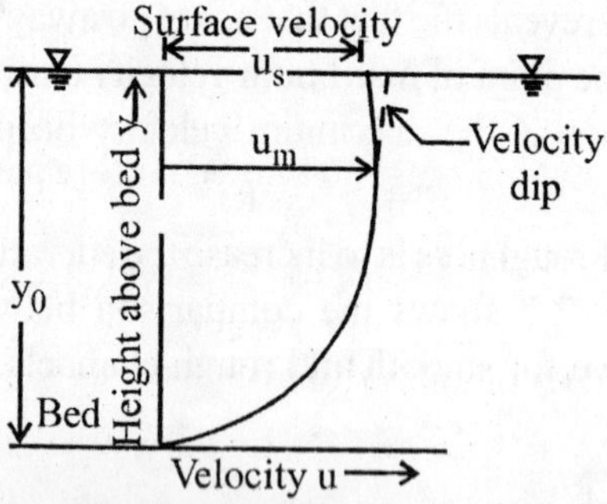

Fig. 2.9: Velocity profile in a narrow channel

2.10 Pressure Distribution in Open Channel Flow

The pressure distribution in a channel section depends upon flow conditions. Lets start with simplest one and then proceed towards more complex one.

(a) *Static condition:* Consider column of liquid having cross section area ΔA. Since liquid column is static or stationary therefore we can say resultant force must be equal to zero. Let P be pressure intensity at bottom of liquid column then force due to pressure at liquid column acting vertically upwards will be $= P\Delta A$.

Also weight of liquid column will be acting downward.

$$\text{Weight} = m \times g = \rho \times (y\,\Delta A) \times g$$

under static condition resultant of all forces will be zero therefore

Pressure force = weight of liquid column

$$P\,\Delta A = \rho \times (y\,\Delta A)\,g$$

$$P = \rho\,g\,y$$

$$\boxed{P = \gamma y} \qquad (\because \gamma\ e\ g)$$

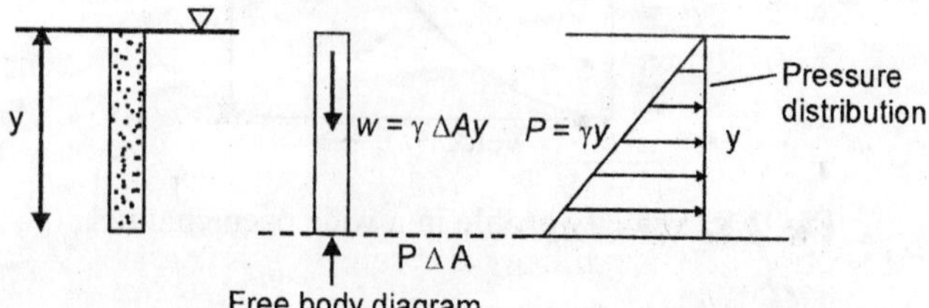

Fig. Pressure distribution in stationary liquid

(*b*) *Pressure distribution in moving liquid:* The pressure distribution given by moving liquid is same as static liquid as long as the streamlines of flowing liquid are parallel and straight

($\therefore$ Normal accelaration $a_n = \frac{v^2}{r} = 0$) where v is the velocity and r is radius of curvature.

For uniform flow streamlines are parallel and straight so radius of curvature is infinity hence normal accelaration is zero.

In general pressure distribution of moving liquid is given by:

$$p = \gamma y \pm \frac{a_n}{g} \gamma y$$

Consider the figure as shown below

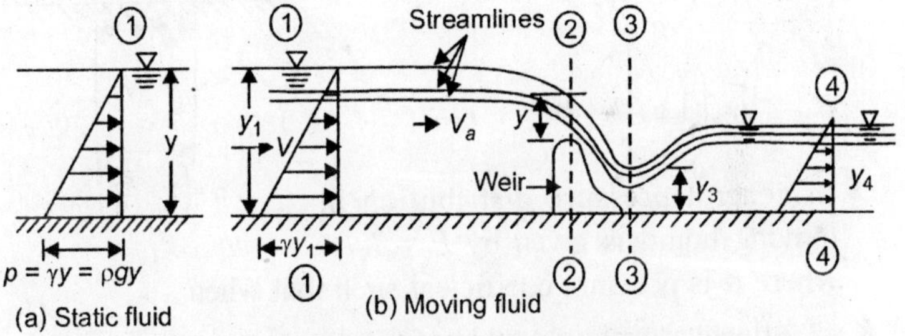

Fig. Prssure distribution in moving liquid

From the figure it is clear that section (1) and section (4) have straight streamlines ($\therefore a_n = 0$)

We know that

$$P = \gamma y \pm \frac{a_n}{g} \gamma y$$

$$\boxed{P = \gamma y}$$ which is same as for static liquid

At section (2) streamlines are convex, so pressure intensity decreases. Which may be written as

$$P_2 = \gamma y_2 - \frac{a_n}{g}\gamma y_2$$

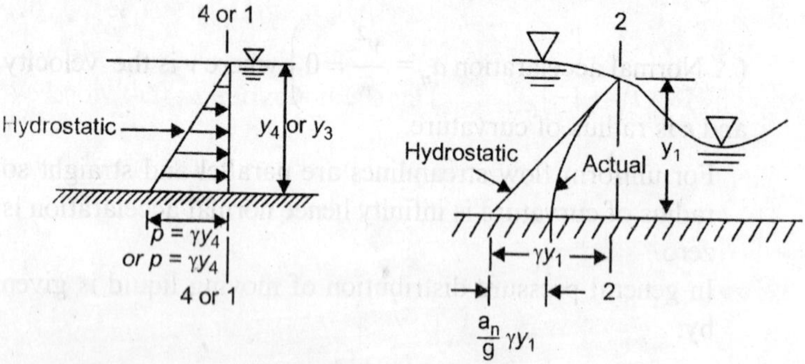

At section (3) streamlines are concave, so pressure intensity increases which may be written as

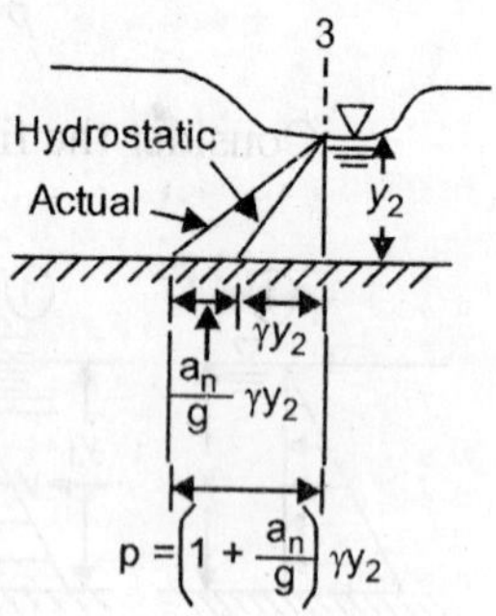

$$P_3 = \gamma y_3 + \frac{a_n}{g}\gamma y_3$$

$$P_3 = \gamma y_3\left(1 + \frac{a_n}{g}\right)$$

In general pressure distribution in moving liquid is given by $P = K\gamma y$, where K is pressure coefficient such that when

$K = 1$, flows are straight and paralel, pressure is hydrostatic

$K > 1$ flow profiles have concave curvature, pressure is more than hydrostatic

$K < 1$ flow profiles have convex curvature pressure is less than hydrostatic

Note in concave surface centrifugal force is in the same direction of gravity force while in convex surface centrifugal force opposing gravity force.

Pressure Distribution on Large Bed Slope

Consider the channel with large slope θ. The depth y is measured vertical from free surface to the channel bottom. Perpendicular distance of bottom from free surface is $d = y \cos\theta$.

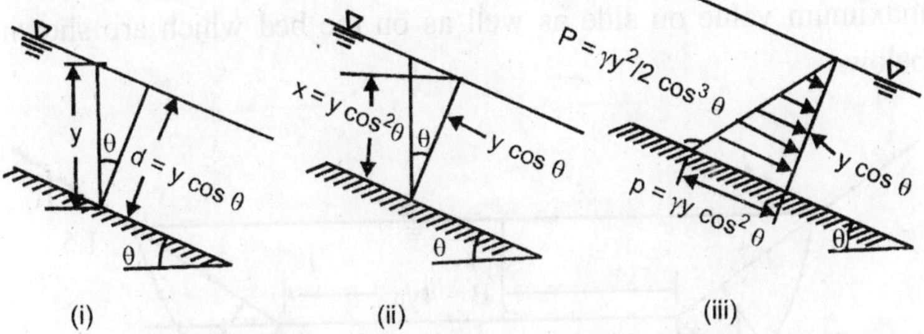

Fig. Effect of bed slope on pressure intensity

The vertical projection of $y \cos\theta$ is x

or $$\frac{x}{y\cos\theta} = \cos\theta \Rightarrow x = y\cos^2\theta$$

The highest intensity of triangular pressure diagram is at the bottom, therefore

$$p = \gamma x$$

or $$p = \gamma y \cos^2\theta$$

The area of the triangular pressure intensity diagram gives the total pressure p/per unit width of channel.

$$p = \frac{1}{2}\gamma y \cos^2\theta \times y\cos\theta$$

$$p = \frac{1}{2}\gamma y^2 \cos^3\theta$$

Which is equation of pressure force per unit width with bigger slope when θ is small, $\cos\theta = 1$ then

$$p = \frac{1}{2}\gamma y^2$$

2.11 Shear Stress Distribution

The average shear stress τ_o on boundary of a channel is given by $\tau_0 = \gamma R S_0$. However, the shear stress is not uniformly distributed over the boundary. It is zero at the intersection of the water surface with the boundary and also at the corners in the boundary. The boundary shear stress τ_0 will have certain maximum value on side as well as on the bed which are shown below.

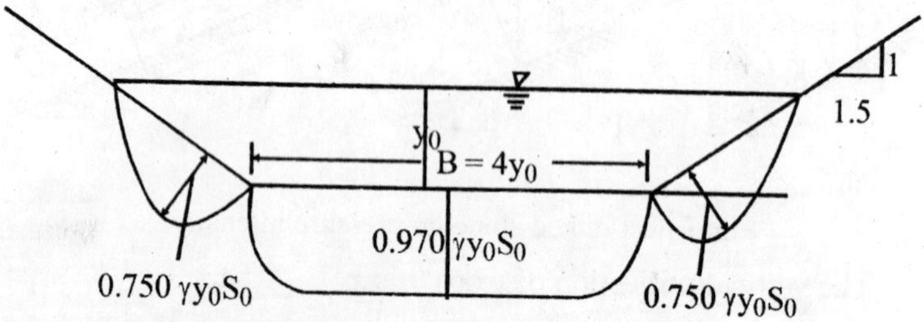

Fig. 2.10: Variation of boundary shear stress in a trapezoidal channel with $B/y_o = 4$ and m = 1.5

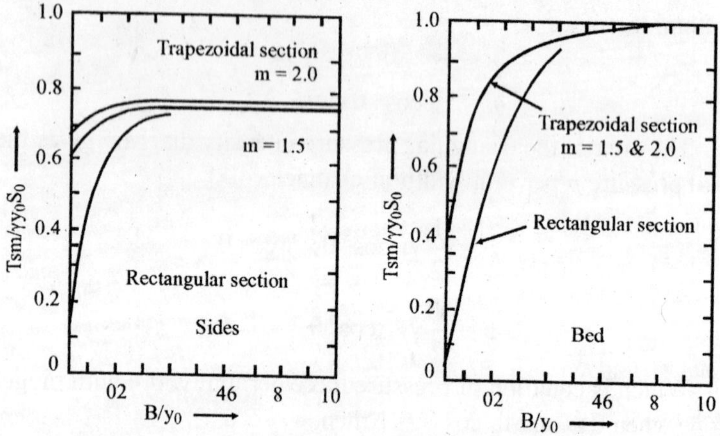

Fig. 2.11: Variation of maximum shear stress on bed and sides of smooth channels

2.12 Most Efficient Channel Section

The most efficient channel section is also called as most economical section or best section. The channel is said to be most efficient channel if it is having minimum wetted perimeter for a given area of flow and also provide maximum discharge and conveyance. Of all the possible open channel section, the semicircular shape has least amount of wetted perimeter for a given area, hence proved to be most economical one.

The relationship between the various geometric elements to form *an* efficient section can be obtained as follows:

(*a*) *Rectangular Section*

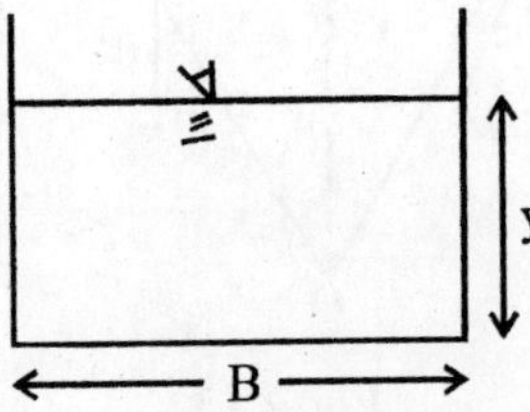

Fig. 2.12

$$\text{Area of flow} = A = By = \text{constant}$$

$$\text{Wetted perimeter} = P = B + 2y$$

$$P = \frac{A}{y} + 2y \qquad \left(\because B = \frac{A}{y}\right)$$

For efficient section P should be minimum, therefore

$$\frac{dp}{dy} = 0$$

$$\frac{dp}{dy} = -\frac{A}{y^2} + 2 = 0$$

which gives

$$A = 2y^2{}_e$$

Also

$$By_e = 2y_e^2 \Rightarrow y_e = \frac{B_e}{2}$$

$$B_e = 2y_e$$

$$R_e = \frac{A_e}{P_e} = \frac{2y_e^2}{2y_e + 2y_e} = \frac{y_e}{2}$$

The suffix '*e*' denotes the Geometric elements of a hydraulically efficient section.

We can say when channel is half square the best section is obtained.

(*b*) *Triangular Section*

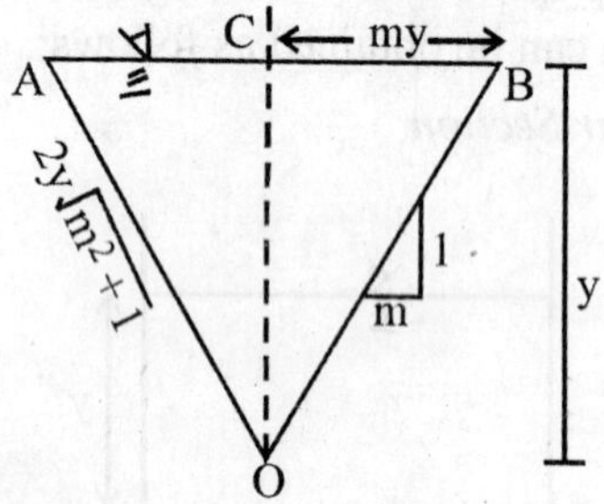

Fig. 2.13

$$CO = y$$

$$\frac{y}{CB} = \frac{1}{m} \Rightarrow CB = my$$

$$\text{Area of triangle} = \frac{1}{2} \times AB \times CO$$

$$A = \frac{1}{2} \times 2my \times y = my^2 = \text{constant}$$

$$A = my^2 \Rightarrow m = \frac{A}{y^2}$$

$$AO = \sqrt{m^2 y^2 + y^2} = y\sqrt{m^2+1}$$

$$P = AO + OB = 2y\sqrt{m^2+1}$$

$$P = 2y\sqrt{\frac{A^2}{y^4}+1} \qquad \left(\because m = \frac{A}{y^2}\right)$$

$$P = 2\left(\frac{A^2}{y^2}+y^2\right)^{1/2}$$

For efficient section P should be minimum, therefore

$$\frac{dp}{dy} = 0$$

$$\frac{1}{2} \times 2\left(\frac{A^2}{y^2} + y^2\right)^{1/2} \times \left(\frac{-2A^2}{y^3} + 2y\right) = 0$$

$$\frac{\left(-2\frac{A^2}{y^3} + 2y\right)}{\sqrt{\frac{A^2}{y^2} + y^2}} = 0$$

or $$-\frac{2A^2}{y^3} + 2y = 0$$

$$A = y_e^2$$

or $$y_e = \sqrt{A_e}$$

We know $$A = my^2$$

therefore $$y_e^2 = my_e^2 \qquad (\because A = y_e^2)$$

$$m = 1$$

We know $$P = 2y\sqrt{m^2 + 1}$$

$$P_e = 2y_e\sqrt{1+1} = 2\sqrt{2}y_e$$

$$R_e = \frac{A_e}{P_e} = \frac{y_e^2}{2\sqrt{2}y_e} = \frac{y_e}{2\sqrt{2}}$$

Thus we can say when section is half square described on diagonal and having equal sloping sides then the best section is obtained.

(c) *Trapezoidal Section*

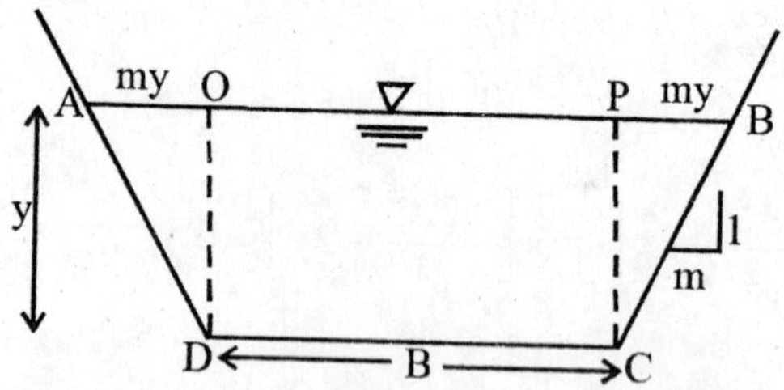

Fig. 2.14

$$\text{Area of ABCD} = \frac{1}{2}[AB + DC] \times OD$$

$$= \frac{1}{2}[(B + 2my) + B] \times y$$

$$A = (B + my)y = \text{constant}$$

$$B = \frac{A}{y} - my$$

wetted perimeter, $P = AD + DC + CB$

$$AD = \sqrt{(my)^2 + y^2} = y\sqrt{m^2 + 1}$$

$$P = y\sqrt{m^2 + 1} + B + y\sqrt{m^2 + 1}$$

$$P = B + 2y\sqrt{m^2 + 1}$$

keeping A and m constant for best section.

$$\frac{dp}{dy} = 0$$

$$\frac{dp}{dy} = -\frac{A}{y^2} - m + 2\sqrt{m^2 + 1} = 0$$

$$A_e = \left(2\sqrt{1 + m^2} - m\right)y_e^2$$

from A_e we get the value of B_e

$$(B + my_e)y_e = \left(2\sqrt{1+m^2} - m\right)y_e^2$$

we get

$$B_e = 2y_e\left(\sqrt{1+m^2} - m\right)$$

$$P_e = B_e + 2y_e\sqrt{m^2+1}$$

Putting value of B_e

$$P_e = 2y_e\left(\sqrt{1+m^2} - m\right) + 2y_e\sqrt{m^2+1}$$

$$P_e = 2y_e\left(2\sqrt{1+m^2} - m\right)$$

$$R_e = \frac{A_e}{P_e} = \frac{\left(2\sqrt{1+m^2} - m\right)y_e^2}{2y_e\left(2\sqrt{1+m^2} - m\right)}$$

$$R_e = \frac{y_e}{2}$$

In above case m is constant if m is allows to vary. The optimum value of m to take P_e most efficient is obtained by putting $\frac{dP_e}{dm} = 0$

$$P_e = 2y_e\left(2\sqrt{1+m^2} - m\right) \qquad ...(2.12)$$

$$A_e = \left(2\sqrt{1+m^2} - m\right)y_e^2 \qquad ...(2.13)$$

from (2.13)

$$y_e = \frac{\sqrt{A_e}}{\sqrt{2\left(\sqrt{1+m^2} - m\right)}} \qquad ...(2.14)$$

Put the value of equation (2.14) in equation (2.12)

$$P_e = \frac{2\sqrt{A_e}}{\sqrt{2\left(\sqrt{1+m^2} - m\right)}} \times \left(2\sqrt{1+m^2} - m\right)$$

$$P_e = 2\sqrt{A\left(2\sqrt{1+m^2}-m\right)} \qquad ...(2.15)$$

Applying condition $\dfrac{dp_e}{dm} = 0$ to equation (2.15)

we get
$$m_{em} = \frac{1}{\sqrt{3}}\cot\theta$$
$$\theta_{em} = 60°$$

where suffic *em* denotes most efficient channel.

$$P_{em} = 2y_{em}\left(2\sqrt{1+\frac{1}{3}}-\frac{1}{\sqrt{3}}\right) = 2\sqrt{3}y_{em}$$

$$B_{em} = 2y_{em}\left(\sqrt{1+\frac{1}{3}}-\frac{1}{\sqrt{3}}\right) = \frac{2}{\sqrt{3}}y_{em}$$

$$A = \left(2\sqrt{1+\frac{1}{3}}-\frac{1}{\sqrt{3}}\right)y_{em}^2 = \sqrt{3}y_{em}^2$$

Proportion of some most efficient sections

Channel Shape	***A***	P_{em}	B_{em}	R_{em}	T_{em}
1. Rectangle (half square)	$2y_{em}^2$	$4y_{em}$	$2y_{em}$	$\frac{y_{em}}{2}$	
2. Trapezoidal (half regular hexagen)	$\sqrt{3}y_{em}^2$	$2\sqrt{3}y_{em}$	$\frac{2}{\sqrt{3}}y_{em}$	$\frac{y_{em}}{2}$	$\frac{4}{\sqrt{3}}y_{em}$
3. Circular (semicircle)	$\frac{\pi}{2}y_{em}^2$	πy_{em}	—	$\frac{y_{em}}{2}$	$2y_{em}$
4. Triangular (vertex angle 90°)	y_{em}^2	$2\sqrt{2}y_{em}$	—	$\frac{y_{em}}{2\sqrt{2}}$	$2y_{em}$

Circular Channel

For circular channel both area and wetted perimeter varies with depth. Thus in case of circular channel two separate condition are taken into account.

(*i*) Condition for maximum discharge through channel.

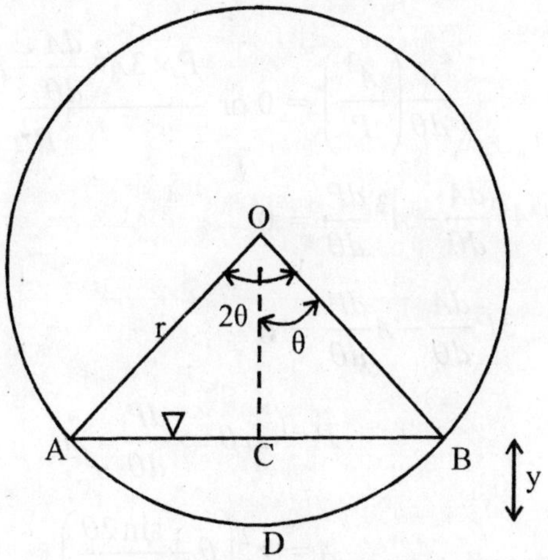

Fig. 2.15

Here $y =$ depth of flow

$r =$ radius of channel

$2\theta =$ angle subtended by water surface AB at the centre in radians.

wetted perimeter = length of arc AB

$$= \frac{2\pi r}{2\pi} \times 2\theta = 2r\theta$$

wetted area = area of $ADBA$

= Area of sector $\Delta ADBO$ – area of $\Delta\ OAB$

$$= \frac{\pi r^2}{2\pi} \times 2\theta - \frac{1}{2} AB \times CO$$

$$= r^2\theta - \frac{1}{2} \times 2BC \times CO$$

$$= r^2\theta - \frac{1}{2}\times 2\times r\sin\theta\times r\cos\theta$$

$$A = r^2\left(\theta - \frac{\sin 2\theta}{2}\right)$$

For maximum discharge condition

$$\frac{d}{d\theta}\left(\frac{A^3}{P}\right) = 0 \text{ or } \frac{P\times 3A^2\dfrac{dA}{d\theta} - A^3\dfrac{dP}{d\theta} = 0}{P^2}$$

$$\Rightarrow \quad 3PA^2\frac{dA}{d\theta} - A^3\frac{dP}{d\theta} = 0$$

or

$$3P\frac{dA}{d\theta} - A\frac{dP}{d\theta} = 0$$

$$P = 2r\theta \Rightarrow \frac{dP}{d\theta} = 2r$$

$$A = r^2\left(\theta - \frac{\sin 2\theta}{2}\right)$$

$$\frac{dA}{d\theta} = r^2(1-\cos 2\theta)$$

Substituting the value of P, A, $\dfrac{dP}{d\theta}$ and $\dfrac{dA}{d\theta}$ in equation

$$3\times 2r\theta\times r^2(1-\cos 2\theta) - r^2\left(\theta - \frac{\sin 2\theta}{2}\right)2r = 0$$

$$6r^3\theta(1-\cos 2\theta) - 2r^3\left(\theta - \frac{\sin 2\theta}{2}\right) = 0$$

Dividing by $2r^3$ we get

$$3\theta(1-\cos 2\theta) - \left(\theta - \frac{\sin 2\theta}{2}\right) = 0$$

$$2\theta - 3\theta\cos\theta + \frac{\sin 2\theta}{2} = 0$$

On solving we get $\quad 2\theta = 308$

$$\theta = \frac{308}{2} = 154°$$

Depth of flow for maximum discharge

$$y = r(1-\cos\theta)$$

$$= r(1-\cos 154°) = 1.9r = 0.95d$$

$$y = 0.95\ d \text{ where } d \text{ is diameter}$$

Hydraulic radius for maximum discharge

$$R = \frac{A}{P} = \frac{r^2\left(\theta - \frac{\sin 2\theta}{2}\right)}{2r\theta} = \frac{r}{2\theta}\left(\theta - \frac{\sin 2\theta}{2}\right)$$

where $\theta = 154° = 154 \times \frac{\pi}{180} = 2.687$ radian

$$R = \frac{r}{2\times 2.687}\left(2.687 - \frac{\sin 308°}{2}\right)$$

$$R = 0.573\ r = 0.29\ d$$

(*b*) Condition for maximum velocity

For maximum velocity $\frac{d}{d\theta}\left(\frac{A}{P}\right) = 0$

where A and P are the function of θ

$$\frac{P\frac{dA}{d\theta} - A\frac{dP}{d\theta}}{P^2} = 0 \Rightarrow P\frac{dA}{d\theta} - A\frac{dP}{d\theta} = 0$$

$$= r^2\left(\theta - \frac{\sin 2\theta}{2}\right)$$

$$\frac{dA}{d\theta} = r^2\left(1 - \frac{\cos 2\theta}{2}\times 2\right)$$

$$= r^2(1-\cos 2\theta)$$

$$P = 2r\theta \Rightarrow \frac{dP}{d\theta} = 2r$$

Substituting value of A, P, $\frac{dA}{d\theta}$ and $\frac{dP}{d\theta}$ in

$$2r\theta\left[r^2(1-\cos 2\theta)\right] - r^2\left(\theta - \frac{\sin 2\theta}{2}\right)2r = 0$$

or $$2r^3\theta(1-\cos 2\theta) - 2r^3\left(\theta - \frac{\sin 2\theta}{2}\right) = 0$$

or $$\theta(1-\cos 2\theta) - \left(\theta - \frac{\sin 2\theta}{2}\right) = 0$$

or $$\theta - \theta\cos 2\theta - \theta + \frac{\sin 2\theta}{2} = 0$$

or $$\theta\cos 2\theta = \frac{\sin 2\theta}{2}$$

or, $$\tan 2\theta = 2\theta$$

Solution gives $2\theta = 257.5°$ or $\theta = 128.75°$.

Depth of flow $= r(1-\cos\theta) = r(1-\cos 128.75°)$.

$$y = 1.62r = 0.81d$$

$$\boxed{y = 0.81\ d}$$

$$R = \frac{A}{P} = \frac{r^2\left(\theta - \frac{\sin 2\theta}{2}\right)}{2r\theta} = \frac{r}{2\theta}\left(\theta - \frac{\sin 2\theta}{2}\right)$$

$$\theta = 128.75° = 128.75 \times \frac{\pi}{180}$$

$$= 2.247 \text{ radians}$$

$$R = \frac{r}{2 \times 2.247}\left(2.247 - \frac{\sin 257.5°}{2}\right)$$

$$R = 0.6086r = 0.305d$$

$$R = \boxed{0.305\ d}$$

Solved Problems

Problem 1: ***A trapezoidal channel is 11.0 m wide and has a side slope of 1.4 horizontal : 1 vertical. The bed slope is 0.0004. The channel is lined with smooth concrete of n = 0.012. Compute the mean velocity and discharge for a depth of flow of 3.0 meters.***

Solution:

Let y_0 = uniform depth flow.

Here B = 11.0 m and side slope m = 1.4

Area $A = (B + my)\, y$

$$= (11.0 + 1.4 \times 3)3.0 = 45.6\text{m}^2$$

wetted perimeter $P = B + 2y\sqrt{m^2 + 1}$

$$= 11.0 + 2 \times 3\sqrt{1.96 + 1}$$

$$= 21.323 \text{ m}$$

Hydraulic radius $R = \dfrac{A}{P} = 2.139$ m

Mean velocity $V = \dfrac{1}{n} R^{2/3}\, S_0^{\,1/2}$

$$= \frac{1}{.012} \times (2.139)^{2/3} (.0004)^{1/2}$$

$$= 2.77 \text{ m/s}$$

Discharge $Q = AV = 126.312\ \text{m}^3/\text{s}$

Problem 2: ***A triangular channel with an apex angle of 75° carries flow of 1.2 m³/sec at a depth of 0.80 meters. If the bed slope is 0.007, find the roughness coefficient of the channel.***

Solution:

y_0 = normal depth = 0.80 m

Area of triangular channel $= \frac{1}{2} \times \text{base} \times \text{height}$

$$= \frac{1}{2} \times 0.80 \times 2 \times 0.8 \tan \frac{75}{2}$$

$$= 0.491 \text{ m}^2$$

wetted perimeter $\quad P = 2 \times 0.8 \times \sec 37.5°$

$$= 2.0168 \text{ m}$$

Hydraulic radius $\quad R = A/P = 0.243 \text{ m}$

Roughness coefficient $\quad n = \dfrac{AR^{2/3} S_0^{\ 1/2}}{Q}$

$$= \frac{(0.491) \times (0.243)^{2/3} (.007)^{1/2}}{1.20}$$

$$= 0.0133.$$

Problem 3*: *For a flow in a rectangular channel of width 5.0 m and a depth of 2.4 m the Darcy-Weisbach friction factor f is estimated to be 0.03. Estimate the values of Chezy's C and Manning's n.

***Solution*:** The coefficient C, n and friction factor f are related as:

$$C = \sqrt{\frac{8g}{f}} = \frac{1}{n} R^{1/6}$$

Area $\quad A = 5.0 \times 2.4 = 12.0 \text{ m}^2$

Wetted perimeter $\quad P = 5.0 + (2 \times 2.40) = 9.8 \text{ m}$

Hydraulic radius $\quad R = A/P = 12/9.8 = 1.224 \text{ m}$

$$C = \sqrt{\frac{8 \times 9.81}{0.03}} = 51.15$$

Also $\quad \frac{1}{n} \cdot (1.224)^{1/6} = C = 51.15$

or $\quad n = 0.0202$

Problem 4: ***A concrete lined trapezoidal channel (n = 0.012) is to have a side slope of 1.0 horizontal: 1 vertical. The bottom slope is to be 0.0004. Find the bottom width of the channel of carry 80 m³/s of discharge at a normal depth of 2.10 m.***

Solution: Let B = bottom width. Here y_0 = normal depth = 2.10 m, m = 1.0

Area $\quad A = (B + 2.10) \times 2.10 \text{ m}^2$

Wetted perimeter $\quad P = \left(B + 2\sqrt{2} \times 2.10\right)$

$$= (B + 5.940) \text{ m}$$

$$\frac{Qn}{\sqrt{S_0}} = \frac{80 \times 0.012}{\sqrt{0.0004}} = 48 = AR^{2/3}$$

$$= \frac{\left[(B + 2.10) \times 2.10\right]^{5/3}}{(B + 5.940)^{2/3}} = 48.0$$

By trial and error $\quad B = 13.97$ m

Problem 5: ***A trapezoidal channel is 11.0 m wide and has a side slope of 1.4 horizontal : 1 vertical. The channel is lined with smooth concrete of n = 0.012. Find the bottom slope necessary to carry only 48 m³/sec of discharge at a depth of 3.0 m.***

Solution:

Area $\quad A = (B + my)\, y$

$$= (11.0 + 1.4 \times 3) \times 3$$

$$= 45.6 \text{ m}^2$$

Wetted perimeter $\quad P = B + 2y\sqrt{m^2 + 1}$

$$= 11.0 + 2 \times 3\sqrt{1.96 + 1}$$

$$= 21.323 \text{ m}$$

Hydraulic radius $\quad R = A/P = 2.139$ m

$$S_0 = \frac{Q^2 n^2}{A^2 \ R^{4/3}}$$

$$= \frac{(48.0)^2 \times (.012)^2}{(45.6)^2 (2.139)^{4/3}}$$

$$= 0.0000579$$

Problem 6: A 4.0 m wide trapezoidal channel having a side slope of 1.5 horizontal : 1 vertical is laid on a slope of 0.00030. The roughness coefficient n = 0.012. Find the normal depth for a discharge of 20 m³/s through this channel.

Solution:

Let y_0 = normal depth

Area $A = (4.0 + 1.5 \times y_0) \times y_0$

Wetted perimeter $P = 4.0 + 2\sqrt{3.25}y_0$

$$R = \frac{A}{P}$$

$$= \frac{(4.0 + 1.5y_0)\, y_0}{4.0 + 2\sqrt{3.25}y_0}$$

$$\frac{Qn}{\sqrt{S_0}} = AR^{2/3}$$

$$\frac{20 \times .012}{\sqrt{.00030}} = \frac{(4.0 + 1.5y_0)^{5/3} \ {y_0}^{5/3}}{(4.0 + 3.606y_0)^{2/3}}$$

Algebrically y_0 can be found from the above equation by trial and error method. The normal depth is found to be 1.844 m.

Problem 7: What size of circular drainage pipe is needed to carry 2.10 m³/s of discharge when flowing half-full. The pipe is laid at a slope of 0.0003 and 'n' for the pipe can be taken as 0.016.

Solution:

Let D = diameter of the pipe

y = depth of flow $D/2$

Area $$A = \frac{1}{2} \times \frac{\pi D^2}{4} = \frac{\pi}{8} D^2$$

Wetted perimeter $$P = \frac{\pi}{2} D$$

Hydraulic radius $R = A/P = D/4$

By Manning's formula, discharge

$$Q = \frac{1}{n} A R^{2/3} S_o^{1/2}$$

$$2.10 = \frac{1}{.016}\left(\frac{\pi}{8} D^2\right)\left(\frac{D}{4}\right)^{2/3} (0.0003)^{1/2}$$

$$D = 2.574 \text{ m}$$

Problem 8: A trapezoidal channel has a bed width of 3.0 m and side slopes of 1:1. The bottom slope of the channel is 0.0030. If a discharge of 25 m³/s passes in this channel at a depth of 1.25 m, estimate the value of Chezy coefficient C.

Solution:

Area $$A = \left[3.0 + (1 \times 1.25) \times 1.25\right]$$

$$= 5.3125 \text{ m}^2$$

Velocity $$V = \frac{25}{5.3125} = 4.71 \text{ m/s}$$

Wetted perimeter $$P = B + 2y\sqrt{m^2 + 1}$$

$$= 3.0 + \left(2 \times 1.25 \times \sqrt{2}\right)$$

$$= 6.5355 \text{ m}$$

$$S_0 = 0.0030$$

Hydraulic Radius $R = \dfrac{A}{P} = \dfrac{5.3125}{6.5355}$

$= 0.8129$ m

By Chezy formula $V = C\sqrt{RS_0}$

$C = 95.37$

Problem 9: ***A rectangular channel 3.5 m wide is laid on a slope of 0.0003. Calculate the normal depth of flow for a discharge of 5.0 m³/s in this channel. The Manning's coefficient 'n' can be taken as 0.01.***

Solution:

Let $y_0 =$ normal depth

Area $A = 3.5 y_0 \text{ m}^2$

Wetted perimeter $P = 3.5 + 2y_0$ m

Hydraulic Radius $R = \dfrac{3.5 y_0}{3.5 + 2y_0}$ m

By Manning's formula $Q = \dfrac{1}{n} A R^{2/3} S_0^{1/2}$

$$5.0 = \frac{1}{0.01} \times \frac{(3.5y_0)(3.5y_0)^{2/3}}{(3.5+2y_0)^{2/3}} \times (0.0003)^{1/2}$$

$$\Rightarrow \quad \frac{(3.5y_0)^{5/3}}{(3.5+2y_0)^{2/3}} = 2.887 \Rightarrow y_0 = 1.080 \text{ m}$$

Problem 10: ***A circular drainage pipe 0.80 m in diameter conveys a discharge at a depth of 0.30 m. If the pipe is laid an a slope of 1 in 800, estimate the discharge. Manning's n = 0.012.***

Solution: Referring to Figure 2.16:

$D = 0.80$ m

$y = 0.30$ m

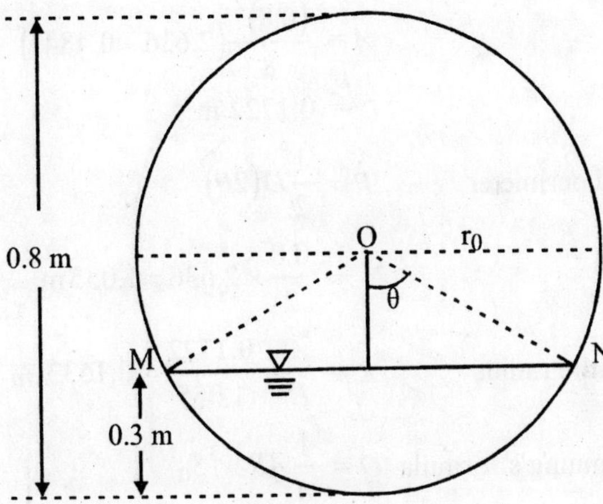

Fig. 2.16

Area of flow section

A = Area of sector

OMN – Area of triangle OMN

$$= \frac{1}{2} r_0^2 .2\theta - \frac{1}{2}(2r_0 \sin\theta) r_0 \cos\theta$$

$$= \frac{D^2}{8}(2\theta - \sin 2\theta)$$

Also, $$\cos\theta = \frac{\left(\frac{D}{2} - y\right)}{\left(\frac{D}{2}\right)} = 1 - \frac{2y}{D}$$

Hence $$2\theta = 2\cos\left(1 - \frac{2y}{D}\right)$$

$$= 2\cos\left(1 - \frac{2\times 0.3}{0.8}\right)$$

$$= 2.636 \text{ rad.}$$

$$\sin 2\theta = 0.4841$$

Area $$A = \frac{(0.8)^2}{8}(2.636 - 0.4841)$$
$$= 0.1722 \text{ m}^2$$

Wetted perimeter $$P = \frac{1}{2}D(2\theta)$$
$$= \frac{0.8}{2} \times 2.636 = 1.055\text{m}$$

Hydraulic radius $$R = \frac{A}{P} = \frac{0.1722}{1.055} = 0.1633 \text{ m}$$

By Manning's formula $$Q = \frac{1}{n}AR^{2/3}S_0^{\ 1/2}$$
$$= \frac{1}{0.012} \times 0.1722 \times (.1633)^{2/3}$$
$$\times \left(\frac{1}{800}\right)^{1/2}$$
$$= 0.152 \text{ m}^3\text{/s}$$

Problem 11: ***A trapezoidal channel is to be designed to convey 75 m³/s of water at a velocity of 3.0 m/s. The bed width to depth ratio is to be 8.0 and the slope are to be 1 horizontal and 1 vertical. It will be lined with a material whose n = 0.03. Calculate the bed width, depth of flow and slope of the channel.***

Solution:

$$m = 1.0, Q = 75 \text{ m}^3\text{/s}$$
$$V = 3.0 \text{ m/s}$$
$$A = \frac{Q}{V} = \frac{75}{3} = 25 \text{ m}^2$$

But $$A = (B + my)y$$
$$= (8y + y)y$$
$$= 9y^2 = 25.0$$
$$y = 1.667 \text{ m}$$

$$B = 8y = 13.33 \text{ m}$$

Wetted perimeter $\quad P = B + 2y\sqrt{1+m^2}$

$$= 13.33 + 2 \times 1.667 \times \sqrt{2}$$

$$= 18.047$$

$$R = \frac{A}{P} = \frac{25}{18.047} = 1.385 \text{ m}$$

By Manning's formula $\quad V = \frac{1}{n} R^{2/3} \sqrt{S_0}$

$$3.0 = \frac{1}{0.03}(1.385)^{2/3} \sqrt{S_0}$$

$$S_0 = 0.00525$$

***Problem 12**: A trapezoidal channel having a cross-sectional area A_1, wetted perimeter P_1, Manning's coefficient n and laid to a slope S, carries a certain discharge Q_1 at a depth of flow equal to d. To increase the discharge the base width of the channel is widened by n. Keeping all other parameters, S, d, side slopes and n same. Prove that*

$$\left[\frac{Q_2}{Q_1}\right]^3 \cdot \left[1 + \frac{x}{P_1}\right]^2 = \left[1 + \frac{xd}{A_1}\right]^5$$

where Q_2 is the new discharge in the channel.

Solution: Let b be the initial base width of the channel and the side slope be z horizontal to 1 vertical. Then

$$A_1 = (b + zd)d$$

$$P_1 = b + 2d\sqrt{1+z^2}$$

when the base width is increased by x, the new cross-sectional area A_2 and wetted perimeter P_2 are given by

$$A_2 = (b + x + zd)d = A_1 + xd \qquad ...(i)$$

$$P_2 = \left(b + x + zd\sqrt{1+z^2}\right) = P_1 + x \qquad ...(ii)$$

Discharge Q_1 is given by

$$Q_1 = \frac{1}{n} A_1 \left(\frac{A_1}{P_1}\right)^{2/3} \sqrt{S_0} \qquad ...(iii)$$

or

$$Q_1 = \frac{1}{n} \frac{A_1^{5/3}}{P_1^{2/3}} \sqrt{S_0}$$

Similarly discharge Q_2 is given by

$$Q_2 = \frac{1}{n} A_2 \left(\frac{A_2}{P_2}\right)^{2/3} \sqrt{S_0}$$

$$Q_2 = \frac{1}{n} \frac{A_2^{5/3}}{P_2^{2/3}} \sqrt{S_0} \qquad ...(vi)$$

From the equations (*iii*) and (*iv*) we get

$$\left[\frac{Q_2}{Q_1}\right]^3 = \left[\frac{A_2}{A_1}\right]^5 \left[\frac{P_1}{P_2}\right]^2$$

By substituting the value of A_2 and P_2 from equations (*i*) and (*ii*) we get

$$\left[\frac{Q_2}{Q_1}\right]^3 = \left[\frac{A_1 + xd}{A_1}\right]^5 \left[\frac{P_1}{P_1 + x}\right]^2$$

or

$$\left[\frac{Q_2}{Q_1}\right]^3 \left[1 + \frac{x}{P_1}\right]^2 = \left[1 + \frac{xd}{A_1}\right]^5$$

Problem* 13: *An earth canal in good condition is 17 m wide at the bottom and has side slope of 2 horizontal to 1 vertical on side slope extend to a height of 7.8 m above the bottom level and the other side extend to an elevation of 1.8 m, then extend to a

flat distance of 150 m and rises vertically. If the slope of the canal is 1 m per 900 m, estimate the discharge when the depth of water is 2.5 m. Assume C (Chezy's) = 55.

Solution:

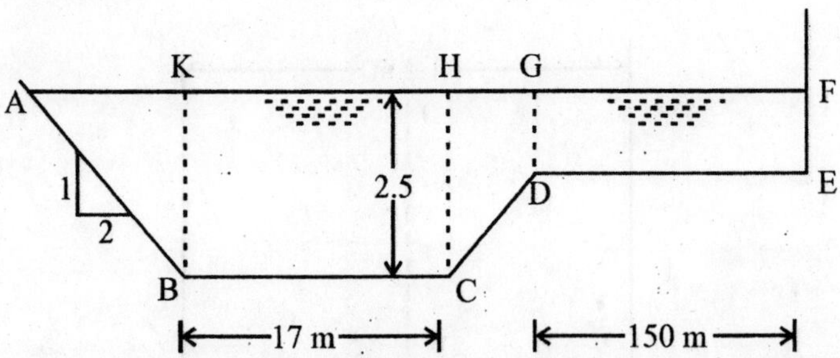

Fig. 2.17

As shown in the above figure the area of flow section of the channel is

$$A = ABK + KBCH + HCDG + GDEF$$

$$= \left(\frac{1}{2} \times 5 \times 2.5\right) + (17 \times 2.5) + \left[\frac{2.5 + 0.7}{2} \times 3.6\right] + (150 \times 0.7)$$

$$= 159.51 \text{ m}^2$$

wetted perimeter

$$P = AB + BC + CD + DE + EF$$

$$= 2.5\sqrt{5} + 17 + 1.8\sqrt{5} + 150 + 0.70$$

$$= 177.31 \text{ m}$$

Thus $$R = A/P = 0.90 \text{ m}$$

Using Chezy's formula

$$Q = AC\sqrt{S_0 \times R}$$

$$= 159.51 \times 55 \times \sqrt{0.90 \times \frac{1}{900}}$$

$$= 277.43 \text{ m}^3/\text{s}$$

***Problem 14**: A channel has a vertical wall 1.2 m apart and a semi-circular invert. If the centreline depth is 0.9 m and the bed slope is 1 in 2000, find the discharge using Chezy's formula C = 50.*

***Solution**:* As shown in the Figure 2.18.

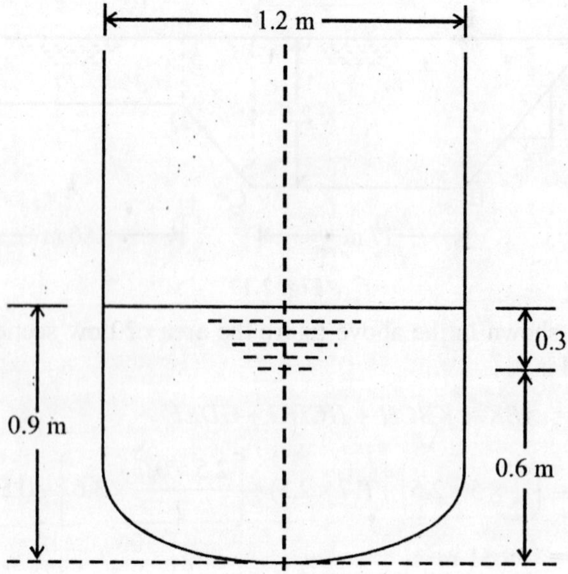

Fig. 2.18

$$A = \left[\frac{\pi \times 0.6^2}{2} + (1.2 \times 0.3)\right]$$

$$= 0.925 \text{ m}^2$$

Wetted perimeter $\quad P = (\pi \times 0.6 + 2 \times 0.3)$

$$= 2.485 \text{ m}$$

Thus, $\quad R = A/P = 0.372$ m

Using Chezy's formula, we get

$$Q = AC\sqrt{RS} = 0.925 \times 50 \times \sqrt{0.372} \times \sqrt{1/2000}$$

$$= 0.6307 \text{ m}^3/\text{s}$$

Problem* 15: *Irrigation channel is to carry a discharge of 18 m³/s with a velocity of 0.9 m/s and a bed slope of 1 in 2500. The side slope are 1 to 1. Find the depth and bottom width. The value of Chezy for this channel for different value of hydraulic radius are tabulated as

Hydraulic radius	***0.7***	***0.8***	***0.9***	***1.0***	***1.1***	***1.2***	***1.3***	***1.4***
Chezy's C	***34***	***35***	***37***	***38***	***39***	***40***	***41***	***41***

***Solution*:**

$$Q = 18 \text{ m}^3/\text{s}$$
$$V = 0.9 \text{ m/s}$$

Area of flow section

$$A = \frac{Q}{V} = \frac{18}{0.9} = 20 \text{ m}^2$$

From Chezy's formula

$$V = C\sqrt{RS_0}$$

By substitution $$0.9 = C\sqrt{R \times \frac{1}{2500}}$$

$$C\sqrt{R} = 45$$

Assuming $$R = 1.0, \text{ then } C = 38$$

$$C\sqrt{R} = 38$$

Assuming $$R = 1.3, \text{ then } C = 41$$

$$C\sqrt{R} = 46.75$$

Assuming $$R = 1.241, \text{ then } C = 40.41$$

$$C\sqrt{R} = 45$$

Which is same as the above

$$R = 1.241$$

We have

$$A = (B+y)y = 20.0 \text{ m}^2 \qquad \text{...}(i)$$

$$R = \frac{20}{B + 2y\sqrt{2}} = 1.241 \qquad \text{...(ii)}$$

from equation (*ii*) we have

$$B = 16.12 - 2.83\,y$$

Substituting the value of B in equation (*i*)

$$(16.12 - 2.83y + y)\,y = 20.0$$

$$16.12y - 1.83y^2 = 20.0$$

$$y^2 - 8.809y + 10.93 = 0$$

$$y = 1.494 \text{ m}$$

Substituting the value of y in equation (*i*) we get

$$B = 11.89 \text{ m}$$

Hence the required dimension of the channel are

$$B = 11.89 \text{ m};\ y = 1.63 \text{ m}$$

Problem 16: ***A lined channel (n = 0.012) is of trapezoidal section with one side vertical and the other on a side slope 1.5 horizontal : 1 vertical. If the channel is to deliver 8.0 m³/s which laid on a slope of 0.00013. Calculate the dimension of the efficient section which require minimum lining. Also calculate the corresponding mean velocity.***

Solution:

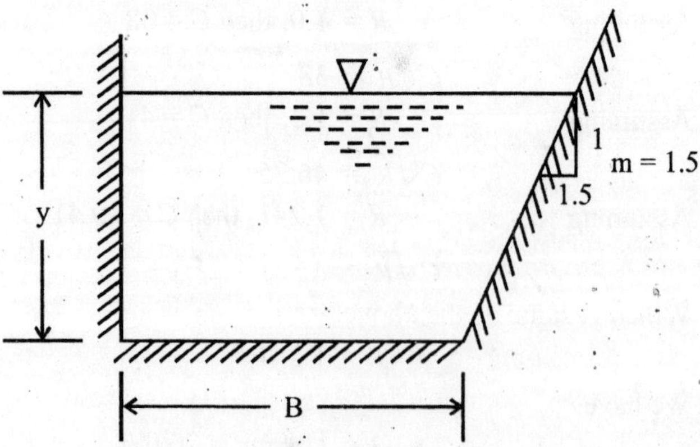

Fig. 2.19

Referring to Fig. 2.19,

$$A = \left(B + \frac{my}{2}\right)y$$

$$B = \frac{A}{y} - \frac{m}{2}$$

wetted perimeter $\quad P = B + y + \sqrt{m^2+1}y$

$$= \frac{A}{y} + \left[1 - \frac{m}{2} + \sqrt{m^2+1}\right]y$$

For minimum perimeter $\frac{dp}{dy} = 0$

$$-\frac{A}{y_e^2} + 1 - \frac{m}{2} + \sqrt{m^2+1} = 0$$

$$A = \left[1 - \frac{m}{2} + \sqrt{m^2+1}\right]y_e^2$$

Putting $m = 1.5$, $A = \left(1 - 0.75 + \sqrt{3.25}\right)y_e^2$

$$= 2.0528\ y_e^2$$

$$B_e = \frac{A}{y_e} - \frac{m}{2}y_e$$

$$= 1.3028\ y_e$$

$$P_e = 1.3028y_e + y_e + \sqrt{3.25}y_e$$

$$= 4.1056\ y_e$$

$$R_e = A_c / P_e = 0.5\ y_e$$

By Manning's formula $Q = \frac{1}{n}R^{2/3}AS_0^{1/2}$

$$\Rightarrow \qquad 8.0 = \frac{1}{0.012}(2.0528y_e^2)(y_e/2)^{2/3}$$

$$\times(0.00013)^{1/2}$$

$$y_e = 2.019$$
$$B_e = 1.3028\, y_e = 2.630$$
$$A_e = 2.0528\, y_e^2 = 8.368$$

Mean velocity $$V = \frac{Q}{A_e}$$
$$= 0.956 \text{ m/s}$$

Problem* 17: *A trapezoidal channel to carry 110 m³/min of water is designed to have a minimum cross-section. Find the bottom width and depth if the bed slope is 1/1500; the side slope are at 45° to the horizontal and Chezy coefficient C = 50.

***Solution*:** For cross-section to be minimum, the cross-section will have the proportion of the efficient section. Thus

$$R = \frac{y_e}{2}$$

and $$B_e + 2my_e = 2y_e\sqrt{m^2+1}$$

$$\text{for } m = 1.0,\ B + 2y_e = 2\sqrt{2}y_e$$

$$Be = 2\sqrt{2}y_e - 2y_e = 0.828y_e$$

Area $$A = (B_e + my_e)y_e = 1.828\, y_e^2$$
$$R = y_e/2$$

Discharge $$Q = \frac{110}{60} = 1.833 \text{ m}^3\text{/s}$$

$$Q = AC\sqrt{RS_0}$$

$$1.833 = 1.828y^2 \times 50 \times \sqrt{0.5y_e \times 1/1500}$$

$$y_e = 1.038 \text{ m}$$

then $$B_e = 0.828 \times 1.038$$
$$= 0.860 \text{ m}$$

Problem* 18: *A trapezoidal channel with side slopes of 2 horizontal : 1 vertical has to be designed to carry 20 m³/s at a slope of 1/6000. Determine the dimension of the efficient section. Assume Manning's n = 0.012.

***Solution*:** For efficient trapezoidal section

$$R = y_e/2$$

$$B + 2my_e = 2y_e\sqrt{m^2 + 1}$$

Here $m = 2.0$ and hence

$$B_e + 4y_e = 2\sqrt{5}y_e$$

$$B_e = 0.4721\ y_e$$

Area $$A = (B_e + my_e)\ y_e$$

$$= (0.4721 + 2)\ y_e^2$$

$$= 2.4721\ y_e^2$$

$$R = \frac{y_e}{2}$$

Discharge by Manning's formula

$$Q = \frac{1}{n} R^{2/3} A S_0^{1/2}$$

$$20 = \frac{1}{0.012}(2.4721 y_e^2)(y_e/2)^{2/3} \times \frac{1}{6000}$$

$$y_e = 2.534\ \text{m}$$

Bottom width $$Be = 2.534 \times 0.4721$$

$$= 1.196\ \text{m}$$

Problem* 19: *Obtain the value of N for (a) a wide rectangular channel, (b) triangualr channels.

***Solution*:**

(*a*) for wide rectangular channel considering unit width $A = y$

We know for wide rectangular channel

$$R = y$$
$$K^2 = C_2 y^N$$
$$K = 1/n\, AR^{2/3}$$
$$= 1/n\,(y).(y)^{2/3}$$
$$K^2 = 1/n^2\,(y^2).(y)^{4/3} = C_2 y^N$$

By equating exponent y on both sides $y^{2+4/3} = y^N \Rightarrow y^{10/3} = y^N$

We get the value of $N = 3.33$

(*b*) For triangular channel of side slope m horizontal : 1 vertical

$$A = my^2,\ P = 2y\sqrt{m^2+1}$$

$$R = \frac{A}{P}$$

$$= \frac{my^2}{2y\sqrt{m^2+1}} = \frac{m}{2\sqrt{m^2+1}}.y$$

$$K^2 = C_2 y^N$$

also
$$K = \frac{1}{n} AR^{2/3}$$

$$K^2 = \frac{1}{n^2}(my^2)^2\left[\frac{m}{2\sqrt{m^2+1}}y\right]^{4/3}$$
$$= C_2 y^N$$

Equating the power of y on both side

$$y^{4+4/3} = y^N$$
$$y^{16/3} = y^N$$

So
$$N = 5.33$$

Problem 19: ***It is required to convey 8 m³/s at a mean velocity of 1 m/s. Calculate the dimensions of the most efficient section of the channel whose shape is***

(a) Rectangular

(b) Triangular

(c) Trapezoidal

(d) Circular

***Solution*:**

Discharge $Q = 8.0 \text{ m}^3/\text{s}$

Velocity $V = 1.0 \text{ m/s}$

Area $A = 8.0/1.0 = 8.0 \text{ m}^2$

(*a*) For the most efficient rectangular channel

$$A = 2y_e^2$$

$$8 = 2y_e^2 \qquad (\therefore y_e = 2.0 \text{ m})$$

Width $B_e = \dfrac{8.0}{2.0} = 4.0m$

perimeter $P_e = (4.0 + 2 \times 2.0) = 8.0 \text{ m}$

(*b*) For the most efficient triangular section $2\theta = 90°$, *i.e.*, side slope = 1.0 horizontal : 1 vertical or $m = 1.0$

$$A = y_e^2$$

$$8 = y_e^2 \qquad (\therefore y_e = 2.828 \text{ m})$$

Perimeter $P_e = 2\sqrt{2}y_e$

$$= 2\sqrt{2} \times 2.828 = 8.0 \text{ m}$$

(*c*) For most efficient trapezoidal section: side slope are inclined at 60° to the horizontal, or $1/\sqrt{3}$ horizontal : 1 vertical, *i.e.*, $m = 1/\sqrt{3}$

$$A = \sqrt{3}\, y_e^2$$

$$8 = \sqrt{3}\, y_e^2$$

$$y_e = 2.149 \text{ m}$$

Bottom width $B_e = \dfrac{2}{\sqrt{3}} y_e$

$$= 2.482 \text{ m}$$

Perimeter $$P = \frac{2}{\sqrt{3}} y$$

$$= 2\sqrt{3} \times 2.149$$

$$= 7.445 \text{ m}$$

(*d*) Circular section : The most efficient circular section is a semi-circle. Hence

$$A = \frac{\pi}{2} y_e^2$$

$$8 = \frac{\pi}{2} y_e^2$$

$$y_e = 2.257 \text{ m}$$

Perimeter $$P_e = \pi y_e = \pi \times 2.257$$

$$= 7.090 \text{ m}$$

***Problem 20:** A compound channel is symmetrical in cross-section and has the following geometric properties.*

Main channel: Trapezoidal cross-section, Bottom width = 15.0 m, Slide Slope = 1.5 H : 1 V, Bank full depth = 3.0 m, Manning's coefficient 0.04 longitudinal slape 0.0008 Flood plains: width = 75 m, side slope = 1.5 H : 1 V, Manning coefficient = 0.06, longitudinal slope 0.0008. Calculate the uniform flow discharges for a flow with total depth of 4.2 m.

Solution: The Fig. 2.20 is given below:

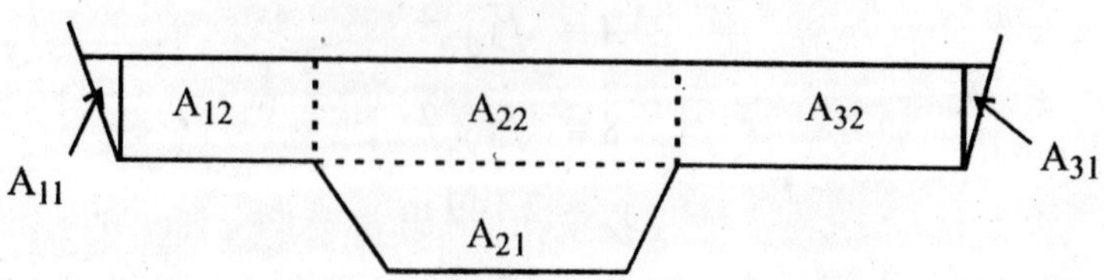

Fig. 2.20

The channel section is considered divided into three sub-section A_1, A_2 and A_3.

Table of Computation of Geometrical Properties *M*

Sub Area	*Area Element*	*Area* (m^2)		*Wetted Perimeter* (*m*)		
A_1	A_{11}	[0.5×1.2×(1.5×1.2)]	1.08	$1.2\times(1+1.5^2)^{0.5}$	2.163	
	A_{12}	75×1.2	90	75	75	
	Total		91.08		77.163	
A_2	A_{21}	[15+(1.5×3.0)]×30	58.5	$15+2\times3\times[1+1.5^2]^{0.5}$	25.817	1.180
	A_{22}	[15+(2×1.5×3)]×1/2	28.8		0	
	Total		87.3		25.817	3.382

Discharge by Manning's formula.

Sub Area A_1

$$Q_1 = \frac{1}{0.06} \times 91.08 \times (1.180)^{2/3} \times (0.0008)^{1/2}$$

$$= 47.94 \text{ m}^3/\text{s}$$

Sub Area A_2

$$Q_2 = \frac{1}{0.04} \times 87.3 \times (3.382)^{2/3} \times (0.0008)^{1/2}$$

$$= 139.085 \text{ m}^3/\text{s}$$

Sub Area A_3

$$Q_3 = Q_1 = 47.94 \text{m}^3/\text{s}$$

Total discharge $= Q_1 + Q_2 + Q_3 = 234.965 \text{ m}^3/\text{s}$

Problem 21: ***Analyse the maximum discharge in a circular channel.***

Solution: Circular channel is channel of second kind.

For maximum discharge

$$\frac{dQ}{dy} = 0$$

Assuming n = constant for all depths, for a constant

So $$\frac{d}{dy}(AR^{2/3}) = 0$$

$$\frac{d}{dy}\left(\frac{A^2}{P^2}\right) = 0 \qquad \left(\because R = \frac{A}{P}\right)$$

For circular channel $A = \dfrac{D^2}{8}(2\theta - \sin 2\theta)$

$P = D\theta$

for maximum discharge

$$\frac{d}{d\theta}\left(\frac{A^5}{P^2}\right) = 0$$

$$5P\left(\frac{dA}{d\theta}\right) - 2A\frac{dp}{d\theta} = 0$$

$$5D\theta.\frac{D^2}{8}(2 - 2\cos 2\theta) - 2.\frac{D^2}{8}(2\theta - \sin 2\theta)D = 0$$

$$3\theta - 5\theta\cos 2\theta + \sin 2\theta = 0$$

The solution of equation is obtained as 150°11'

$$\frac{y_0}{D} = \frac{1-\cos\theta}{2} = 0.938$$

Depth of flow for max discharge $= y_0 = 0.938D$

At $\dfrac{y_0}{D} = 0.935,\qquad \dfrac{AR^{2/3}}{D^{8/3}} = 0.3353$

Also when $\dfrac{y_0}{D} = 1.0 \qquad \dfrac{AR^{2/3}}{D^{8/3}} = 0.3117$

Hence if Q_f = Discharge with $y_0 = D$, *i.e.*, the pipe running just full, and

Q_m = maximum discharge then

$$\frac{Q_m}{Q_f} = \frac{0.3353}{0.3117} = 1.0757$$

Thus maximum discharge will be 7.6 percent more than the pipe full discharge.

Review Questions

1. What is uniform flow in open channel?
2. Derive Chezy's equation or channel of constant velocity for uniform flow condition in open channel.
3. On what factors the Manning coefficient '*n*' depends?
4. Draw velocity profile for the following:
   (*a*) Rectangular Channel
   (*b*) Trapezoidal Channel
   (*c*) Triangular Channel
   (*d*) Circular Channel
   (*e*) Wide and Narrow Open Channel
5. Draw velocity distribution in horizontal and vertical plane for rectangular channel.
6. Draw the velocity distribution for the following:
   (*a*) Velocity Distribution in Rough Channel
   (*b*) Velocity Distribution in Smooth Channel
7. Draw shear stress variation on bed and sides for rectangular as well as trapezoidal channel?
8. What do you mean by equivalent roughness?
9. Define the following:
   (*a*) Channel of First Kind
   (*b*) Channel of Second Kind
   (*c*) Conveyance of Channel
   (*d*) Most efficient Channel
   (*e*) Second Hydraulic exponent N
   (*f*) Compound Channels
10. Find the discharge in the following channels with a bed slope of 0.0004 and $n = 0.012$.

(a) Rectangular, $B = 2$ m, $y_0 = 1.8$ m.

(b) Trapezoidal, $B = 3.0$ m, $m = 2.1$, $y_0 = 1.10$ m.

(c) Triangular, $m = 1.5$, $y_0 = 2.5$ m.

11. A concrete lined trapezoidal channel ($n = 0.012$) is 7.0 m wide and has a side slope of 1H : 1V. The longitudinal slope is 0.008. Estimate the normal depth.

12. A rectangular channel is to be laid on a slope of 0.0002. The sides will be of smooth concrete ($n = 0.014$). What width of channel is necessary to carry a discharge of 8.0 m³/s with a normal depth of 2.4 m?

13. A rectangular channel having a width of 3.0 m and a slope of 0.0001 was guaged to determine its roughness coefficient. Given Q = 16 m³/s and $y = 2.0$ m, estimate value of 'n'.

14. A trapezoidal channel of 2.0 m bed width and side slope of 1H : 1V carries a full supply of 8.0 m³/s at a depth of 2.5 m. What would be the discharge at half of full supply depth *i.e.*, at 1.5 m? What would be the depth at half of full supply discharge?

15. Show that the normal depth in a hydraulically most efficient trapezoidal is given by

$$y_{em} = 0.9678\left(\frac{Qn}{\sqrt{S_0}}\right)^{3/8}$$

16. Show that the flow rate in a most efficient trapezoidal section is given by

$$Q = \frac{0.7435}{n} B^{8/3} \sqrt{S_0}$$

17. For a triangular cross-sectional canal, show that cross-section is most efficient when the sides make an angle of 45° with the vertical. Determine the hydrualic radius in term of depth of flow for such a channel.

18. Show that a hydraulically efficient parabolic section $(T = R\sqrt{y})$ will have the following relationship between hydraulic radius R_e, top width T_e and depth Y_e:

$$\left[\text{Take } P = T + \frac{8}{3}y^2/T\right]$$

$$R_e = \frac{y_e}{2} \text{ and } T_e = \sqrt{8}\,y_e$$

19. For the compound channel shown below estimate the discharge for a depth of flow of (*i*) 1.5 m and (*ii*) 2.5 m

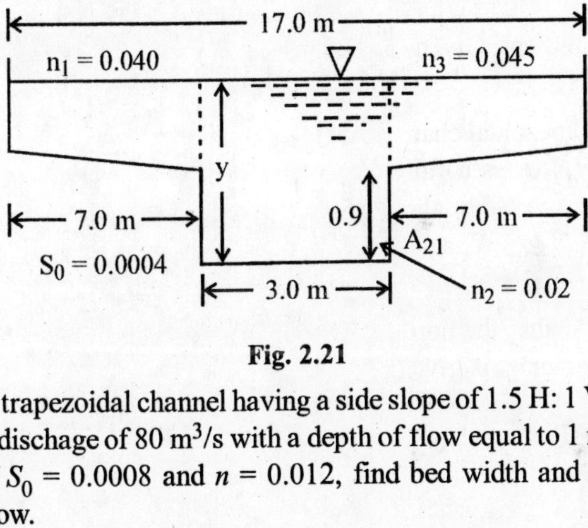

Fig. 2.21

20. A trapezoidal channel having a side slope of 1.5 H: 1 V carries a dischage of 80 m^3/s with a depth of flow equal to 1 m width. If S_0 = 0.0008 and n = 0.012, find bed width and depth of flow.
21. A rectangular channel (n = 0.014) carries a flow of 15 m^3/s with a depth of flow equal to the width of the channel. If S_0 = 0.0003, find the Froude number of the flow.
22. Show that the normal depth in a triangular channel of side slope m horizontal and 1 vertical in given by

$$y_0 = 1.1892\left[\frac{Qn}{\sqrt{S_0}}\right]^{3/8}\left[\frac{m^2+1}{m^5}\right]^{1/8}$$

23. Show that a hydraulically efficient triangular channel section has $R_e = \frac{y_e}{2\sqrt{2}}$.

24. A lined channel ($n = 0.012$) is of trapezoidal section with one side vertical and other side on a slope of 1H : IV. If the canal has to deliver 13 m^3/s when laid on a slope of 0.0003. Calculate the dimensions of the efficient section which required minimum of lining.

Unit 3

Gradually Varied Flow

3.1 INTRODUCTION

Uniform flow, discussed in the previous chapter, occurs where the depth, slope, cross-section and velocity remain constant over given length of channel. However, there are several disturbances that may cause the flow depth and velocity to change from one section to the other. Such flow is said to be varied or non-uniform flow.

Gradually varied flow is one of the category of varied or non-uniform flow in which the change in depth of flow takes place gradually in a long length of the channel.

3.2 EQUATION OF GRADUALLY VARIED FLOW

The following assumptions are made for analysing gradually varied flow:

1. The velocity distribution is fairly uniform over a cross-section.
2. The kinetic energy correction factor and momentum correction factor can be taken as unity.
3. The bed slope of channel is very small.
4. The pressure distribution over the channel section is hydrostatic.
5. The roughness coefficient is constant for the length of the channel and it does not depend upon depth of flow.
6. The Chezy and Manning equations are equally applicable to gradually varied flow.

3.3 DIFFERENTIAL EQUATION OF GVF

Let us consider gradually varied flow in a channel of small slope. The total energy is given by expression

$$H = Z + E = z + y + \frac{v^2}{2g} \quad \text{...(3.1)}$$

where H = total energy, E = Specific energy

Z = Datum head

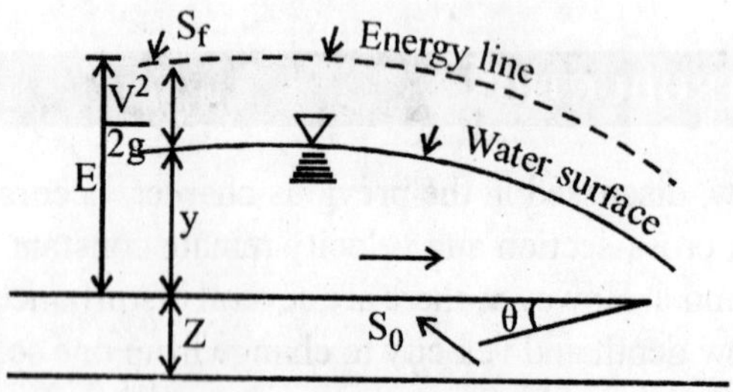

Fig. 3.1: Sketch of GVF

As we know from the definition of GVF that water surface varied in the longitudinal direction, therefore the depth of flow and total energy are function of x. Differentiating equation 3.1 with respect to x.

$$\frac{dH}{dx} = \frac{dz}{dx} + \frac{dE}{dx} \quad \text{...(3.2)}$$

$$\frac{dH}{dx} = \frac{dz}{dx} + \frac{dy}{dx} + \frac{d}{dx}\left(\frac{V^2}{2g}\right) \quad \text{...(3.3)}$$

$$\left(\because E = y + \frac{V^2}{2g}\right)$$

In equation (3.3), $\frac{dH}{dx}$ represents the energy slope, denoting it by S_f,

$\frac{dz}{dx}$ denotes the bottom slope, denoting it by S_0, $\frac{dy}{dx}$ represents water surface slope relative to bottom of the channel.

As we know total energy of the flow always decrease in the direction of motion. So we can take

$$\frac{dH}{dx} = -S_f \qquad ...(3.4)$$

Also, it is common to consider the channel slope with bed elevation decreasing in the downstream. So we can take,

$$\frac{dz}{dx} = -S_0 \qquad ...(3.5)$$

We can write $\frac{d}{dx}\left(\frac{V^2}{2g}\right)$ as

$$\frac{d}{dx}\left(\frac{V^2}{2g}\right) = \frac{d}{dy}\left(\frac{Q^2}{2gA^2}\right)\frac{dy}{dx} \qquad \left(\because V = \frac{Q}{A}\right)$$

$$= -\frac{Q^2}{gA^3}\frac{dA}{dy}.\frac{dy}{dx}$$

We know $\frac{dA}{dy} = T$ (Top width)

$$\frac{d}{dx}\left(\frac{V^2}{2g}\right) = \left(-\frac{Q^2T}{gA^3}\right)\frac{dy}{dx} \qquad ...(3.6)$$

Equation 3.3 can now be written as

$$-S_f = -S_0 + \frac{dy}{dx} - \left(\frac{Q^2T}{gA^3}\right)\frac{dy}{dx}$$

From this we can say

$$\frac{dy}{dx} = \frac{S_0 - S_f}{1 - \frac{Q^2T}{gA^3}} \qquad ...(3.7)$$

Equation 3.7 is basic differential equation of GVF also known as dynamic equation of GVF.

Equation 3.7 can be written in other form also we know

$$Q = K\sqrt{S}$$

where K = conveyance at any depth y

We can write
$$K = \frac{Q}{\sqrt{S_f}} \quad ...(3.8)$$

(By assumption of GVF)

Also
$$K_0 = \frac{Q}{\sqrt{S_0}} \quad ...(3.9) \text{ (Uniform flow)}$$

where K_0 = conveyance corresponding to normal depth.

from equation 3.8 and 3.9, we conclude

$$\frac{S_f}{S_0} = \frac{K_0^2}{K^2} \quad ...(3.10)$$

In the same manner if Z = section factor at depth y and Z_c = section factor at critical depth, y_c.

$$Z^2 = \frac{A^3}{T} \quad ...(3.11)$$

Also
$$Z_c^2 = \frac{A_c^3}{T_c} = \frac{Q^2}{g} \quad ...(3.12) \left(\because \frac{Q^2T}{gA^3} = 1\right)$$

Dividing equation (3.12) by equation (3.11)

$$\frac{Q^2T}{gA^3} = \frac{Z_c^2}{Z^2} \quad ...(3.13)$$

Equation (3.7) can be written as

$$\frac{dy}{dx} = S_0 \frac{\left(1 - \frac{S_f}{S_0}\right)}{\left(1 - \frac{Q^2T}{gA^3}\right)} \quad ...(3.14)$$

Putting the value of equations (3.1) and (3.13) in equation (3.14) we get

$$\frac{dy}{dx} = S_0 \cdot \frac{1-\left(\frac{K_0}{K}\right)^2}{1-\left(\frac{Z_c}{Z}\right)^2} \qquad ...(3.15)$$

If Q_n represents normal discharge at depth y and Q_c denotes critical discharge at a same depth.

$$Q_n = K\sqrt{S_0} \qquad ...(3.16)$$

$$Q_c = Z\sqrt{g} \qquad ...(3.17)$$

Also from equation (3.9), $Q = K_0\sqrt{S_0}$ and from equation (3.12)

$$Z_c^2 = \frac{Q^2}{g} \Rightarrow Q = Z_c\sqrt{g}$$

Dividing equation (3.9) by (3.16)

$$\frac{Q}{Q_n} = \frac{K_0}{K}$$

$$\left(\frac{Q}{Q_n}\right)^2 = \left(\frac{K_0}{K}\right)^2 \qquad ...(3.18)$$

Dividing equation (3.12) by (3.17)

$$\frac{Q}{Q_c} = \frac{Z_c}{Z}$$

$$\left(\frac{Q}{Q_c}\right)^2 = \left(\frac{Z_c}{Z}\right)^2 \qquad ...(3.19)$$

Putting value of equation, (3.18) and (3.19) in equation (3.15) we get

$$\frac{dy}{dx} = S_0 \cdot \frac{1-\left(\frac{Q}{Q_n}\right)^2}{1-\left(\frac{Q}{Q_c}\right)^2} \qquad ...(3.20)$$

Also equation (3.2) can be written as

$$\frac{dE}{dx} = S_0 - S_f \qquad ...(3.21)$$

$$\left(\because S_0 = \frac{dz}{dx}, S_f = \frac{dH}{dx}\right)$$

Equation (3.21) is called as differential energy equation.

For rectangular channel equation (3.7) can be written as

$$\frac{dy}{dx} = \frac{S_0 - S_f}{1 - F^2} \qquad ...(3.22)$$

where F = Froude number

Since for rectangular section

$$\frac{Q^2T}{gA^3} = \frac{V^2B}{gA}$$

$$\Rightarrow \quad \frac{V^2}{gy} = \left(\frac{V}{\sqrt{gy}}\right)^2 = (F)^2$$

3.4 Limitation of Gradually Varied Flow

Following are the limitations of GVF:

1. Roughness coefficient is not constant for length of channel.
2. The pressure distribution may not be hydrostatic over entire channel section.
3. The energy and momentum correction factor are not unity.
4. The flow may not be always steady.
5. The velocity distribution may not be uniform over or cross-section.

3.5 Flow Classification and Surface Profiles

We have discussed the flow classification in chapter first. Here we are interested only in GVF and its surface profiles. Before discussing the surface profiles, it is necessary to understand the concept of critical depth and normal depth.

The critical depth depends on the cross-sectional shape and discharge while normal depth depends on cross-section shape, discharge, boundary roughness and bed slope. To understand the relationship between the normal depth and critical depth, consider the case of laboratory flume with adjustable bed slope in which constant discharge flows through it. (Assuming rectangular cross-section and Mannings, *n*). If we start from very small bed slope (almost horizontal), the normal depth would be very large. The critical depth, however, will not depend on the bed slope as we have assumed constant discharge and can be computed using $y_c = \left(\frac{q^2}{g}\right)^{1/3}$ (as the channel is rectangular). Since the normal depth is very large, it is reasonable to say that normal depth (y_0) is greater than critical depth (y_c). Now, if we go on increasing the bed slope, the normal depth will keep on decreasing and stage would be reached when the normal depth becomes equal to the critical depth. This bed slope is called critical bed slope (S_c). Further increase in bed slope will cause the normal depth to become smaller than the critical depth. A bed slope smaller than critical slope $(S_0 < S_c)$ is called mild slope $(y_0 > y_c)$ and that larger than critical slope $S_0 < S_c$ is called steep slope $(y_0 < y_c)$. Most natural channels will have mild slope but steep slopes may be encountered in short reaches. There are two other classifications of bed slope, which are not very common. These are horizontal slope $(S_0 = 0)$ and adverse slope $(S_0 < 0)$, which do not have normal depth. (Normal depth is infinite for horizontal slopes and complex number for adverse slopes). These two slopes are called non sustaining slopes as the driving force is absent (the component of weight in the flow direction being zero for horizontal slope and negative for adverse slope channels) implies that uniform flow cannot be sustained.

The commonly used two character classification of the GVF profile uses the first letter of the slope (A, H, M, C, S for adverse, horizontal, mild, critical and steep slopes) as its first character. The second character indicates the relationship of the flow depth in comparison with normal and critical depths, if uniform flow occurs in the channel, the water surface will be parallel to the bed at a distance (measured perpendicular to the bed) of y_0. This line is called normal depth line and is represented by NDL (_._._). Similarly, if we draw line parallel to the bed at a distance of y_c, it can be called as critical depth line and is represented by CDL (.....). These two lines divide the space above the channel into three zones (Irrespective Wether the NDL is above CDL or below it). The space above the top most line is called *zone* 1. The space between two lines is called *zone* 2 and the space between channel bed and lower most line is called *zone* 3. The zone number is used as the second character in the classification of GVF profiles leading to 15 possible combination (A_1, A_2, A_3, H_1, H_2, H_3, C_1, C_2, C_3, S_1, S_2, S_3, M_1, M_2, M_3), out of these A_1, H_1 and C_2 profiles do not exist. A_1 and H_1 do not exist because there is no zone 1 on adverse and horizontal slopes, which have no normal depth, similarly in C_1 the NDL, and CDL concide and zone 2 collapses into a line.

Table 3.1: Classification of Channels

S. No	*Channel Category*	*Symbol*	*Characteris-ticcondition*	*Remark*
1.	Mild slope	M	$y_0 > y_c$	Subcritical flow at normal depth
2.	Steep slope	S	$y_c > y_0$	Super critical flow at normal depth
3.	Critical slope	C	$y_c = y_0$	Critical flow at normal depth

4.	Horizontal slope	H	$S_0 = 0$	Cannot sustain unifor-m flow
5.	Adverse slope	A	$S_0 < 0$	Cannot sustain unif-orm flow

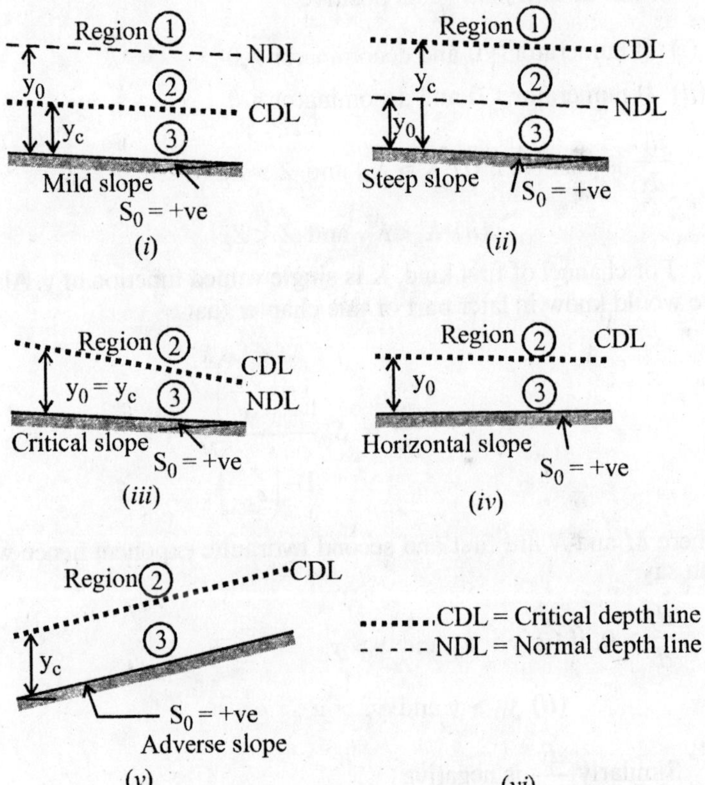

Fig. 3.2: Regions of flow profiles

The water surface profile will depend upon channel category and region of flow. The nature of GVF profile that whether the profile will have an increasing or decreasing water depth in the direction of flow will depend upon dy/dx being positive or negative.

We know
$$\frac{dy}{dx} = S\frac{1-\left(\frac{K_0}{K}\right)^2}{1-\left(\frac{Z_c}{Z}\right)^2}$$

It can be said that $\frac{dy}{dx}$ is positive.

(*i*) If numerator > 0 and denominator > 0

(*ii*) If numerator < 0 and denominator < 0

$\frac{dy}{dx}$ is positive if (*i*) $K > K_0$ and $Z > Z_c$ or

(*ii*) $K < K_0$ and $Z < Z_c$

For channel of first kind, *K* is single valued function of *y*. Also we would know in later part of this chapter that

$$\frac{dy}{dx} = S_0\frac{1-\left(\frac{y_0}{y}\right)^N}{1-\left(\frac{y_c}{y}\right)^M}$$

where *M* and *N* are first and second hydraulic exponent hence we can say

$\frac{dy}{dx} > 0$, if (*i*) $y > y_0$ and $y > y_c$

(*ii*) $y_0 > y$ and $y_c > y$

Similarly $\frac{dy}{dx}$ is negative

(*i*) $y_0 > y$ and $y_c > y$ or $y_0 > y > y_c$

(*ii*) $y_0 > y$ and $y_c > y$ or $y_c > y > y_0$

3.5.1 Salient Features of the Surface Profile

We know that gradually varied equation for channel is given by

$$\frac{dy}{dx} = \frac{S_0 - S_f}{1 - \frac{Q^2 T}{gA^3}}$$

Also
$$\frac{dy}{dx} = S_0 \frac{\left[1 - \left(\frac{y_0}{y}\right)^N\right]}{\left[1 - \left(\frac{y_c}{y}\right)^M\right]}$$

From these two equation we draw following conclusion.

1. **Near Normal depth line:** As $y \to y_0$, $\frac{dy}{dx} \to 0$, *i.e.*, the water surface approached normal depth line asympotically.
2. **Near critical depth line:** If $y \to y_c$, $\frac{dy}{dx} \to \infty$, *i.e.*, the water surface meets the critical depth line vertically.
3. At larger depth $y \to \infty$, $\frac{dy}{dx} \to S_0$, *i.e.*, the water surface meets a very large depth as a horizontal asymptote.
4. **Slope of water surface:** A positive slope of water surface (back water curve) is obtained when both numerator and denominator in GVF equation are of same sign and draw down curve (negative *dy*/*dx*) is obtained when numerator and denominator of GVF equation are of opposite sign. It is evident from Figure 3.3 that all the curve in region 1 have positive slopes. These are known as backwater curves. Similarly, all the curves in region 2 have negative slopes are referred as draw down curve.

Table 3.2: Types of GVF Profiles

Channel	*Region*	*Condition*	*Type*
Mild slope	1	$y > y_0 > y_c$	M_1
	2	$y_0 > y > y_c$	M_2
	3	$y_0 > y_c > y$	M_3
Steep slope	1	$y > y_c > y_0$	S_1
	2	$y_c > y > y_0$	S_2
	3	$y_c > y_0 > y$	S_3
Critical slope	1	$y > y_0 = y_c$	C_1
	3	$y < y_0 = y_c$	C_3
Horizontal slope	2	$y > y_c$	H_2
	3	$y < y_c$	H_3
Adverse slope	2	$y > y_c$	A_2
	3	$y < y_c$	A_3

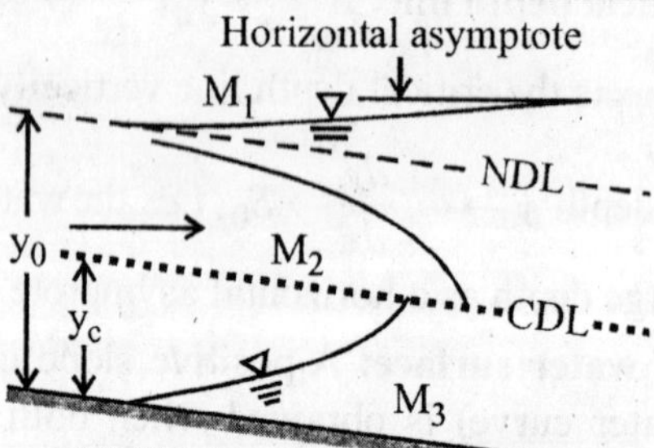

(*a*) Mild slope

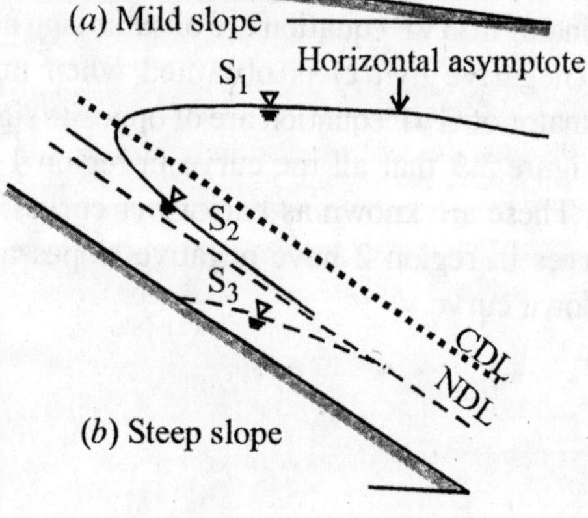

(*b*) Steep slope

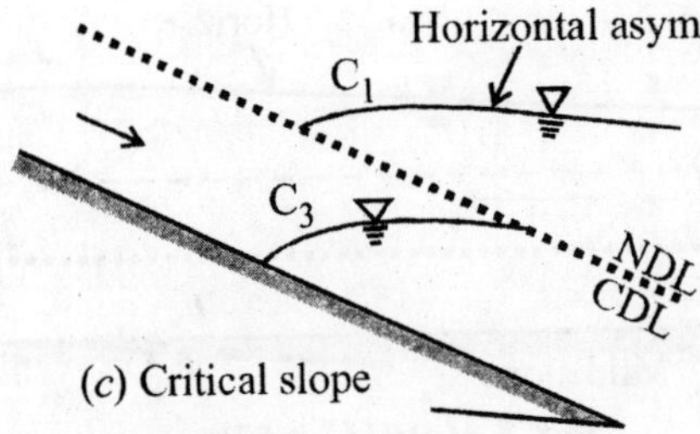

(*c*) Critical slope

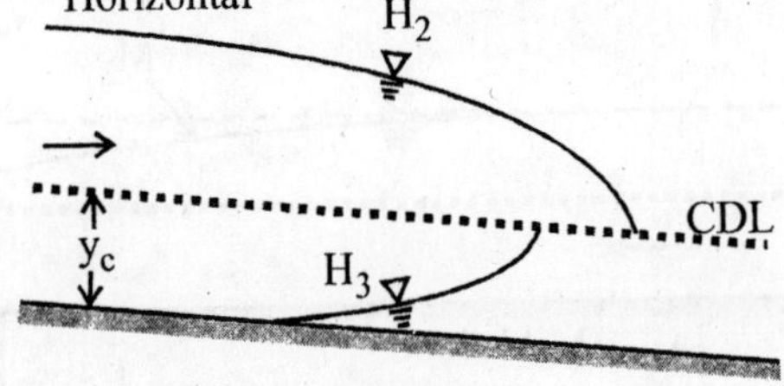

(*d*) Horizontal slope

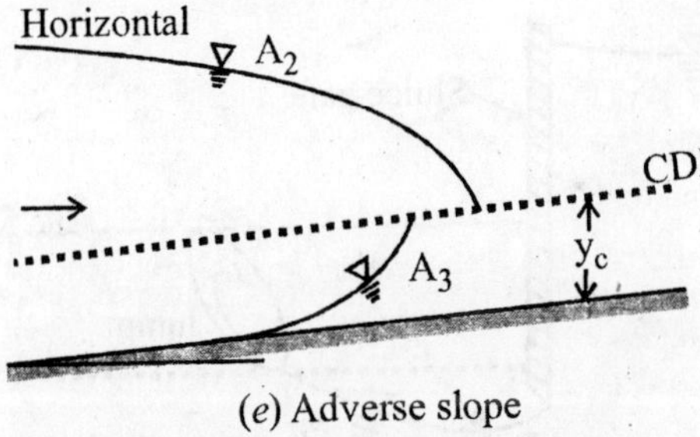

(*e*) Adverse slope

Fig. 3.3: Various GVF profiles

3.5.2 Examples of Flow Profiles

(*a*) M-profiles:

(*i*) M_1 profiles are obtained by obstruction to flow, such as weirs, dams, control structures and bends.

(*ii*) M_2 profiles occurs at a sudden drop in bed of channel, at construction type of transition and the canal outlet into pools.

(*iii*) M_3 profiles occurs where supercritical stream enter mild slope channel. The flow from sluice gate to a mild slope is a example of M_3 profile.

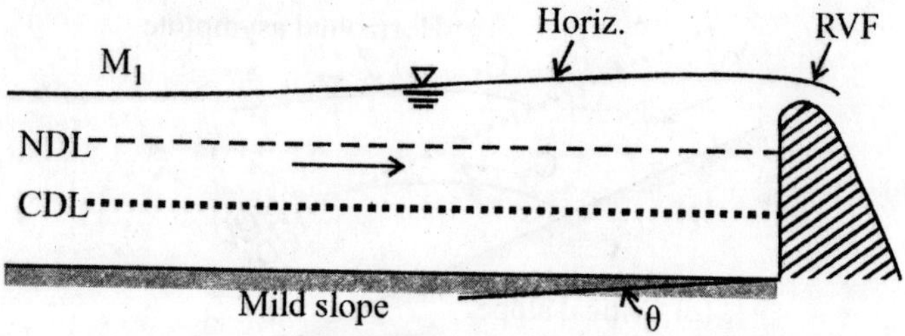

Fig. 3.4(*a*): M_1 profile

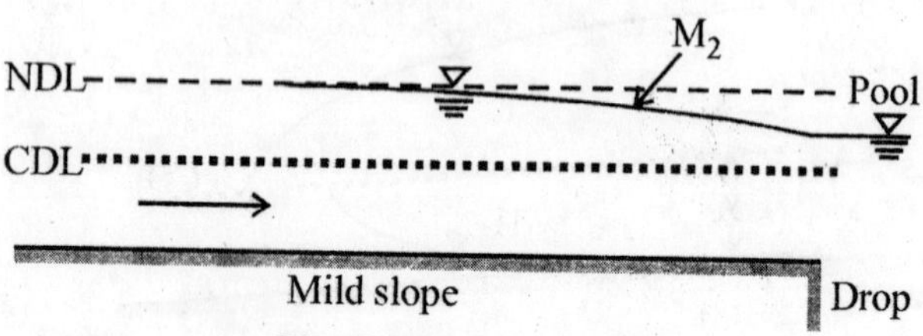

Fig. 3.4(*b*): M_2 profile

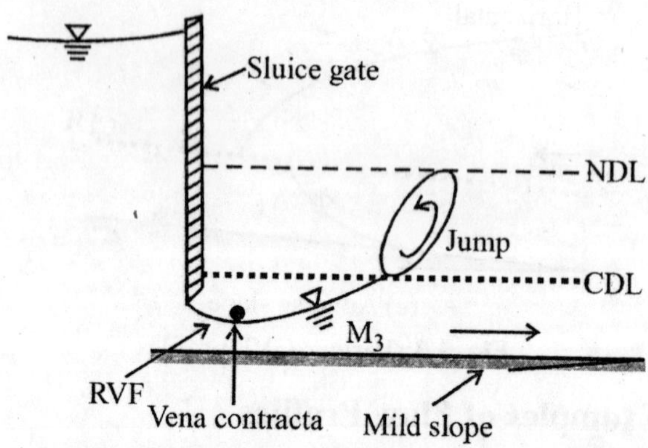

Fig. 3.4(*c*): M_3 profile

(*b*) S-profiles:

(*i*) **S_1 profiles:** It is obtained when the flow is from steep channel slope ended by deep pool created by dam or weir.

(*ii*) **S_2 profiles:** It is obtained at the entrance portion of a steep slope channel leading from reservoir.

(*iii*) **S_3 profiles:** It is obtained when a free flow occurs from sluice gate with steep slope at downstream. It can also be obtained when a flow occurs from a steeper slope to a less steep slope.

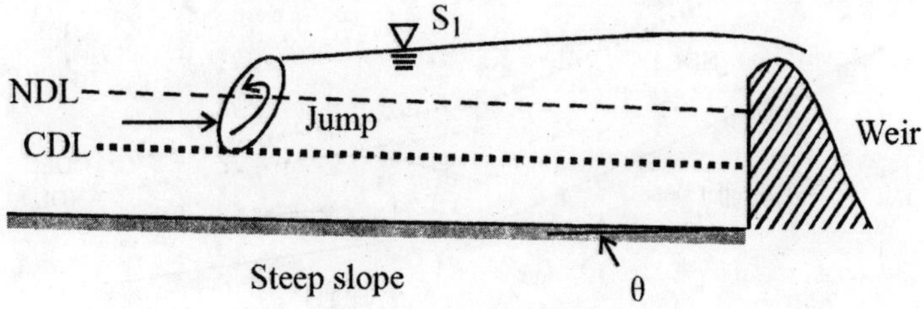

Fig. 3.4(*d*): S_1 profile

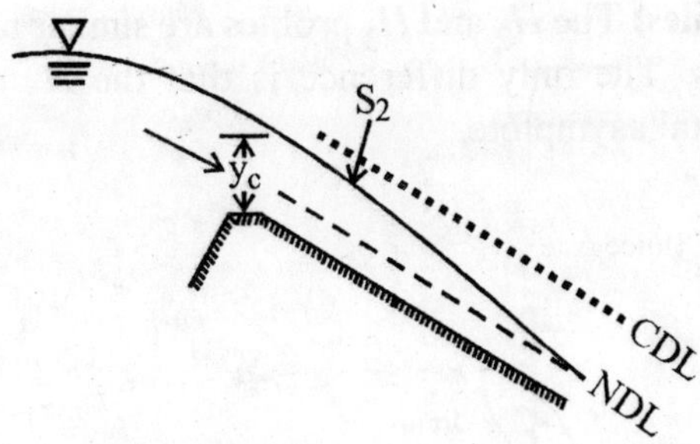

Fig. 3.4(*e*): S_2 profile

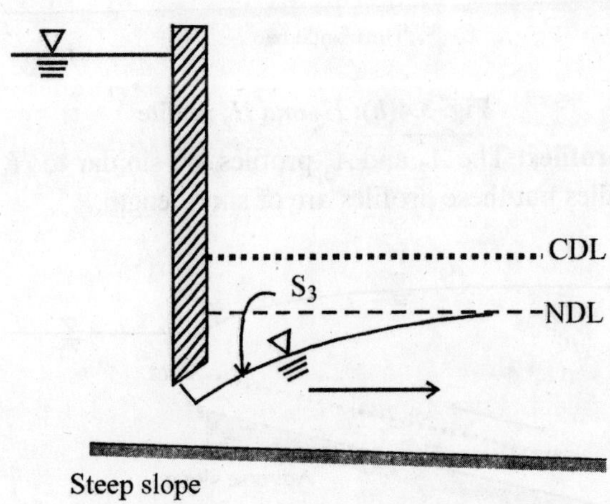

Fig. 3.4(*f*): S_3 profile

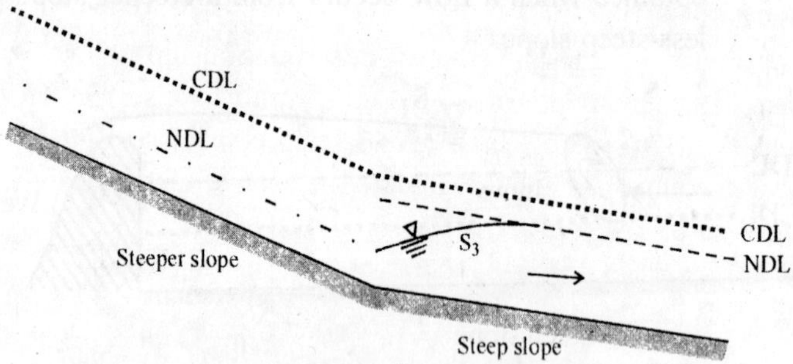

Fig. 3.4(*g*): S_3 profile

(*c*) **H-profiles:** The H_2 and H_3 profiles are similar to M_2 and M_3 profiles. The only difference is that the H_2 curve has a horizontal asymptote.

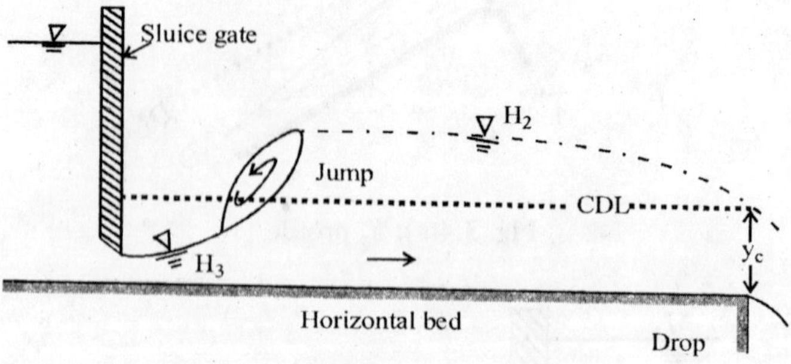

Fig. 3.4(*h*): H_2 *and* H_2 profile

(*d*) **A-profiles:** The A_2 and A_3 profiles are similar to H_2 and H_3 profiles but these profiles are of short length.

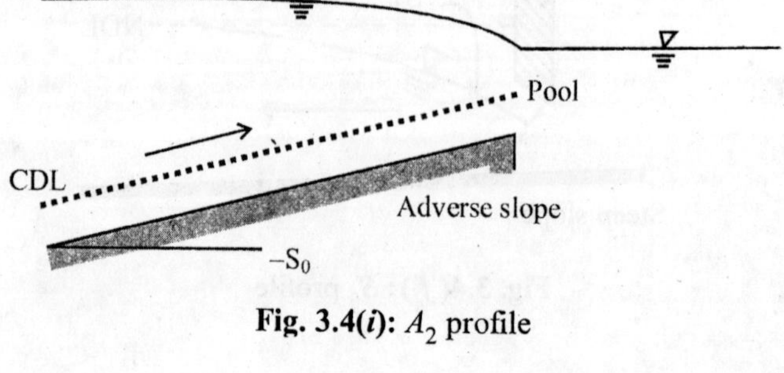

Fig. 3.4(*i*): A_2 profile

3.5.3 Control Section

It is defined as a section in which a fixed relationship exists between the discharge and depth of flow. Weirs, spillways, sluice gates are examples of structure which give rise to control sections. A few typical examples are given in figures the control section is indicated by dark point in the fig. 3.5.

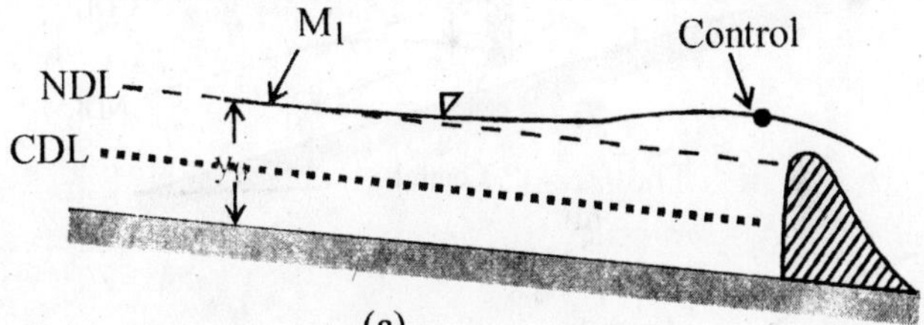

(a)

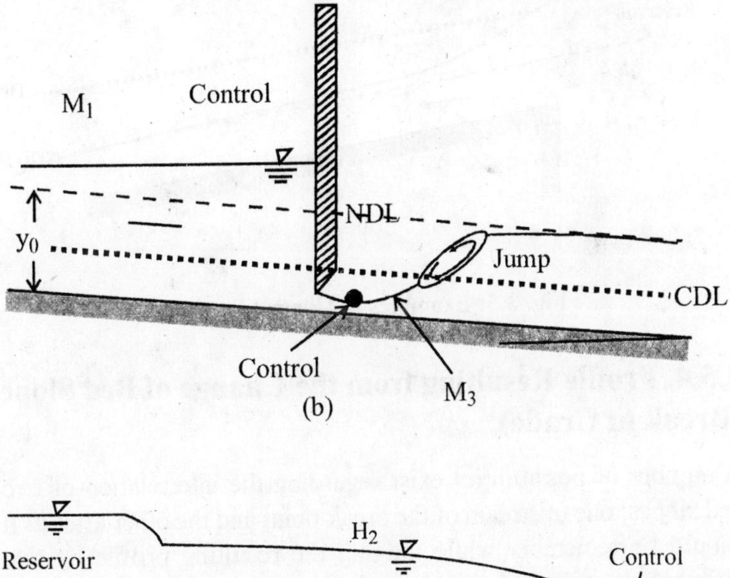

(b)

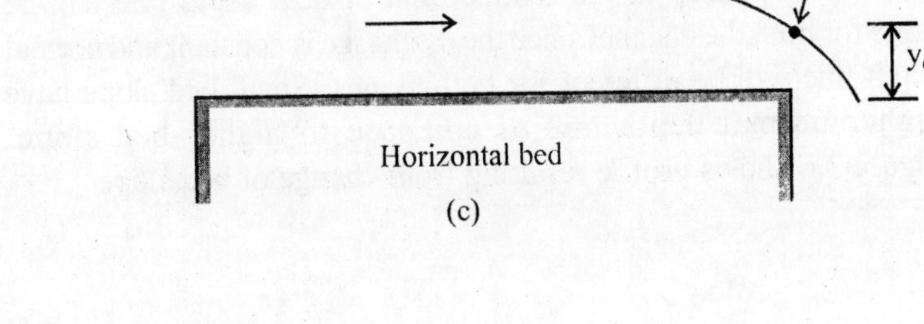

(c)

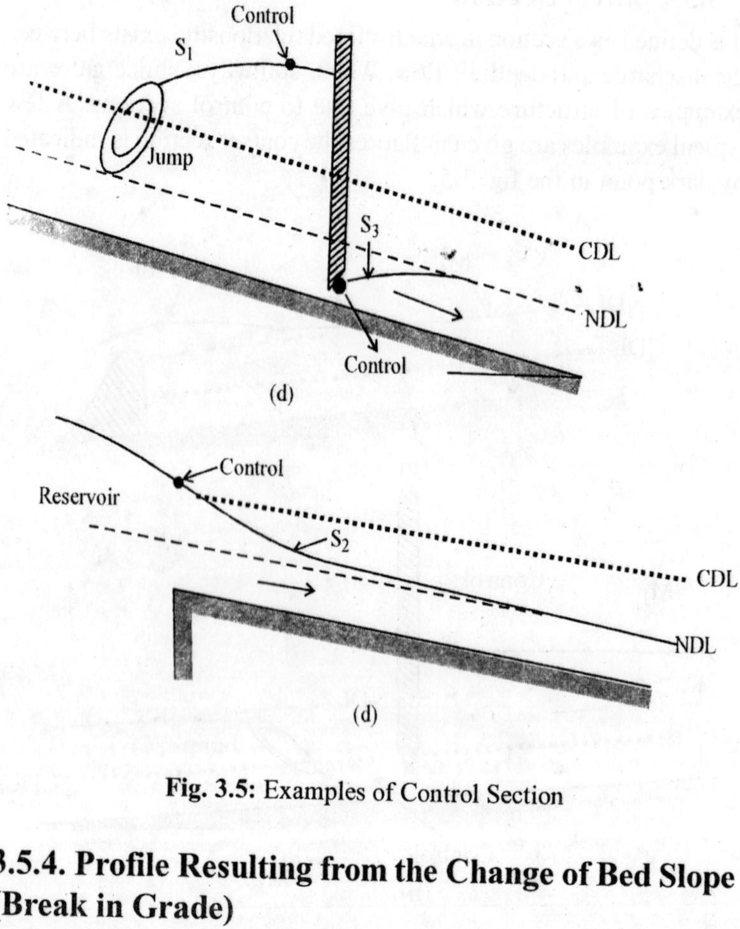

Fig. 3.5: Examples of Control Section

3.5.4. Profile Resulting from the Change of Bed Slope (Break in Grade)

A number of possibilities exist regarding the interrelation of two bed slopes; one upstream of the break point and the other after it. It should be remember while judging the resulting profile by two different slope bed of the channel that critical depth line will be same for both the channel since the discharge is constant and normal depth line will be different for both slopes. Small bed slope have higher normal depth line as compare to higher bed slope. Figure 3.6 shows profile resulting from change of bed slope.

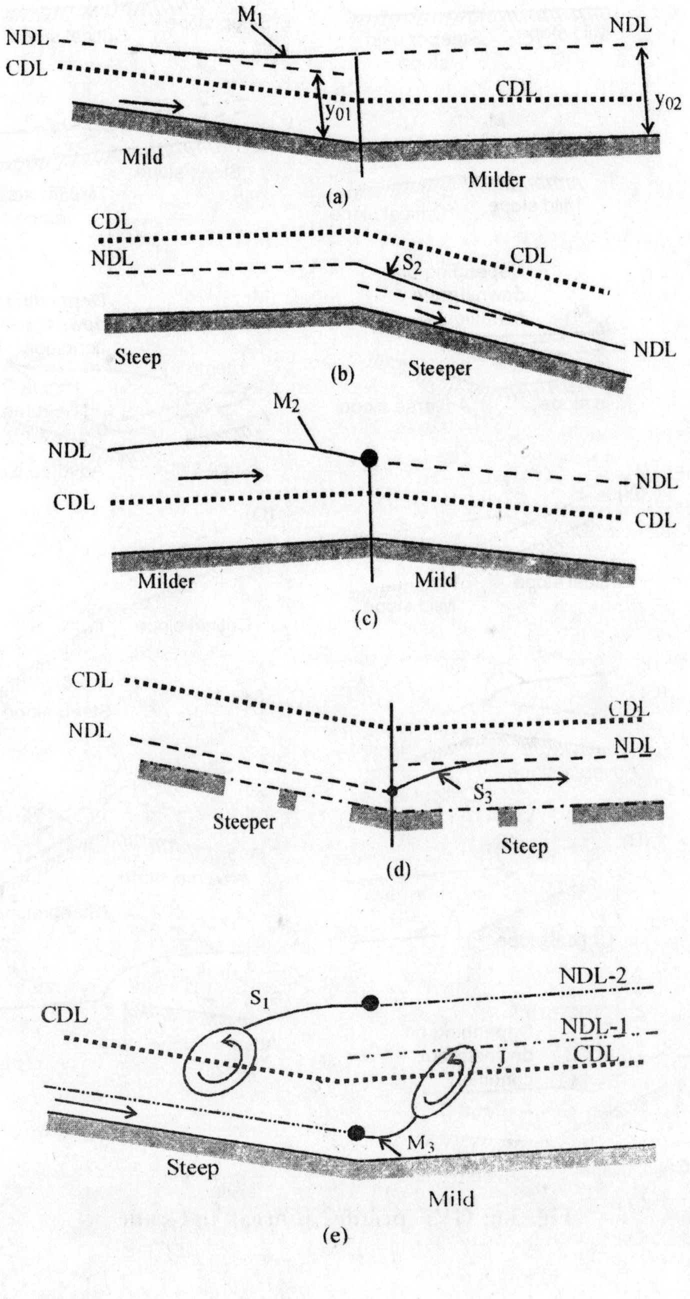

M1
NDL
NDL
CDL
CDL
y01
y02
Mild
Milder
(a)
CDL
NDL
S2
CDL
NDL
Steep
Steeper
(b)
M2
NDL
NDL
CDL
CDL
Milder
Mild
(c)
CDL
NDL
CDL
NDL
S3
Steeper
Steep
(d)
NDL-2
S1
CDL
NDL-1
J
CDL
M3
Steep
Mild
(e)

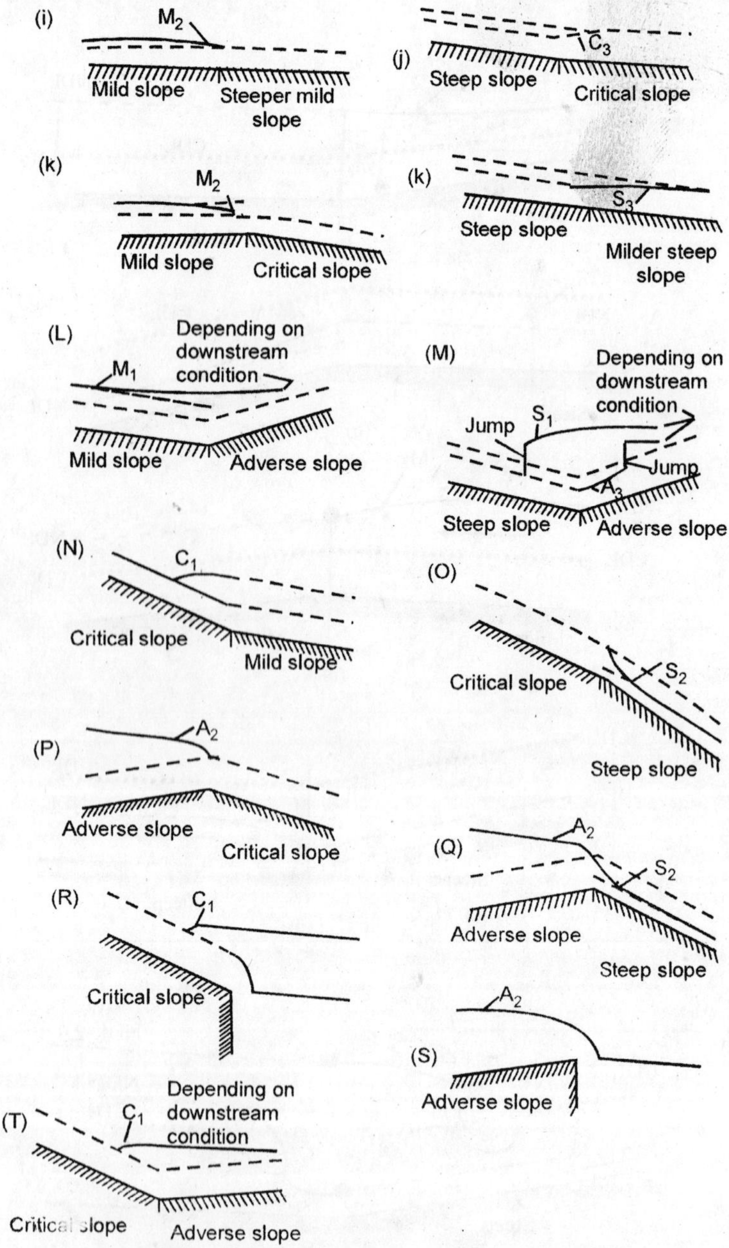

Fig. 3.6: GVF profiles at break in Grade

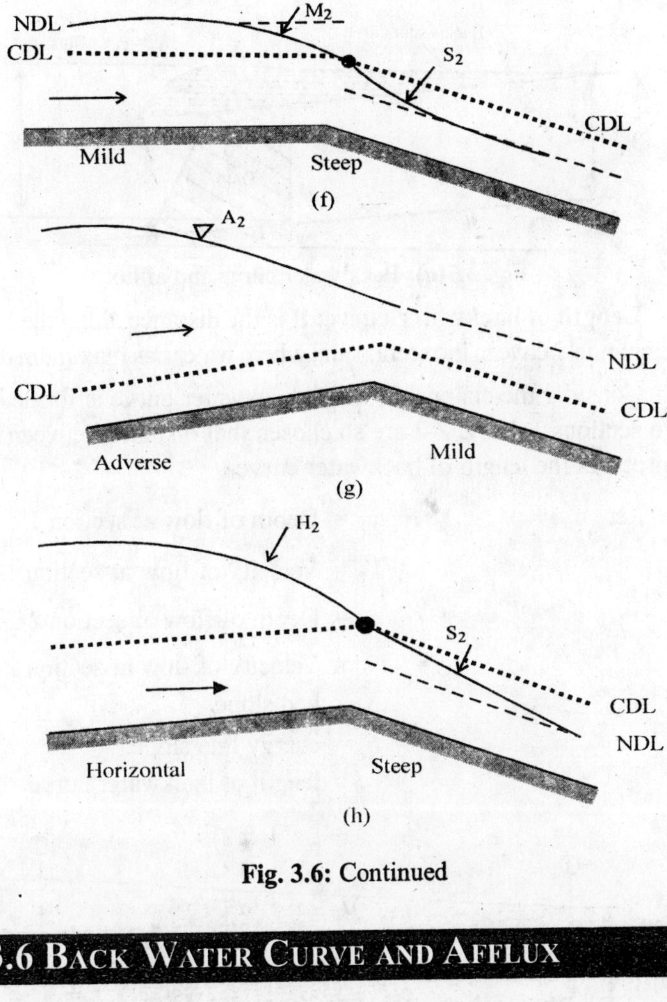

Fig. 3.6: Continued

3.6 Back Water Curve and Afflux

When an obstruction like a dam, weir, *etc.*, comes across the channel width the water level rises and it has maximum depth from the bed at some section. If y_1 is the depth of water at the point, where water start rising up and y_2 is the maximum height of rising water from the bed then this increase in bed $(y_2 - y_1)$is called as afflux and the curved surface of the liquid when its concavity upwards, is known as back water curve.

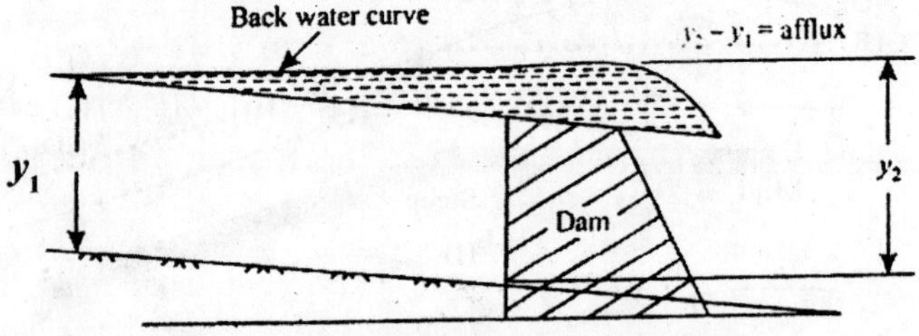

Fig. 3.7(*a*): Back water curve and afflux

Length of backwater curve: It is the distance along the bed of the channel between the section and where water has maximum depth.

Consider the channel in which backwater curve is formed. Let two sections 1-1 and 2-2 are so chosen that distance between them represents the length of backwater curve.

Let
y_1 = Depth of flow at section 1-1
V_1 = Velocity of flow at section 1-1
y_2 = Depth of flow at section 2-2
V_2 = Velocity of flow at section 2-2
S_o = bed slope,
S_e = energy line slope.
l = length of backwater curve

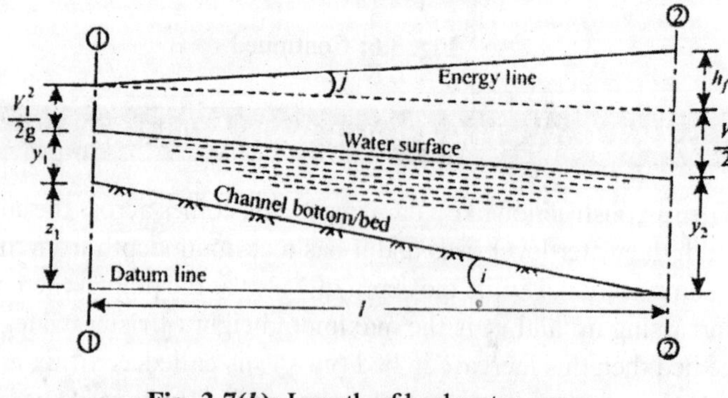

Fig. 3.7(*b*): Length of backwater curve

Applying Bernoullis equation at the two sections with channel bed 2-2 as the datum for potential head, we have

$$Z_1 + y_1 + \frac{V_1^2}{2g} = y_2 + \frac{V_2^2}{2g} + h_f \qquad (\because Z_2 = 0)$$

where $\qquad h_f$ = loss of head due to friction = $S_e l$

$$Z_1 = S_0 l$$

$$S_o l + y_1 + \frac{V_1^2}{2g} = y_2 + \frac{V_2^2}{2g} + S_e l$$

$$S_o l - S_e l = \left(y_2 + \frac{V_2^2}{2g}\right) - \left(y_1 + \frac{V_1^2}{2g}\right)$$

$$S_o l - S_e l = E_2 - E_1$$

$$l = \frac{E_2 - E_1}{S_o - S_e}$$

where E_1 and E_2 are specific energies. The value of S_e is determined by Manning's formula or Chezy's formula corresponding to flow condition at mean/average depth of flow.

3.7 COMPUTATIONS FOR GRADUALLY VARIED FLOW PROFILE

The various available procedures for computing GVF profile can be classified as:

1. Analytical Method as Integration Method
2. Numerical Method
3. Graphical Method

Out of these numerical method is most extensively used. It is the only method available to solve practical problems in natural channels. Analytical Method is essentially of academic interest. While Graphical method is practically obsolute and is seldom used.

3.7.1 Analytical Method

As we have already discussed that differential equation of GVF for prismoidal channel is given by

$$\frac{dy}{dx} = S_0 \cdot \frac{1-\left(K_o^2 / K^2\right)}{1-\left(Z_C{}^2 / Z^2\right)} = F(y) \qquad \text{(a)}$$

Which is non-linear, first order differential equation let it is required to find $y = f(x)$ in the depth range y_1 to y_2 the following assumption are made.

1. Conveyance at any depth y is given by

$$K^2 = C_2 y^N \qquad \text{...(3.23)}$$

At depth y_0 $\quad K_0{}^2 = C_2 y_0{}^N \qquad \text{...(3.24)}$

This implies C_2 and N are constant

2. The section factor Z is given by

At any depth y $\quad Z^2 = C_1 y^M \qquad \text{...(3.25)}$

At critical depth Yc $\quad Z_C{}^2 = C_1 Y_c{}^M \qquad \text{...(3.26)}$

This implies coefficient C_1 and the first hydraulic exponent M are constant

Substituting the value of equations (3.23), (3.24), (3.25) and (3.26) in equation (*a*), we get

$$\frac{dy}{dx} = S_0 \frac{1-\left(y_0/y\right)^N}{1-\left(y_C/y\right)^M} \qquad \text{...(3.27)}$$

Let us consider $u = y/y_0 \Rightarrow dy = y_0 du$

Put $u = y/y_0$ and $dy = y_0 du$ in equation (3.27)

$$\frac{du}{dx} = \frac{S_o}{y_0}\left[\frac{1-\frac{1}{u^N}}{1-\left(\frac{Y_C{}^M}{y_0{}^M}\right)\frac{1}{u^M}}\right]$$

therefore, $dx = \frac{y_0}{S_o}\left[1-\frac{1}{1-u^N}+\left(\frac{y_C}{y_0}\right)^M \frac{u^{N-M}}{1-u^N}\right]du$

On integration, we get

$$x = \frac{y_0}{S_o}\left[u - \int_0^u \frac{du}{1-u^N} + \left(\frac{y_C}{y_0}\right)^M \int_0^u \frac{u^{N-M}}{1-u^N}du\right] + \text{constant} \qquad ...(3.28)$$

Take $\int_0^u \frac{du}{1-u^N} = F(u,N)$

Also take $V = u^{\frac{N}{J}}$ where $J = \frac{N}{N-M+1}$

$$dV = \frac{N}{J}u^{\frac{N}{J}-1}du$$

$$= (N-M+1)u^{N-M}du$$

$$\int_0^u \frac{u^{N-M}}{1-u^N}du = \frac{1}{N-M+1}\int_0^v \frac{dV}{1-V^J}$$

$$= \frac{J}{N}F(V,J)$$

Here we notice that $F(V, J)$ is the same function as $f(u, N)$ with, u and N replaced by V and J respectively.

So equation (3.28) can be written as

$$x = \frac{y_0}{S_0}\left[u - F(u,N) + \left(\frac{y_c}{y_0}\right)^M \frac{J}{N}F(V,J)\right] \qquad ...(3.29)$$

The function $f(u, N)$ is called as varied flow function.

For a wide rectangular channel if the Chezy formula with C = constant is used, the hydraulic exponent will be $M = 3.0$ and $N = 3.0$. Putting these value in equation (3.29), we get

$$x = \frac{y_0}{S_o}\left[u - 1 - \left(\frac{y_c}{y_0}\right)^3 F(u,3)\right] + \text{constant} \qquad ...(3.30)$$

Since $$J = \frac{N}{N-M+1} = \frac{3}{3-3+1} = 3$$

where $F(u, 3) = \int_0^u \frac{du}{1-u^3}$

$F(u, 3)$ is called Bresse's function.

3.8 Direct Integration or Analytical Method for Circular Channel

Keifer and Chu's Method

Let Q be the actual discharge in circular channel of diameter D and bed shope S_o then

$$Q = K\sqrt{S_f} \qquad ...(3.31)$$

$$Q = K_o\sqrt{S_o} \qquad ...(3.32)$$

Where K and K_o are conveyance at depths y and y_0 (normal depth) respectively.

Let Q_D = hypothetical discharge corresponding to uniform flow with channel flowing full.

$$Q_D = K_D\sqrt{S_o} \qquad ...(3.33)$$

where K_D = conveyance at depth D

We know $K = \frac{1}{n}AR^{2/3}$. Also we know $= R = \frac{A}{P}$

$$R = \frac{\pi D^2/4}{\pi D} = \frac{D}{4}$$

$$K_D = \frac{1}{n}\left(\pi \frac{D^2}{4}\right)\left(\frac{D}{4}\right)^{2/3} \qquad ...(3.34)$$

$$\left(\frac{K_o}{K}\right)^2 = \left(\frac{K_o}{K_D}\right)^2 \left(\frac{K_D}{K}\right)^2 \qquad ...(3.35)$$

from equations (3.32) and (3.33) we get

$$\left(\frac{K_o}{K_D}\right)^2 = \left(\frac{Q}{Q_D}\right)^2 = Q_r^{\,2} \qquad ...(3.36)$$

we know

$$K = \frac{1}{n} A R^{\frac{2}{3}} \qquad ...(3.37)$$

from equations (3.34) and (3.37) we get

$$\left(\frac{K_D}{K}\right)^2 = \frac{\left(\pi \frac{D^2}{4}\right)\left(\frac{D}{4}\right)^{2/3}}{AR^{2/3}} = f_1\left(\frac{y}{D}\right) \qquad ...(3.38)$$

So we can write equation (3.35) as

$$\left(\frac{K_o}{K}\right)^2 = Q_r^2 f_1(y/D) \qquad ...(3.39)$$

we know

$$\frac{dy}{dx} = S_o \frac{\left(1 - \left(\frac{K_o}{K}\right)^2\right)}{1 - \frac{Q^2 T}{gA^3}}$$

we can write it as

$$\frac{dy}{dx} = S_o \frac{\left(1 - Q_r^2 f_1(y/D)\right)}{1 - \frac{Q^2 T}{gA^3}}$$

Since $\left(\frac{K_o}{K}\right)^2 = Q_r^{\,2} f_1\left(\frac{y}{D}\right)$

Also

$$\frac{Q^2 T}{gA^3} = \frac{Q^2}{g} \cdot \frac{1}{D^5} \frac{T/D}{A^3/D^6} = \frac{Q^2}{gD^5} f_2\left(\frac{y}{D}\right)$$

take $y/D = (\eta)$

$$dy = D\,d\eta$$

$$\frac{d\eta}{dx} = \frac{S_o}{D}\left[\frac{1 - Q_r^2 f_1(\eta)}{1 - \frac{Q^2}{g_D{}^5} f_2(\eta)}\right] \qquad ...(3.40)$$

$$dx = \frac{D}{S_o}\left[\frac{d\eta}{1 - Q_r^2 f_1(\eta)} - \frac{Q^2}{g_D{}^5} \frac{f_2(\eta)\,d\eta}{1 - Q_r^2 f_1(\eta)}\right]$$

on integration we get

$$x = \frac{D}{S_o}\left[\int_0^{\eta} \frac{d\eta}{1 - Q_r^2 f_1(\eta)} - \frac{Q^2}{g_D{}^5} \int_0^{\eta} \frac{f_2(\eta)\,d\eta}{1 - Q_r^2 f_1(\eta)} + \text{constant}\right]$$

or $x = \frac{D}{S_o}\left[I_1 - \frac{Q^2}{g_D{}^5} I^2\right] + \text{constant}$ ...(3.41) where I_1 and I_2 are known as Keifer and Chu's function.

$$I_1 = \int_0^{\eta} \frac{d\eta}{1 - Q_r^2 f_1(\eta)} = I_1(Q_r, \eta)$$

$$I_2 = \int_0^{\eta} \frac{f_2(\eta)\, d\eta}{1 - Q_r^2 f_1(\eta)} = I_2(Q_r, \eta)$$

3.8.1. Numerical Method

The numerical method is broadly classified into two categories.

(*i*) **Simple Numerical Method:** They were developed for simple calculation where computation can be done by hand. Two commanly used simple numerical methods are

Direct Step Method: This method is suitable for prismatic channels.

We know that $$\frac{dE}{dx} = S_o - S_f$$

It can also be written in finite difference form

$$\frac{\Delta E}{\Delta x} = S_o - \overline{S}_f \qquad \text{...(3.42)}$$

where $\overline{S}_f$ is the average friction slope.

$$\Delta x = \frac{\Delta E}{S_o - \overline{S}_f}$$

Taking two sections 1 and 2.

$$x_2 - x_1 = \frac{E_2 - E_1}{S_o - \frac{1}{2}\left(S_{f_1} - S_{f_2}\right)} \qquad \text{...(3.43)}$$

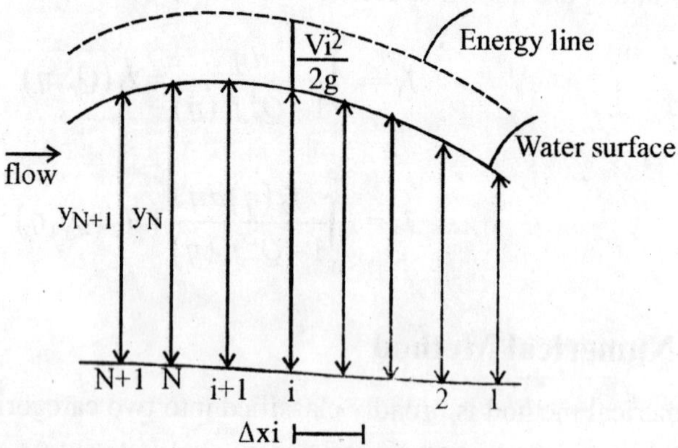

Fig. 3.8 Sketch for direct step method

Consider the above water surface profile and it is divided into *N* parts and each part of different depth.

$$\Delta E = \Delta\left(y + \frac{v^2}{2g}\right) = \Delta\left(y + \frac{Q^2}{2gA^2}\right)$$

$$\Delta E = E_{i+1} - E_i = \left(y_{i+1} + \frac{Q^2}{2gA_{i+1}^2}\right) - \left(y_i + \frac{Q^2}{2gA_i^2}\right)$$

$$\bar{S}_f = \frac{1}{2}\left(S_{f_{i+1}} - S_{f_i}\right) = \frac{n^2Q^2}{2}\left[A_{i+1}^{1/2} R_{i+1}^{4/3} + A_i^{1/2} R_i^{4/3}\right]$$

For general

$$\Delta x_i = \frac{E_{i+1} - E_i}{S_0 - \bar{S}_f} \qquad ...(3.44)$$

Standard Step Method

It is used for solving natural channel problems we know that in natural channel the cross-sectional shape are likely to vary from section to section.

Thus the problem of computation of GVF profile can be solved as follows.

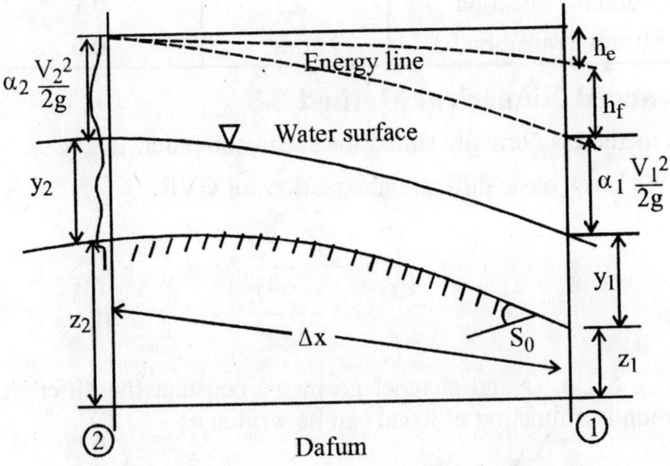

Fig. 3.9: Sketch for the standard step method

Consider sections 1 and 2 with 1 downstream of section 2 at a distance Δx.

Equating total energy at sections 1 and 2.

$$z_2 + y_2 + \alpha_2 \frac{v_2^2}{2g} = z_1 + y_1 + \alpha_1 \frac{v_1^2}{2g} + h_f + h_e$$

h_f = friction loss $\quad h_e$ = eddy loss.

α_1 and α_2 = kinetic energy correction.

The h_f can be calculated as

$$h_f = \bar{S}_f \Delta x = \frac{1}{2}\left(S_{f_1} + S_{f_2}\right)\Delta x$$

$$S_f = \frac{n^2 Q^2}{A^2 R^{4/3}}$$

$$h_e = C_e \left[\frac{\alpha_1 v_1^2 - \alpha_2 v_2^2}{2g}\right]$$

Value of C_e	*Expansion*	*Contraction*
Prismatic Channel	0	0
Gradual Transition	0.3	0.1
Abrupt Transition	0.8	0.6

Advanced Numerical Method

This method is normally suited for use in computer.

We know basic differential equation for GVF.

$$\frac{dy}{dx} = \frac{S_o - S_f}{1 - \frac{Q^2 T}{gA^3}}$$

keeping S_o, n, Q and channel goemetry constant the differential equation is a function of y and can be written as

$$\frac{dy}{dx} = f(y) \text{ or } f(y) = \frac{S_o - S_f}{1 - \frac{Q^2 T}{gA^3}}$$

using notation
$$y_i = y(x_i) \text{ means } y \text{ at } x_i$$
$$y_{i+1} = y(x_i + 1) \text{ means } y \text{ at } x_{i+1}$$
$$x_{i+1} = x_i + \Delta x$$

The various numerical method are as follows:

(*a*) Standard Fourth Order Runge-Kutta Method (SRK)

$$y_{i+1} = y_i + \frac{1}{6}(k_1 + 2k_2 + 2k_3 + k_4)$$
$$k_1 = \Delta x\, f(y_i)$$
$$k_2 = \Delta x\, f\left(y_i + \frac{k_1}{2}\right)$$
$$k_3 = \Delta x\, f\left(y_i + \frac{k_2}{2}\right)$$

$$k_4 = \Delta x\, f\left(y_i + k_3\right)$$

(b) Kutta Merson Method (KM)

$$y_{i+1} = y_i + \frac{1}{2}\left(k + 4k_4 + k_5\right)$$

$$k_1 = \frac{1}{3}\Delta x\, f\left(y_i\right)$$

$$k_2 = \frac{1}{3}\Delta x\, f\left(y_i + k_1\right)$$

$$k_3 = \frac{1}{3}\Delta x\, f\left(y_i + \frac{k_1}{2} + \frac{k_2}{2}\right)$$

$$k_4 = \frac{1}{3}\Delta x\, f\left(y_i + \frac{3}{8}k_1 + \frac{9}{8}k_3\right)$$

$$k_5 = \frac{1}{3}\Delta x\, f\left(y_i + \frac{3}{2}k_1 - \frac{9}{2}k_3 + 6k_4\right)$$

(c) Trapezoidal Method (TRAP): This is Iterative method with

$$y_{i+1} = y_i + \frac{1}{2}\Delta x\left\{F\left(y_i\right) + F\left(y_{i+1}\right)\right\}$$

The calculation starts with the assumption of $F\left(y_{i+1}\right) = F\left(y_i\right)$ in the right hand side of equation: The value of y_{i+1} is evaluated from above equation and substituted in $\frac{dy}{dx} = f(y)$ to get $F\left(y_{i+1}\right)$. The process is repeated. Thus rth iteration will have

$$y_{i+1}^{r} = y_i + \frac{1}{2}\Delta x\left\{F\left(y_i\right) + F\left(y_{i+1}\right)^{r-1}\right\}$$

3.8.2 Graphical Method

We know that $$\frac{dy}{dx} = \frac{S_o - S_f}{1 - \frac{Q^2 T}{gA^3}}$$

It can be written as

$$\frac{dy}{dx} = \frac{S_o - \dfrac{\eta^2 Q^2}{A^2 R^{4/3}}}{1 - \dfrac{Q^2 T}{g A^3}}$$

or

$$x_{1,2} = \int_{y_1}^{y_2} \frac{1 - \dfrac{Q^2 T}{g A^3}}{S_o - \dfrac{\eta^2 Q^2}{A^2 R^{4/3}}} dy$$

Where $x_{1,2}$ represents the distance between two sections where the flow depths are y_1 and y_2. The integral can be obtained graphically by plotting the variation of the function within the Integral *i.e.*, plotting $\frac{dy}{dx}$ verse y and by calculating area enclosed gives the required result.

3.9 Transitions of Subcritical and Supercritical Flow

A Channel Transition is actually the local change in cross section in terms of area or shape producing variation from one uniform state to another. A transition may also involve a change in flow direction. These transitions are generally short but their effects are felt for considerable distance upstream and downstream. Bends, expansion and contraction are the examples of channel transitions Now we will discuss one by one both cases of transitions.

(a) Subcritical flow transitions: It is provided to enable smooth change from the channel of one dimension to another. Either an expanding or a contracting transition is provided depending upon the relative sizes of the two channels. In case the channel is flumed over certain length for reason of economy, contracting and expanding transitions need to be provided, one at each end of the flumed portion. These transitions are usually gradual there by leading to small energy losses.

(i) *Contracting transition:* The energy loss in contracting flows is generally small as compared to that expanding flows. The energy loss in contractions is due to expansions of flow from vena contracta section to the section where the flow reaches itself to the side walls of the downstream section. The energy loss in the contraction may be expressed $C_L \frac{V^2}{2g}$. In which V is the velocity in the smaller channel and C_L is the loss coefficient depend up on type of the contraction and width–depth ratio.

(ii) **Abrupt Expansions:** The expression for energy loss can be derived on the basis of following assumptions:

(a) $h_1 = h_2$

(b) $E_1 = E_2$

(c) $M_2 = M_3$

(d) Boundary friction is negligiable

(e) F_1 is small enough. for higher power say F_1^4 can be neglected.

The energy loss can be given by

$$E_L = \frac{V_1^2}{2g}\left(1 - \frac{B_1}{B_2}\right)^2 \left[1 + \frac{2F_1^2\left(\frac{B_1}{B_2}\right)^3}{(1 - B_1/B_2)}\right] \quad \text{(a)}$$

The square bracket is negligible if $F_1 \geq 0.5$ and $\frac{B_2}{B_1} \leq 1.5$ then equation a reduces to

$$E_L = \frac{v_1^2}{2g}\left(1 - \frac{B_1}{B_2}\right)^2 \qquad ...(b)$$

Scientist Abbott & Kline found that the length of the eddy on both walls is same so long as $\frac{B_2}{B_1} \leq 1.5$. At larger values of $\frac{B_2}{B_1}$, the eddy on one side becomes larger than on the other and centre line of the channel no longer remain the line of the maximum velocity

(iii) *Gradual expansions:* It is provided in any major structure to ensure speedy recovery of the normal velocity distribution and to minimise loss. The recommended length of transition generally varies from $2(B_2 - B_1)$ to $2.5(B_2 - B_1)$

A common type of transition used in India is that proposed by Mitra. Based on the assumption of constant depth and constant $\frac{dv}{dx}$ along the length, Mitra obtained the following expression for width B_x at any distance x from the beginning of the transition:

$$B_x = \frac{B_1 B_2 L}{LB_2 - (B_2 - B_1)x}$$

Where L is eddy length.

(b) *Supercritical flow transition:* The formation of a standing wave in supercritical flow is well understood by considering the movement of small disturbance in a stationary liquid at a velocity V greater than speed of natural wave velocity C. When the particle is at any point P_1, the disturbance travels outwards from the point same velocity in all directions. In other words, at any subsequent instant the wavefront may be described by a circle around P_1. Similar fronts are generated as the particle reaches its successive positions P_2, P_3 etc. Since $V > C$, the partical is always ahead of wane front. When the particle is at P_n, the wave front can be enveloped by the tanget O_1P_n and this itself is a wave front.

$$\sin\beta = \frac{O_1P_1}{P_1P_n} = \frac{C\cdot dt}{V\cdot dt} = \frac{C}{V} = \frac{1}{F}$$

A' similar front called shock waves occurs in the case of compressible gases a supersonic speeds and sin β is then equal to $\frac{1}{M}$, M is the Mach number when the liquid is moving at velocity V and particle is stationary, stationary oblique front will be obtained.

Transitions for which the approach flow is supercritical offer a design challenge because of the existence and propagation of standing wave fronts. The reason for the occurrence of standing wave fronts in supercritical flow can be visualized from the viewpoint of an observer riding on a small particle or disturbance moving at a constant speed, V, in still water. At each instant of time, the disturbance, P, sends a circular wave front outward that moves at a speed equal to the wave celerity, c. If $V < c$, as in Figure, the wave fronts outdistance the moving disturbance, so that no pileup or addition of wave front occurs. On the other hand, for cases of $V > c$,

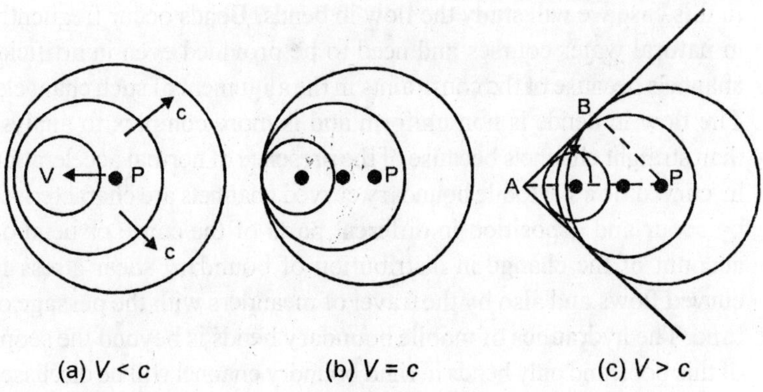

(a) $V < c$ (b) $V = c$ (c) $V > c$

Fig. 3.10 Movement of point disturbance, p at speed V in still water (C = Wave celerity)

the disturbance, *P*, moves faster than the wave fronts. The result is an accumulation of wave fronts, the outer locus of which forms a straight line at an angle β to the path of the disurbance that can be defined as a standing wave front. In between these two extremes, $V = c$, and the standing wave front is perpendicular to the path of the disturbance, *P*. If the disturbance is infinitesimal, so that $c = (gy)^{0.5}$, where y = flow depth, then obviously, V/c is the Froude number.

The supercritical case can be analyzed in more detail to determine the angle β. In the time t_1, the distance moved by point *P* to point *A* is Vt_1 while, at the same time, the initial wave front grows from *P* to point *B* over a radial distance given by ct_1. Then. $\sin \beta = 1/F$, where F is the Froude number. If the viewpoint of the observer is changed, the fluid moves at a speed *V* in the supercritical case and any boundary irregularity, such as that caused by a change in wall direction in the contraction gives rise to a standing wave front at an angle, β_1, relative to the original flow direction. This analysis indicates that larger Froude numbers result in smaller angles of deflection of the standing wave front, but the possibility of a finite height of the standing wave front is not considered. In this circumstance, the analysis must be modified.

3.10 Flow in Curved Channels

In this case we will study the flow in bends. Bends occur frequently in natural water courses and need to be provided even in artificial channels because of the constraints in the alignment of such channels. The flow in bends is non uniform and is more complex to analyse than straight channels because of the presence of normal acceleration in curved flows. Mobile boundary curved channels are characterised by scour and deposition in different parts of the curve or bend on account of the change in distribution of boundary shear stress in curved flows and also by the travel of meanders with the passage of time. The hydraulics of mobile boundary bends is beyond the scope of this book and only bends in rigid boundry channel will be discussed in detain in 7th unit.

Solved Problems

***Problem 1:** A rectangular channel with a bottom width of 4.0 m and a bottom slope of 0.0008 has a discharge of 1.50 m³/s. In gradually varied flow in this channel, the depth at certain location is found to be 0.30 m. Assuming n = 0.016, determine the type of GVF profile.*

Solution: (*a*)

$$Q = \frac{1}{n} AR^{2/3} S_o^{1/2}$$

$$1.50 = \frac{1}{0.016} \times By_o \times \frac{(By_0)^{2/3}}{(B+2y_0)^{2/3}} \times \sqrt{.0008}$$

$$\therefore \qquad \frac{(By_0)^{5/3}}{(B+2y_0)^{2/3}} = 0.848528157$$

$$\frac{(4y_0)^{5/3}}{(4+2y_0)^{2/3}} = 0.848528157$$

By trial–and–error method, we get

$$y_0 = 0.426 \text{ m}$$

(*b*) Critical depth

$$q = Q/B = \frac{1.5}{4.0} = 0.375 \text{ m}^3\text{/s/m}$$

$$y_c = \left(q^2/g\right)^{1/3} = \left((0.375)^2 / 9.81\right)^{1/3} = 0.243 \text{m}$$

(*c*) Type of profile

Since $y_0 > y_c$; the channel is a mild-slope channe . Also, given $y = 0.30$ m is such that

$$y_0 > y > y_c$$

As such profile is of the M_2 type.

Problem 2: Water flowing in triangular channel of side slope 1 horizontal 1 vertical and longitudinal slope of 0.002. Determine whether the channel is mild, steep or critical when a discharge of 0.2 m^3/s flow through it. Assume Manning's n = 0.015. For what range of depth will the flow be type 1, 2 or 3?

Solution: For a depth of flow of y in the channel

$$A = y^2$$

$$T = 2y$$

$$R = y^2/2\sqrt{2}y = \frac{1}{2\sqrt{2}}y$$

Critical depth y_c:

$$\frac{Q^2}{g} = \frac{Ac^3}{T_c} = \frac{y_c^6}{2y_c} = \frac{y_c^5}{2}$$

$$y_c = \left(\frac{2Q^2}{g}\right)^{1/5} = \left(\frac{2\times(0.2)^2}{9.81}\right)^{1/5} = 0.382\,\text{m}$$

Normal depth y_0:

$$Q = \frac{1}{n}AR^{2/3}\,So^{1/2}$$

$$= \frac{1}{n}(y_0)^2\left(\frac{y_0}{2\sqrt{2}}\right)^{2/3} S_o^{1/2}$$

$$y_0^{8/3} = \frac{nQ}{0.5\,S_o^{1/2}} = \frac{0.015\times 0.2}{0.5\times(0.002)^{1/2}} = 0.13416$$

$$y_0 = 0.471\text{ m}$$

Since $y_0 > y_c$, the channel is a mild slope channel for this discharge. If y is the depth of flow

For M_1 curve	$y > 0.471$ m
For M_2 curve	$0.471 > y > 0.382$ m
For M_3 curve	$y < 0.382$ m

Problem 3: ***A wide rectangular channel has a Manning's coefficient of 0.018. For a discharge intensity of 1.5 m³/s/m. Identify the possible type of gradually varied profile (GVF) produced in the following break in the grade of the channel.***

(a) $S_{o1} = 0.0003$ ***and*** $S_{o2} = 0.016$

(b) $S_{o1} = 0.015$ ***and*** $S_{o2} = 0.0003$

Solution: Discharge intensity $q = 1.5\ m^3/s/m$

Critical depth, $y_c = \left(q^2/g\right)^{1/3} = \left(\frac{(1.5)^2}{9.81}\right)^{1/3}$

$= 0.612$ m

Normal depth y_0 = For wide rectangular channel

$R = y_0$

$$q = \frac{1}{n} y_0 \times y_0^{2/3} \times S_o^{1/2}$$

$$y_0 = \left[\frac{nq}{\sqrt{S_o}}\right]^{3/5} = \left[\frac{0.018 \times 1.5}{\sqrt{S_o}}\right]^{3/5}$$

Slope	y_0
0.0003	1.305 m
0.0003	1.305 m
0.015	0.404 m
0.016	0.396 m

Case	y_c	y_{01}	y_{02}	Type of grade charge
(a)	0.612	1.305	0.396	Mild to Steep
(b)	0.612	0.404	1.305	Steep to Mild

Case (*a*): Possible type of profile are M_2 and S_2

Case (*b*):

(*i*) M_3 curve and a jump or

(*ii*) Jump and S_1 curve

(*iii*) In a limiting case of (*i*) and (*ii*) the jump may occur of the break in grade and there will be no GVF profile in a such case.

Problem 4: Show that differential equation gradually varied flow in a rectangular channel of variable width B can be expressed as

$$\frac{dy}{dx} = \frac{S_o - S_f + \left(\frac{Q^2 y}{gA^3} \frac{dB}{dx}\right)}{1 - \frac{Q^2 y}{gA^3}}$$

Solution: We know gradually varied flow equation

$$\frac{dH}{dx} = \frac{dz}{dx} + \frac{dy}{dx} + \frac{d}{dx}\left(\frac{V^2}{2g}\right) \qquad ...(1)$$

But we know that $\frac{dH}{dx} = -S_f, \frac{dz}{dx} = -S_o$

Putting these value in equation (1)

$$-S_f = -S_o + \frac{dy}{dx} + \frac{d}{dx}\left(\frac{Q^2}{2gA^2}\right)$$

$$S_o - S_f = \frac{dy}{dx} + \frac{d}{dx}\left(\frac{Q^2}{2g \times y^2 B^2}\right)$$

$$S_o - S_f = \frac{dy}{dx} + \frac{d}{dy}\left(\frac{Q^2}{2gy^2 B^2}\right) \times \frac{dy}{dx} + \frac{d}{dB}\left(\frac{Q^2}{2gy^2 B^2}\right)\frac{dB}{dx}$$

$$S_o - S_f = \frac{dy}{dx} - \frac{Q^2 B}{gA^3}\frac{dy}{dx} - \frac{Q^2 y}{gA^3}\frac{dB}{dx}$$

on rearranging, we get.

$$\frac{dy}{dx} = \frac{S_o - S_f + \frac{Q^2 y}{gA^3}\frac{dB}{dx}}{1 - \frac{Q^2 B}{gA^3}}$$

Problem 5: Show that for horizontal, frictionless rectangular channel of varying width B, the above relation reduces to

$$\left(1 - F^2\right)\frac{dy}{dx} - F^2\left(\frac{y}{B}\right)\frac{dB}{dx} = 0$$

where F = Froude Number.

Solution: From above equation proved in previous question,

$$\frac{dy}{dx} = \frac{S_o - S_f + \frac{Q^2 y}{gA^3}\ \frac{dB}{dx}}{1 - \frac{Q^2 B}{gA^3}}$$

As, in case if horizontal, frictionless rectangular channel, $S_o = 0$ and $S_f = 0$, we get

$$\frac{dy}{dx} = \frac{\frac{Q^2 y}{gA^3}\ \frac{dB}{dx}}{1 - \frac{Q^2 B}{gA^3}} \qquad ...(1)$$

We know that,

Froude number, $F = \dfrac{V}{\sqrt{g\dfrac{A}{T}}}$

or, $F^2 = \dfrac{V^2}{g \times y}$ [In case of rectangular channel $\dfrac{A}{T} = y$]

or, $$F^2 = \frac{Q^2}{gy^3B^2}$$

or, $$F^2 = \frac{q^2}{gy^3} \qquad \text{(where } q = Q/B\text{)}$$

or, $$F^2 = \frac{Q^2}{gB^2y^3} = \frac{Q^2B}{g\left(y^3B^3\right)}$$

$\Rightarrow$ $$F^2 = \frac{Q^2B}{gA^3} \qquad ...(2)$$

Putting these value in equation (1), we get

$$\frac{dy}{dx} = \frac{F^2\left(\frac{y}{B}\right)\frac{dB}{dx}}{1-F^2}$$

$$\left(1-F^2\right)\frac{dy}{dx} - F^2\left(\frac{y}{B}\right)\frac{dB}{dx} = 0$$

where F = Froude number

Problem 6: Show that for a wide rectangular channel the slope is mild or steep according to S_o being less or greater than

$$\left(\frac{n^2 g^{10/9}}{q^{2/9}}\right)$$

Solution: (*a*) To find normal depth

$$Q = \frac{A}{n}R^{2/3}\ S_o^{1/3} \qquad ...(1)$$

For wide rectangular channel $R = y_0$

$$Q = \frac{B \times y_0}{n} \times y_0^{2/3} \times S_o^{1/2}$$

$$q = \frac{y_0^{5/3}}{n} S_o^{1/2}$$

$$\Rightarrow \qquad y_0 = \left(\frac{n^2 q^2}{S_o}\right)^{3/10} \qquad ...(2)$$

(*b*) To find critical depth

$$y_c = \left(\frac{q^2}{g}\right)^{1/3}$$

For mild slope

$$y_0 > y_c$$

$$\left(\frac{n^2 q^2}{S_o}\right)^{3/10} > \left(\frac{q^2}{g}\right)^{1/3}$$

$$S_o < \left(\frac{n^2 g^{10/9}}{q^{2/9}}\right)$$

Similarly for steep channel

$$y_o < y_c$$

$$\left(\frac{n^2 q^2}{S_o}\right)^{3/10} < \left(\frac{q^2}{g}\right)^{1/3}$$

$$\Rightarrow \qquad S_o > \left(\frac{n^2 g^{10/9}}{q^{2/9}}\right)$$

Problem 7: For a wide rectangular channel, if the Manning's formula is used, the differential equation of GVF become

$$\frac{dy}{dx} = S_o \left[\frac{1-(y_0/y)^{3.33}}{1-(y_c/y)^3}\right]$$

Solution: The form of basic differential equation is given by

$$\frac{dy}{dx} = \frac{S_o - S_f}{1 - \dfrac{Q^2T}{gA^3}}$$

$$= S_o \left[\frac{1 - \dfrac{S_f}{S_o}}{1 - \dfrac{Q^2T}{gA^3}} \right]$$

$$\frac{dy}{dx} = S_o \left[\frac{1 - \left(\dfrac{K_o}{K}\right)^2}{1 - \left(\dfrac{Z_c}{Z}\right)^2} \right] \quad ...(1)$$

where $K =$ convenyance of one channel

$$K = \frac{1}{n} A R^{2/3}$$

$$K = \frac{1}{n} B y^{5/3}$$

Similarly for K_o

$$K_o = \frac{1}{n} B\, y_o^{5/3}, \text{ Hence } \frac{K_o}{K} = \left(\frac{y_0}{y}\right)^{5/3} \quad ...(2)$$

Now, we know that

$$Z^2 = C_1 y^M$$

where $M =$ First hydraulic exponent

for rectangular channel $M = 3$.

Hence $Z^2 = C_1 y^3$

$$Z_c^2 = C_1 y_c^3$$

$$\left(\frac{Z_c}{Z}\right)^2 = \left(\frac{y_c}{y}\right)^3 \quad ...(3)$$

Putting these value is equation (1), we get

$$\frac{dy}{dx} = S_o\left[\frac{1-\left(\frac{y_o}{y}\right)^{10/3}}{1-\left(\frac{y_c}{y}\right)^3}\right]$$

$$\frac{dy}{dx} = S_o\left[\frac{1-\left(\frac{y_o}{y}\right)^{3.333}}{1-\left(\frac{y_c}{y}\right)^3}\right]$$

Problem 8: For a very wide rectangular channel, if Chezy formula is used, the differential equation of GVF is given by

$$\frac{dy}{dx} = S_o\left[\frac{1-(y_o/y)^3}{1-(y_c/y)^3}\right]$$

Solution: The basic GVF equation is given by

$$\frac{dy}{dx} = \frac{S_o - S_f}{1-\frac{Q^2T}{gA^3}}$$

$$\frac{dy}{dx} = S_o\left[\frac{1-S_f/S_o}{1-\frac{Q^2T}{gA^3}}\right] \qquad ...(1)$$

By assumption (2) of GVF

$$Q = CA\sqrt{RS_f} \qquad ...(2)$$

(for wide rectangular channel $R = y$)

$$Q = C \times By\sqrt{yS_f} \qquad ...(3)$$

For normal discharge $Q = C \times By_o \times \sqrt{y_o S_o}$ ...(4)

Dividing equations (3) and (4), we get

$$\frac{S_f}{S_o} = \left(\frac{y_o}{y}\right)^3 \qquad ...(5)$$

$$\frac{Q^2 T}{gA^3} = \frac{(y_c)^3}{(y)^3} \qquad ...(6)$$

[because $\frac{Q^2}{gB^2} = y_c^3$]

Putting these value in equation (1), we get

$$\frac{dy}{dx} = S_o\left[\frac{1-\left(\frac{y_o}{y}\right)^3}{1-\left(\frac{y_c}{y}\right)^3}\right]$$

Problem 9: For wide horizontal channel, by assuming the Zy coefficient C = constant, show that the gradually varied flow can be expressed as

$$x = \frac{C^2}{g}\left[y - \frac{y^4}{4y_C^3}\right] + \textbf{constant}$$

Solution: The basic differential equation of GVF is

$$\frac{dy}{dx} = \frac{S_o - S_f}{1 - \frac{Q^2 T}{gA^3}} \qquad ...(1)$$

For a wide rectangular channel, by considering unit width, $T = 1$, $A = y$ and $Q = q = Vy$, $R = y$

Also by Chezy formula

$$V = C\sqrt{RS_f} = C\sqrt{yS_f}$$

$$S_f = \frac{V^2}{C^2 y} = \frac{q^2}{C^2 y^3}$$

For a horizontal channel $S_o = 0$

Hence, with these for a wide horizontal channel, equation (1) becomes

$$\frac{dy}{dx} = \frac{-S_f}{1 - \frac{Q^2 T}{g A^3}} = \frac{-q^2 / C^2 y^3}{1 - \frac{q^2}{g} \frac{1}{y^3}}$$

But $$q^2/g = y_C^3$$

Hence $$\frac{dy}{dx} = \frac{-g/C^2 \left(y_C/y \right)^3}{1 - \left(y_C/y \right)^3}$$

$$dx = \left[\frac{1}{-\frac{g}{C^2}\left(y_C/y \right)^3} + \frac{\left(y_C/y \right)^3}{\frac{g}{C^2}\left(y_C/y \right)^3} \right] dy$$

$$x = \left[-\frac{C^2}{4g} \frac{y^4}{y_C^3} + \frac{C^2}{g} y \right] + \text{constant}$$

$$x = \frac{C^2}{g} \left[y - \frac{y^4}{4 y_C^3} \right] + \text{constant}$$

Problem 10: In a gradually varied flow in a rectangular channel of bottom width 3.0 m the discharge is 8 m³/s and the depth of flow changes from 1.4 m at section M to 1.05 m at section N. Calculate the average energy slope between these two section. Assume n = 0.018.

Solution: In gradually varied flow the Manning's formula is written for any section as

$$V = \frac{1}{n} R^{2/3} S_f^{1/2}$$

where S_f = energy slope at that section

Hence
$$S_f = \frac{n^2 V^2}{R^{4/3}}$$

Average energy slope between two section

$$\overline{S_f} = \frac{S_{f_1} + S_{f_2}}{2}$$

The calculation as shown in table.

Property	***Section M***	***Section N***
A	$3.0 \times 1.4 = 4.2\text{m}^2$	$3 \times 1.05 = 3.15\text{ m}^2$
B	$3.0 + (2 \times 1.4) = 5.8\text{m}$	$3.0 + 2 \times 1.05 = 5.10\text{ m}$
R	$4.2/5.8 = 0.724\text{m}$	$3.15/5.10 = 0.617\text{ m}$
V	$8.0/4.2 = 1.9048\text{m/s}$	$8.0/3.5 = 2.5398\text{ m/s}$
S_f	$\frac{(0.018)^2 \times (1.9048)^2}{(0.724)^{4/3}}$ $= 1.8077 \times 10^{-3}$	$\frac{(0.018)^2 \times (2.5397)^2}{(0.6176)^{4/3}}$ $= 3.973 \times 10^{-3}$

$$\overline{S_f} = \frac{1.8077 \times 10^{-3} + 3.9730 \times 10^{-3}}{2}$$

$$= 2.890 \times 10^{-3}$$

Problem* 11: (*a*) *Intergrate the differential equation of GVF for a horizontal channel to get the profile equation as

$$x = \frac{y_C}{S_C}\left[\frac{\left(\frac{y}{y_C}\right)^{N-M+1}}{N-M+1} - \frac{\left(\frac{y}{y_C}\right)^{N+1}}{N+1}\right] + \textbf{constant}$$

where S_C ***= critical slope.***

(b) By using the result of part (a) above develop an equation for GVF profile in a wide, rectangular, horizontal channel.

Solution:

$$\frac{dy}{dx} = \frac{S_o - S_f}{1 - \dfrac{Q^2 T}{gA^3}} \qquad ...(1)$$

For a horizontal channel, $S_o = 0$

Also,
$$Sf = \frac{Q^2}{K^2} = \frac{K_C^2 S_C}{K^2} = \left(\frac{y_C}{y}\right)^N S_C$$

Further,
$$\frac{Q^2}{g} = Z_C^2 \text{ and } \frac{A^3}{T} = Z^2$$

Thus
$$\frac{Q^2 T}{gA^3} = \left(\frac{y_C}{y}\right)^M$$

Equation (1), now written as

$$\frac{dy}{dx} = \frac{-\left(\frac{y_C}{y}\right)^N S_C}{1 - \left(\frac{y_C}{y}\right)^M}$$

$$-dx.\, S_C = \left[1 - \left(\frac{y_C}{y}\right)^M\right] \frac{1}{\left(\frac{y_C}{y}\right)^N} dy$$

$$-dx.\, S_C = \left[\left(\frac{y}{y_C}\right)^N - \left(\frac{y}{y_C}\right)^{N-M}\right] dy$$

$$\therefore -S_C.\, x = \left[\frac{\left(\frac{y}{y_C}\right)^{N+1}}{N+1} - \frac{\left(\frac{y}{y_C}\right)^{N-M+1}}{N-M+1}\right] + \text{constant}$$

$$\therefore x = \frac{y_C}{S_C}\left[\frac{\left(\frac{y}{y_C}\right)^{N-M+1}}{N-M+1} - \frac{\left(\frac{y}{y_C}\right)^{N+1}}{N+1}\right] + \text{constant}$$

(*b*) For a wide rectangular channel,

$$S_C = \frac{n^2 g}{y_C^{1/3}}$$

We know that $N = 10/3$ and $M = 3$, for rectangular channel substituting for S_C, N and M in result of part (*a*)

$$\therefore \quad x = \frac{y_C}{S_C}\left[\frac{\left(y/y_C\right)^{N-M+1}}{N-M+1} - \frac{\left(y/y_C\right)^{N+1}}{N+1}\right] + \text{constant}$$

We get for a wide rectangular, horizontal channel, the GVF profiles as

$$\therefore \quad x = \frac{y_C^{4/3}}{n^2 g}\left[\frac{3}{4}\left(y/y_C\right)^{4/3} - \frac{3}{13}\left(y/y_C\right)^{13/3}\right] + \text{constant}$$

Problem 12: ***A wide rectangular channel of slope 0.0003 and n = 0.02 connects two reservoir 1.5k m apart. The upstream reservoir level is constant at an elevation of 104.00 m and the elevation of canal at the intake is 101.00 m. The intake is free and the loss of energy at the intake can be neglected. (a) What should be the downstream level to cause uniform flow in the entire length of the channel?***

Solution:

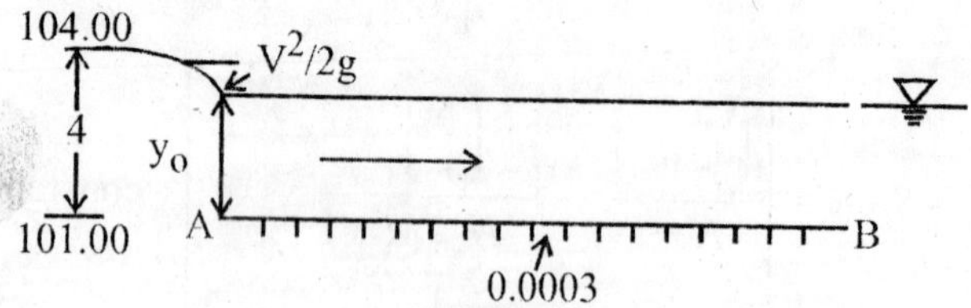

Fig. 3.11

As given in Fig. 3.11. Neglecting losses at the entry.

$$H = y_0 + \frac{V_0^2}{2g}$$

As the channel is wide,

$$V_0 = \frac{1}{n} y_0^{2/3} S_0^{1/2}$$

$$y_0 = H - \frac{1}{2g}\left[\frac{1}{n} y_0^{2/3} S_0^{1/2}\right]^2$$

$$y_0 = 3.0 - \frac{1}{2 \times 9.81}\left[\frac{1}{(0.02)^2} y_0^{4/3} \times 0.0003\right]$$

Solving by trial-and-error,

$$y_0 = 2.70 \text{ m}$$

Bed unit at downstream end of the channel *i.e.*, at the section B.

$$= 101.00 - (0.003 \times 1500)$$

$$= 100.55 \text{ m}$$

Down stream pool elevation.

$$= 100.55 + 2.70$$

$$= 103.25 \text{ m.}$$

Problem* 13: *In a rectangular channel 12 m wide and 3.6 m deep water is flowing with a velocity of 1.2 m/s. The bed slope of the channel is 1/4000. If flow of water through the channel is regulated in such a way that energy line is having a slope of 0.00004. Find the rate of change of depth of water in the channel?

***Solution*:**

Width of the Channel $b = 12$ m

Depth of Channel $y = 3.6$ m

Velocity of flow $v = 1.2$ m/s

Bed slope $S_F = 1/4000 = 0.00025$

Slope of energy line $S_e = 0.00004$

$$\frac{dy}{dx} = ?$$

$$\frac{dy}{dx} = \frac{S_0 - S_f}{1 - \frac{V^2}{gy}}$$

$$\frac{dy}{dx} = \frac{0.00025 - 0.00004}{\left(1 - \frac{1.2^2}{9.81 \times 3.6}\right)}$$

$$\frac{dy}{dx} = 0.0002189$$

Problem 14: ***A weir is installed across a rectangular open channel thereby raising the flow depth from 1.5 m in a normal flow to 2.5 m of the weir. The width of the channel is 10 m and it is laid to a slope of 1/10000 find an approximate length of back water curve considering the average velocity, average depth and average slope midway between the two sections. Take the value of mannings rugosity coefficient equal to 0.02.***

Solution:

Upstream section 1 – 1

Width of the channel, $b_1 = 10$ m

$y_1 = 1.5$ m

Area of flow $A_1 = b_1 y_1 = 10 \times 1.5 = 15$ m2

Wetted perimeter $P_1 = b_1 + 2y_1 = 10 + 2 \times 1.5 = 13$ m

Hydraulic radius $R_1 = \frac{A_1}{P_1} = \frac{15}{13} = 1.154\,\text{m}$

Chezy's constant $C_1 = \frac{1}{n} R^{1/6} = \frac{1}{0.02} \times (1.154)^{1/6}$

$C_1 = 51.2$

Velocity of flow $V_1 = C_1 \sqrt{RS_0}$

$$= 51.2 \sqrt{1.154 \times \frac{1}{10000}}$$

$$= 0.55 \text{ m/s}$$

$$E_1 = y_1 + \frac{V_1^2}{2g} = 1.5 + \frac{0.55^2}{2 \times 9.81} = 1.515\,\text{m}$$

Downstream Section 2 – 2

$b_2 = 10$ m

$$y_2 = 2.5 \text{ m}$$

$$A_2 = b_2 y_2 = 10 \times 2.5 = 25 \text{ m}^2$$

Wetted perimeter $P_2 = b_2 + 2y_2 = 10 + 2 \times 2.5 = 15$ m

$$R_2 = \frac{A_2}{P_2} = \frac{25}{15} = 1.667 \text{ m}$$

$$A_1 V_1 = A_2 V_2$$

$$V_2 = \frac{A_1 V_1}{A_2} = \frac{15 \times 0.55}{25} = 0.33 \text{ m/s}$$

Specific energy $E_2 = y_2 + \frac{V_2^2}{2g}$

$$= 2.5 + \frac{0.33^2}{2 \times 9.81} = 2.505 \text{ m}$$

The value of S_f (energy line slope) is calculated by Chezy's formula corresponding to flow condition at the average depth of flow.

Average depth of flow $= \frac{y_1 + y_2}{2} = y = \frac{1.5 + 2.5}{2} = 2$ m

Area of flow; $A = b_y = 10 \times 2 = 20$ m2

wetted perimeter $P = b + 2y = 10 + 2 \times 2 = 14$ m2

Hydraulic radius $R = \frac{A}{P} = \frac{20}{14} = 1.428$ m

$$AV = A_1 V_1$$

$$V = \frac{A_1 V_1}{A} = \frac{15 \times 0.55}{20} = 0.41?5 \text{ m/s}$$

Chezy's constant $C = \frac{1}{n} R^{\frac{1}{6}} = \frac{1}{0.02} \times (1.428)^{\frac{1}{6}} = 53.06$

Velocity, $V = C\sqrt{RS_f}$

$$0.4125 = 53.06\sqrt{1.428 \times S_f}$$

$$S_f = 0.0000423$$

Length of backwater curve

$$L = \frac{E_2 - E_1}{S_0 - S_f} = \frac{2.505 - 1.515}{0.0001 - 0.0000423}$$

$$= 17157 \text{ m or } 17.157 \text{ km}$$

Problem 15: Show that the gradually varied flow equation for flow in a rectangular channel of variable width B may be expressed as

$$\frac{dy}{dx} = \frac{S_0 - S_f + \dfrac{V^2}{gB}\dfrac{dB}{dx}}{1 - F^2}$$

Solution: We know that

$$H = z + y + \frac{V^2}{2g}$$

differentiating with respect to x

$$\frac{dH}{dx} = \frac{dz}{dx} + \frac{dy}{dx} + \frac{d}{dx}\left(\frac{V^2}{2g}\right)$$

$$-S_f = -S_0 + \frac{dy}{dx} + \frac{d}{dx}\left(\frac{Q^2}{2gA^2}\right)$$

$$S_0 - S_f = \frac{dy}{dx} + \frac{d}{dx}\left(\frac{Q^2}{2gB^2y^2}\right)$$

$$S_0 - S_f = \frac{dy}{dx} + \frac{d}{dx}\left(\frac{Q^2}{2gy^2B^2}\right)\frac{dy}{dx} + \frac{d}{dB}\frac{Q^2}{2gy^2B^2}\frac{dB}{dx}$$

$$S_0 - S_f = \frac{dy}{dx} - \frac{Q^2B}{gA^3}\frac{dy}{dx} - \frac{Q^2y}{gA^3}\frac{dB}{dx}$$

or rearranging, we get

$$\frac{dy}{dx} = \frac{S_0 - S_f + \dfrac{Q^2 y}{gA^3}\dfrac{dB}{dx}}{1 - \dfrac{Q^2 B}{gA^3}} \qquad \ldots(a)$$

put $\quad \dfrac{Q^2 y}{gA^3} = \dfrac{A^2 V^2 y}{gA^3} = \dfrac{V^2}{gB}$ and $\dfrac{Q^2 B}{gA^3}$ can be written as

$$\frac{A^2 V^2 B}{gA^3} = \frac{V^2}{gy}$$

where $\dfrac{V^2}{gy} = F^2$ in case of rectangular channel put all the value in equation (a) we get

$$\frac{dy}{dx} = \frac{S_0 - S_f + \dfrac{V^2}{gB}\dfrac{dB}{dx}}{1 - F^2}$$

Problem 16: ***A frictionless liquid flows along a flat, open channel of variable breadth B subject to the influence of gravity g, obtain expression***

$$\frac{dy}{dx} = -\frac{\dfrac{dB}{B}}{1 - \left(\dfrac{gy}{V^2}\right)},$$ ***where y is the depth of the liquid and V is the mean velocity of flow.***

Solution: We know that

$$H = z + y + \frac{V^2}{2g}$$

Differentiating with respect to x we get

$$\frac{dH}{dx} = \frac{dz}{dx} + \frac{dy}{dx} + \frac{d}{dx}\left(\frac{V^2}{2g}\right)$$

$$-S_f = -S_0 + \frac{dy}{dx} + \frac{V}{g}\frac{dV}{dx}$$

Since in a frictionless liquid there is no energy loss, the total energy line is horizontal

therefore = $S_0 = S_f = 0$

or
$$0 = \frac{dy}{dx} + \frac{V}{g}\frac{dv}{dx}$$

$$\frac{dy}{dx} = \frac{V^2}{g}\left(\frac{1}{y}\frac{dy}{dx} + \frac{1}{B}\frac{dB}{dx}\right)$$

$$\frac{dy}{dx}\left(1 - \frac{V^2}{gy}\right) = \frac{V^2}{gy}\frac{dB}{dx}$$

or
$$\frac{dy}{dx} = -\left(\frac{\frac{dB}{B}}{1 - \frac{gy}{V^2}}\right) \quad \textbf{Proved.}$$

Problem 17: *Show that the following relationship in GUV infrictionless rectangular channel is given by*

$$x = \frac{y}{S_0}\left[1 + \frac{1}{2}\left(\frac{y_c}{y}\right)^3\right] + \text{Const.}$$

Solution: The General equation for GVF is given by:

$$\frac{dy}{dx} = \frac{S_0 - S_f}{1 - \frac{Q^2 T}{gA^3}} \quad \ldots(a)$$

As channel is frictionles, $\therefore S_f = 0$

Also since channel is rectangular

$$\frac{Q^2T}{gA^3} = \frac{Q^2B}{gB^3y^3} = \left(\frac{q^2}{g}\right)\frac{1}{y^3} = \left(\frac{y_c}{y}\right)^3$$

As we know $\left(\dfrac{q^2}{g} = y_c^3\right)$

Now equation a reduces to

$$\frac{dy}{dx} = \frac{S_0}{1-\left(\dfrac{y_c}{y}\right)^3}$$

$$1-\left(\frac{y_c}{y}\right)^3 dy = S_0\, dx$$

Assume channel as rectangular channel

The continuity equation is given by

$Q = ByV$, Since width varies with x differentiating continuity equation wrt x

$$\frac{dQ}{dx} = By\frac{dV}{dx} + BV\frac{dy}{dx} + Vy\frac{dB}{dx}$$

Dividing throughtout ByV and for constant discharage, $\dfrac{dQ}{dx} = 0$

we get

$$0 = \frac{1}{y}\frac{dy}{dx} + \frac{1}{V}\frac{dV}{dx} + \frac{1}{B}\frac{dB}{dx} \qquad \ldots(a)$$

as we know $\dfrac{dy}{dx} = -\dfrac{V}{g}\dfrac{dV}{dx}$ $\qquad \ldots(b)$

From equation (a) we get

$$-\frac{dV}{dx} = V\left[\frac{1}{y}\frac{dy}{dx} + \frac{1}{B}\frac{dB}{dx}\right]$$

Put in (b) we get

or $$S_0 x = \left(y - \frac{y_c^3}{= 2y^2} \right)$$

or $$x = \frac{y}{S_0}\left[1 + \frac{1}{2}\left(\frac{y_c}{y}\right)^3\right] + \text{Const}$$

Problem 18: The normal depth of flow in a channel having the cross-section shown in Fig. 3.12 is 1.5 m. The channel slope is 1 × 10^{-4} and n 0.020. A weir at the downstream end of the chanel raises the water depth to 3.0 m immediately behind the weir. Obtain the surface profile upstream of the weir.

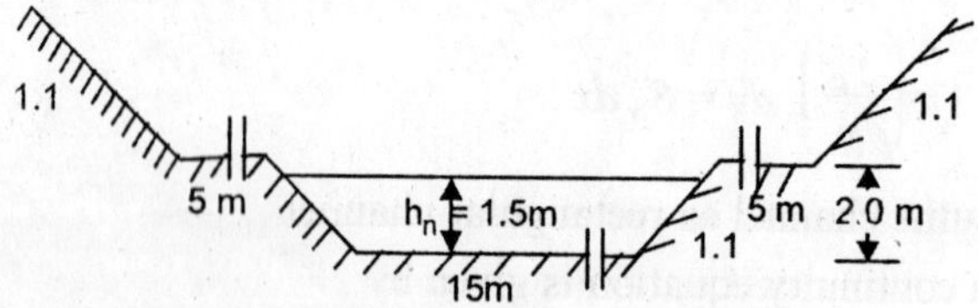

Fig. 3.12: Cross-sectional shape of channel (example 6.1)

Solution: By Manning's formula

$$U = \frac{1}{n} R^{2/3} S^{1/2}$$

$$= \frac{1}{0.02}\left[\frac{(15+1.5)1.5}{15+2\sqrt{2}\times 1.5}\right]^{2/3} \times (10^{-5})^{1/2}$$

$$= 0.591 \text{ m/s}$$

$$Q = A\,U$$

$$= (15 + 1.5)\ 1.5 \times 0.591$$

$$= 14.63 \text{ m}^3\text{/s}$$

$$\frac{U}{\sqrt{gD}} = \frac{0.591}{\sqrt{9.8 \times \frac{(15+1.5)1.5}{15+2\times 1.5}}} = 0.161 < 1.0$$

Hence the flow is subcritical, i.e. the slope is mild. Since $h > h_n > h_c$, the profile formed upstream of the weir is an M_1 profile. The depth of 3.0 m upstream of the weir acts as the control and calculations may be made up to 1.01 h_n, i.e. 1.515 m as shown in Table.

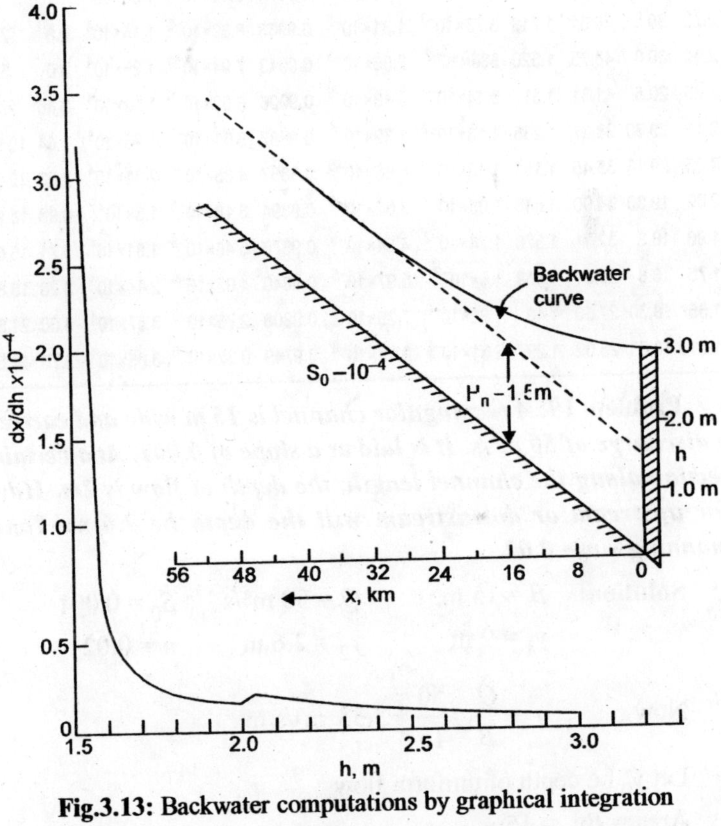

Fig.3.13: Backwater computations by graphical integration

The curve of h versus dx/dh shows a kink near $h = 2.0$ m because of the sudden increase in channel width at this elevation. The distance between any two stations as well as the distance from the weir are entered in the last two columns. The backwater curve plotted using these results is shown in the inset of Figure. A depth one per cent greater than the normal depth is seen to be attained at about 45 km from the weir. This length is often thought of as the length of the M_1 curve.

Computations by Graphical Integration Method

h, m	T m	A_2 m	R, m	$\frac{Q^2T}{gA^3}$	$\frac{Q^2n^2}{A^2R^{4/3}}$	$1-\frac{Q^2T}{A^2R^{4/3}}$	$S_0-\frac{Q^2n^2}{A^2R^{4/2}}$	$\frac{dx}{dh}$	Dx, km	x km
3.0	31.0	64.0	1.911	2.58×10^{-2}	8.83×10^{-6}	0.9974	9.12×10^{-4}	1.09×10^{4}		
2.75	30.5	56.31	1.718	3.73×10^{-3}	1.31×10^{-5}	0.9963	8.69×10^{-5}	1.15×10^{4}	2.8	2.8
2.50	30.0	48.75	1.520	566×10^{-2}	2.06×10^{-5}	0.9943	7.94×10^{-5}	1.25×10^{4}	3.0	5.8
2.25	29.5	41.31	1.317	9.14×10^{-3}	3.48×10^{-5}	0.9906	6.52×10^{-5}	1.52×10^{4}	3.46	9.26
2.15	29.30	38.37	1.235	1.13×10^{-2}	4.39×10^{-5}	0.9887	5.61×10^{-5}	1.76×10^{4}	1.64	10.90
2.05	29.10	35.45	1.151	1.43×10^{-2}	5.65×10^{-5}	0.9857	4.35×10^{-5}	2.30×10^{4}	2.03	12.93
2.00	19.00	34.00	1.646	1.06×10^{-2}	3.82×10^{-5}	0.9894	6.18×10^{-5}	1.6×10^{4}	0.98	13.91
1.90	18.8	32.11	1.576	1.24×10^{-2}	4.54×10^{-5}	0.9876	5.46×10^{-5}	1.81×10^{4}	1.71	15.62
1.75	18.8	29.31	1.469	1.6×10^{-2}	5.97×10^{-5}	0.9840	4.03×10^{-5}	2.44×10^{4}	3.20	18.82
1.65	18.30	27.50	1.40	1.92×10^{-2}	7.25×10^{-5}	0.9808	2.75×10^{-5}	3.57×10^{4}	3.00	21.82
1.515	18.03	25.02	1.297	2.51×10^{-2}	9.68×10^{-5}	0.9749	0.32×10^{-5}	3.05×10^{4}	22.70	44.52

Problem 19: ***A rectangular channel is 15 m wide and carries a discharge of 50 m^3/s. It is laid at a slope of 0.001. At a certain section along the channel length, the depth of flow is 2m. How far upstream or downstream will the depth be 2.6 m. Take manning's n = 0.02.***

Solution: $B = 15$ m, $\quad Q = 50$ m³/s $\quad S_0 = 0.001$

$y_1 = 2$ m $\quad y_2 = 2.6$ m $\quad n = 0.02$

Now $\quad q = \dfrac{Q}{B} = \dfrac{50}{15} = 3.33$ m³/s/m

Let y_0 be depth of uniform flow

Area $= By_0 = 15y_0$

Wetted perimeter $= P = B + 2y_0 = 15 + 2y_0$

Hydraulic radius $R = \dfrac{A}{P} = \dfrac{15y_0}{15+2y_0}$

Discharge $\quad Q = A \times V$

$$= A \times \frac{1}{n} \times R^{2/3} S_0^{1/2}$$

$$50 = 15y_0 \times \left[\frac{1}{0.02} \times \left(\frac{15y_0}{15+2y_0}\right)^{2/3} \times 0.001\right]^{1/2}$$

$$50 = 7.5\ y_0 \times \left(\frac{15y_0}{15+2y_0}\right)^{2/3}$$

$$1.09 = \frac{y_0^{5/3}}{(15+2y_0)^{2/3}}$$

Using calculator by trial and error,

$$y_0 = 3.6$$

Note: Make above equation in calculator try different value of y_0 by assuming $y_0 = 1, 2, 3, 4$ etc.

The value of y_0 which in very near to 1.09 is taken as y_0. That why this method called trial and error method.

$$\text{Now critical depth } y_c = \left(\frac{q^2}{g}\right)^{1/3}$$

$$y_c = \left(\frac{3.33^2}{9.8}\right)^{1/3} = 1.04 \text{ m}$$

Since $y_0 > y_c$, the slope is mild. Slope. Also the value of $y = 2$ to 2.6 therefore $y_0 > y > y_c$, the profile is M_2 profile for computation of M_2 profile consider three step with depth increment = $\frac{2.6-2}{3} = 0.2$ m.

The length between depth of flow 2m to a section where depth is 2.6 m is 1887.69.

Important notes: $\bar{S}_f$ is calculated by taking average of two successive S_f as calculated by column (8) for example,

Table Problem 19

1	2	3	4	5	6	7	8	9	10	11	12	13
Depth y (m)	Area $A = By$ $= 15y$	Perimeter $P = B + 2y$ $P = 15+2y$	Hydraulic *radius* $R = A/P$	velocity $V = Q/A$ $= 50/A$	$\frac{V^2}{2g}$	$E = y + \frac{V^2}{2g}$	$S_f = \frac{n^2V^2}{R^{4/3}}$	$\bar{S}_f$	$S_0 - \bar{S}_f$	$\Delta E = E_2 - E_1$	$dx = \frac{\Delta E}{S_0 - \bar{S}_f}$	Σx
2	30	19	1.578	1.66	0.14	2.14	6.04×10^{-4}					
								5.28×10^{-4}	-4.28×10^{-4}	0.18	420.56	
2.2	33	19.4	1.701	1.51	0.12	2.32	4.52×10^{-4}					
								4×10^{-4}	-3×10^{-4}	0.17	566.66	
2.4	36	19.8	1.818	1.38	009	2.49	3.48×10^{-4}					
								3.11×10^{-4}	-2.11×10^{-4}	0.19	900.47	
2.6	39	20.2	1.930	1.28	0.08	2.68	2.74×10^{-4}					
											Total	1887.69

$$\overline{S}_f = \frac{S_{f_2} + S_{f_2}}{2}$$

$$= \left(\frac{6.04 + 4.52}{2}\right) \times 10^{-4} = 5.28 \times 10^{-4}$$

Similarly we can calculate further value of $\overline{S}_f$.

Difference of specific energy ΔE is calculatd by taking the difference of two successive specific energy for example as in above case from table

$$\Delta E = E_2 - E_1 = 2.32 - 2.14 = 0.18$$

Similarly all other value can be calculated also Σx can be calculated out by taking sum of all Δx. Therefore from above problem

$$\Sigma x = 420.56 + 56.66 + 900.47$$
$$= 1887.69 \text{ m}$$

Problem 20: ***A rectangular channel 12 m wide carries a discharge of 30 m³/s. It is laid at a shope of 0.001. If at a section in this channel the depth is 1.6 m how far upstream or downstream from the section will the depth be 2.0 m. Take Manning's n = 0.015.***

Solution: $B = 12$ m, $Q = 30$ m³/s, $n = 0.015$

$S_0 = 0.001$, $y_1 = 1.6$ m, $y_2 = 2$ m

$$q = \frac{Q}{B} = \frac{30}{12} = 2.5 \text{ m}^3\text{/s/m}$$

Let y_0 = normal depth

$$\text{Area} = By_0 = 12y_0$$

$$P = B + 2y_0 = 12 + 2y_0$$

$$R = \frac{A}{P} = \frac{12y_0}{12 + 2y_0}$$

$$Q = AV = 12y_0\left[\frac{1}{n} \cdot R^{2/3} S_0^{1/2}\right]$$

$$30 = 12y_0\left[\frac{1}{0.015}\cdot\left(\frac{12y_0}{12+2y_0}\right)^{2/3}\cdot 0.0001^{1/2}\right]$$

$$3.75 = y_0\left(\frac{12y_0}{12+2y_0}\right)^{2/3}$$

$$3.75 = y_0 \times \left(\frac{y_0}{1+\dfrac{y_0}{6}}\right)^{2/3}$$

$$3.75 = \frac{y_0^{5/3}}{\left(1+\dfrac{y_0}{6}\right)^{2/3}}$$

By calculator using trial and error

$$y_0 = 2.55 \text{ m}$$

$$\text{Now } y_c = \left(\frac{q^2}{g}\right)^{1/3} = \left(\frac{2.5}{9.8}\right)^{1/3} = 0.63 \text{ m}$$

Since $y_0 > y_c'$, the profile is mild.

Also $y_0 > y > y_c$, as the depth 1.6m to 2 m his between y_0 and y_c hence surface profile is M_2 curve.

Computation of M_2 profile:

Consider 4 step with depth increment

$$= \frac{2-1.6}{4} = 0.1$$

Hence the required distance between two section is 2072.16 m.

For Table, see the next page.

Problem 21: ***A wide rectangular channel Conney's a discharge of 4 m^3/s/m with bed slope of 0.0001, n = 0.02. If the depth of***

Table Problem 20

1	2	3	4	5	6	7	8	9	10	11	12	13
y	$A = By$ $= 12y$	$P = B + 2y$ $P = 12+2y$	$R = A/P$	$V = Q/A$ $= 30/A$	$\frac{V^2}{2g}$	$E = y + \frac{V^2}{2g}$	$S_f = \frac{n^2V^2}{R^{4/3}}$	$\bar{S}_f$	$S_0 - \bar{S}_f$	$\Delta E =$ $E_2 - E_1$	$dx = \frac{\Delta E}{S_0 - \bar{S}_f}$	Σdx
1.6	19.2	15.2	1.26	1.56	0.124	1.72	4.02×10^{-4}					
								3.68×10^{-4}	-2.68×10^{-4}	0.09	335.82	
1.7	20.4	15.4	1.32	1.47	0.11	1.81	3.34×10^{-4}					
								3.07×10^{-4}	-2.07×10^{-4}	0.08	386.47	
1.8	21.6	15.6	1.38	1.38	0.098	1.89	2.8×10^{-4}					
								2.5×10^{-4}	-1.55×10^{-4}	0.009	580.64	
1.9	22.8	15.8	1.44	1.31	0.088	1.98	2.3×10^{-4}					
								2.17×10^{-4}	-1.17×10^{-4}	0.09	769.23	
2	24	16	1.5	1.25	0.079	2.07	2.05×10^{-4}					
											Σdx	2072.16m

section is 4m determine how far upstream or downstream of the sectin the depth would be within 10% of normal depth. Take 2 step.

Solution: q =4 m3/s/m $\qquad S_0 = 0.0001$

$y_1 = 3.5 \qquad n = 0.02$

For wide rectangular channel:

$$R = y_0$$

$$q = \frac{Q}{B} = \frac{AV}{B} = y_0\left[\frac{1}{n}R^{2/3}S_0^{1/2}\right]$$

$$4 = y_0 \times \left[\frac{1}{0.02} \times (y_0)^{2/3} \cdot (0.0001)^{1/2}\right]$$

$$B = y_0^{5/3}$$

$$y_0 = 3.48 \text{ m}$$

$$y_c = \left(\frac{q^2}{g}\right)^{1/3} = \left(\frac{4^2}{9.8}\right)^{1/3} = 1.17 \text{ m}$$

$y_0 > y_c$, hence Mild slope.

Since $y = 4$ m is greater than y_0 and y_c the profile is

$$y > y_0 > y_c \qquad (M_1 \text{ Profile})$$

In case of M1 profile depth within 10% of mean depth means 10% more than normal depth.

$$y = 1.1\, y_0 = 1.1 \times 3.48 = 3.83$$

For computation of M_1 profile

Consider two steps with depth increment

$$\frac{4 - 3.83}{2} = 0.085$$

Depth will be within 10% of normal depth $a \times a$ distance of 5092.72 m on upstream of sections.

Table Problem 21

1	2	3	4	5	6	7	8	9	10
Depth y	velocity $V = Q/A$	$\frac{V^2}{2g}$	$E = y + \frac{V^2}{2g}$	$S_f = \frac{n^2V^2}{R^{4/3}}$	$\bar{S}_f$	$S_0 - \bar{S}_f$	$\Delta E = E_2 - E_1$	$dx = \frac{\Delta E}{S_0 - \bar{S}_f}$	Σx
3.83	1.04	0.055	3.885	0.72×10^{-4}					
					0.695×10^{-4}	3.05×10^{-5}	0.083	2721.3	
3.915	1.02	0.053	3.968	0.67×10^{-4}					
					0.65×10^{-4}	3.5×10^{-5}	0.083	2371.42	
4	1	0.051	4.051	0.63×10^{-4}					
								Σx	5092.72 m

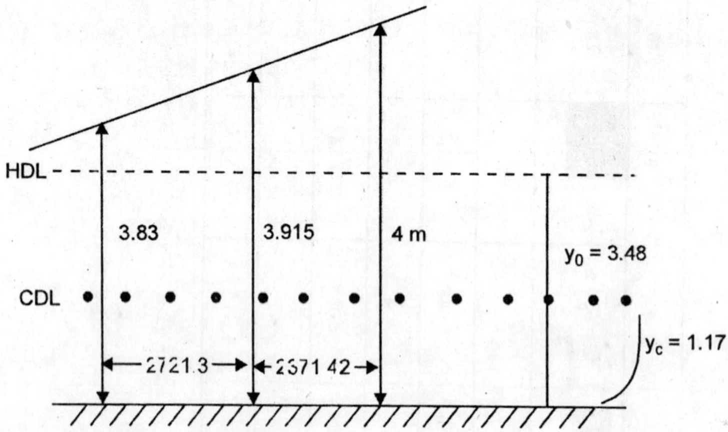

Problem 22: ***A river bed 80 m wide and 2m deep has an average bed slope of 0.0005. Estimate the length of GVF profile produced by a low dam which raises the water surfaces just upstream if it by 1.5 m. Take manning's constant n = 0.035.***

Solution: $B = 80$ m $y_0 = 2$ m, $S_0 = 0.0005$ $n = 0.035$

$$q = \frac{Q}{B} = y_0 \times \left(\frac{1}{n} R^{2/3} S_0^{1/2}\right)$$

for river $\boxed{R = y_0}$

$$q = \frac{1}{n} y_0 \times y_0^{2/3} S_0^{1/2}$$

$$q = \frac{1}{n} y_0^{5/3} \times (0.0005)^{1/2}$$

$$= \frac{1}{0.035} 2^{5/3} \times (0.0005)^{1/2}$$

$$q = 2.03 \text{ m}^3/\text{s/m}$$

$$y_c = \left(\frac{q^2}{g}\right)^{1/3} = \left(\frac{2.03^2}{9.8}\right)^{1/3} = 0.14 \text{ m}$$

$$y = y_0 + 1.5 = 2 + 1.5 = 3.5$$

Table Problem 22

1	2	3	4	5	6	7	8	9
Depth y	velocity $V = Q/A$	$\frac{V^2}{2g}$	$E = y + \frac{V^2}{2g}$	$S_f = \frac{n^2V^2}{R^{4/3}}$	$\bar{S}_f$	$S_0 - \bar{S}_f$	$\Delta E = E_2 - E_1$	$dx = \frac{\Delta E}{S_0 - \bar{S}_f}$
2	1.015	0.05	2.05	0.5×10^{-4}				
					0.36×10^{-3}	1.35×10^{-4}	0.484	3585.18
2.5	0.812	0.034	2.534	0.23×10^{-3}				
					0.18×10^{-3}	3.2×10^{-4}	0.489	1528.125
3	0.67	0.023	3.023	0.13×10–3				
					0.1×10^{-3}	4×10^{-4}	0.494	1235
3.5	0.58	0.017	3.517	0.077×10^{-3}				
							Σx	6348.3

Since $y > y_0 > y_c$, the GVF profile is M_1 curve with the depth of 3.5 m at the low dam as the control.

Computation of M_1 profile:

$$\text{Consider three step : } \frac{35-2}{3} = 0.5\text{ m}$$

Problem 23: ***A trapezoidal channel having b = 20 ft, z == 2, S_0 = 0.0016, and n = 0.025 carries a discharge of 400 cfs. Compute the backwater profile created by a dam which backs up the water to a depth of 5ft. immediately behind the dam. The upstream end of the profile is assumed at a depth equal to 1% greater than the normal depth. The energy coefficient α = 1.10.***

Solution: The critical and normal depths are found to be y_c - 2.22 ft and y_n = 3.36 ft, respectively. Since y_n is greater than y_c and the flow starts with a depth greater than y_n, the flow profile of the M_1 type. The section factor $Z_c = 400/\sqrt{g/\alpha} = 74.0$ and the conveyance $K_n = 400/\sqrt{0.0016} = 10{,}000$.

For simplicity, the channel bottom at the site of the dam is chosen as the origin and the *x* value in the upstream direction is taken as positive. The computation of *dx*/*dy* is given in Table for various values of *y* varying from 5 ft. to 1% greater than y_n or 3.40 ft. For instance, when *y* = 5.00 ft, the value in other columns of the table are

$$T = 40.00\text{ ft}$$

$$A = 150.00\text{ ft}^2$$

$$R = 3.54\text{ ft}$$

$$R^{2/3} = 3.323$$

$$K = \frac{1.49AR^{2/3}}{n} = \frac{1.49 \times 150.00 \times 2.323}{0.025} = 20{,}800$$

$$Z = \sqrt{\frac{A^3}{T}} = \sqrt{\frac{150.00^3}{40{,}000}} = 290.2$$

Table: Computation of the flow profile for example by Graphical Integration

$Q = 400$ cfs $n = 0.025$ $S_0 = 0.0016$ $y_c = 2.22$ ft $y_n = 3.36$ ft $\alpha = 1.10$

y	T	A	R	$R^{2/3}$	K	Z	dx/dy	DA	x
5.00	40.00	150.00	3.54	2.323	20,800	290.2	760		
4.80	39.20	142.08	3.43	2.274	19,234	270.4	792	155	155
4.60	38.40	134.32	3.31	2.221	17,770	251.5	836	163	318
4.40	37.60	126.72	3.19	2.167	16,360	232.3	913	175	493
4.20	36.80	119.28	3.08	2.117	15,050	214.5	1,000	191	684
4.00	36.00	112.00	2.96	2.062	13.750	197.5	1,140	214	898
3.80	35.20	104.88	2.84	2.006	12,550	181.0	1,130	257	1,155
3.70	34.80	101.38	2.77	1.972	11,910	173.0	1,750	159	1,314
3.60	34.40	97.92	2.71	1.944	11,350	165.0	2,260	201	1,515
3.55	34.20	96.21	2.68	1,929	11,060	161.1	2,770	126	1,641
3.50	34.00	94.50	2.65	1.916	10,800	157.3	3,480	156	1,797
3.47	33.88	93.48	2.63	1.904	10,600	155.2	4,520	120	1,917
3.44	33.76	92.45	2.61	1,894	10,440	153.0	5,990	158	2,075
3.42	33.68	91.80	2.60	1.890	10,340	151.7	7,930	139	2,214
3.40	33.60	91.12	2.59	1.886	10,230	150.0	10,760	187	2,401
3.36	33.44	89.78	2.56	1.872	10,000	147.0			

$$\frac{dx}{dy} = \frac{1}{S_0}\frac{1-(Z_c/Z)^2}{(K_n/K)^2}$$

$$= \frac{1}{0.0016}\frac{1-(74.0/290.2)^3}{1-(10{,}000/20{,}800)^2} = 760$$

Value of y are then plotted against the corresponding value of dx/dy (Fig.). The increments in area ΔA are plainmetered and the table (the cumulative value of ΔA should give the length x of the flow profile. Finally, the backwater profile is obtained by plotting y against x).

It should be noted that, when the depth approaches the normal depth,the incremental area varies so greatly with the change in y value that it becomes difficult to plainmeter. In such a case, the area may be computed by assuming it is a trapezoid. For instance, the incremental area between $y = 3.42$ and 3.40 is $\Delta A = (7{,}930 + 10{,}760)$ $3.42 - 3.40)/2 = 187$.

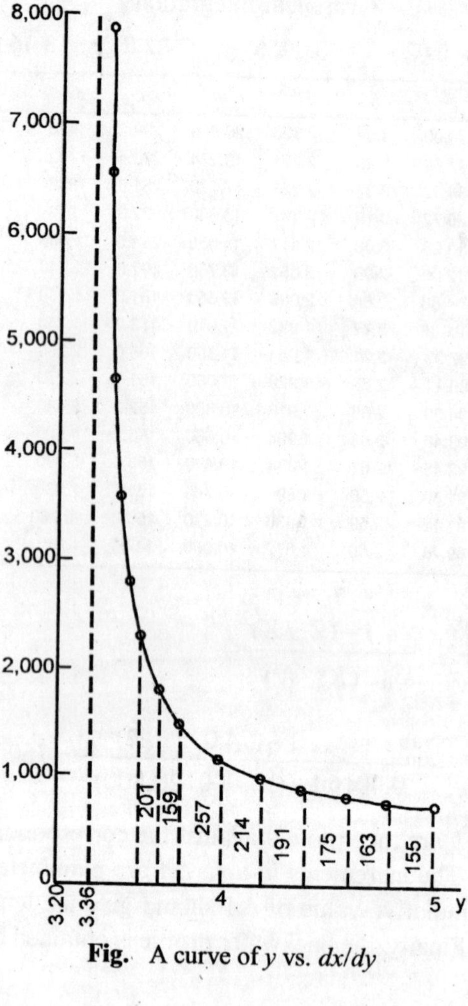

Fig. A curve of y vs. dx/dy

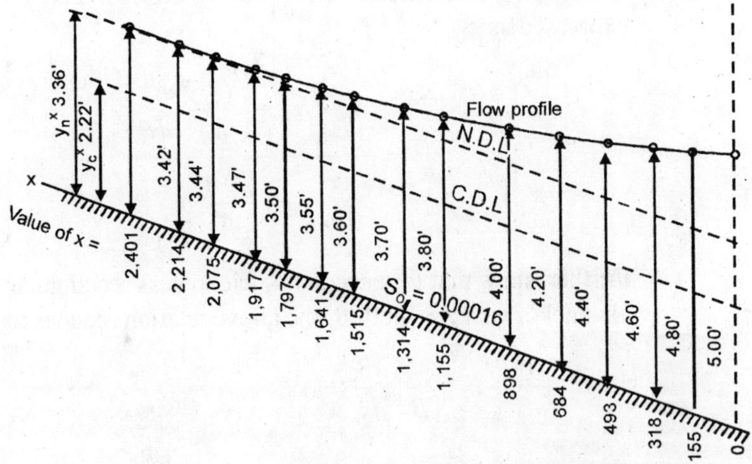

Fig. An M_1 flow profile computed by the graphical-integration method

Review Questions

1. Discuss the Assumption of Gradually veried flow?
2. Prove that dynamic equation of GVF is given by:

$$\frac{dy}{dx} = \frac{S_o - S_f}{1 - \dfrac{Q^2 T}{gA^3}}$$

3. What are the limitations of GVF?
4. Describe flow classification and its surface profiles in GVF?
5. Write short notes on the following:
   (*a*) Integration of varied flow equation by analytical method.
   (*b*) Integration of varied flow equation by graphical method.
   (*c*) Integration of varied flow equation by numerical method.
   (*d*) Flow in channel of non-linear alignment.
6. (*i*) Show that differential equation of gradually varied flow

in a retangular channel of variable width B can be expressed as:

$$\frac{dy}{dx} = \frac{S_o - S_f + \left(\frac{Q^2 y}{gA^3}\frac{dB}{dx}\right)}{1 - \frac{Q^2 B}{gA^3}}$$

(*ii*) Further, show that for horizontal, frictionless rectangular channel of varying width B, the above relation reduces to

$$\left(1 - F^2\right)\frac{dy}{dx} - F^2\left(\frac{y}{B}\right)\frac{dB}{dx} = 0$$

7. Using basic differential equation, show that $\frac{dB}{dx}$ is positive for S_1, M_3 and S_3 profiles.
8. Show that for a wide rectangular channel the slope is mild or steep according to S_o being less than or greater than.

$$\frac{\eta^2 g^{10/9}}{q^{2/9}}$$

9. Sketch Possible GVF profiles in the following serial arrangement of channels and controls. The flow is from left to right.
   (*a*) steep-horizontal-mild slope.
   (*b*) mild-sluice gate-steep-horizontal-sudden drop.
   (*c*) steep-steeper-mild-milder slope.
   (*d*) free intake-steep-sluice gate-mild slope.
   (*e*) steep-mild-sluice gate-mild-sudden drop.
   (*f*) Sluice gate-adverse-horizontal-steep slope.
10. Show that the basic differential equation in terms of conveyance and section factors can be written as:

$$\frac{dy}{dx} = S_o \frac{1-\left(\frac{K_0}{K}\right)^2}{1-\left(\frac{Z_c}{Z}\right)^2}$$

11. Show that for wide rectangular channel, if the Manning's formula is used, the differential equation of GVF is given by:

$$\frac{dy}{dx} = S_o \frac{1-\left(\frac{y_0}{y}\right)^{3.33}}{1-\left(\frac{y_c}{y}\right)^3}$$

12. Show that for very wide rectangular channel, if Chezy formula is used, the differential equation of GVF is given by:

$$\frac{dy}{dx} = S_o \frac{1-\left(\frac{y_0}{y}\right)^3}{1-\left(\frac{y_c}{y}\right)^3}$$

13. Show that for a horizontal channel, by assuming Chezy C = constant, the GVF profile is given by

$$x = \frac{C^2}{g}\left[y - \frac{y^4}{4y_c^3}\right] + \text{constant}$$

14. Establish that the GVF profile in a frictionless rectangular channel is given by

$$x = \frac{y_0}{S_o}\left[1+\frac{1}{2}\left(\frac{y_c}{y}\right)^3\right] + \text{constant}$$

15. A rectangular channel with a bottom width of 4 m and a bottom slope of 0.0008 has a discharge of 1.5 m³/s. In gradually varied flow in this channel, the depth at a certain location is found to be 0.30 m. Assuming $n = 0.016$, determine the type of GVF profile?

16. In a rectangular channel 12 m and 3.6 m deep water is flowing with a velocity of 1.2 m/s. The bed slope of a channel is 1 in 4000 if flow of water through the channel is regulated in such a way that energy line having a slope of 0.004. Find the rate of change of depth of water in the channel?
17. A weir is installed a cross a rectagular open channel thereby raising the flow depth from 1.5 m in a normal flow to 2.5 m at the weir. The width of channel is 10 m and it is laid to a slope of 1 in 1000. Find appoximate length of back water curve considering the average velocity, average depth and average depth mid way between the two sections? Take the value of Mannings rugosity co-efficent equal to 0.02.

Unit 4

Rapidly Varied Flow RVF

4.1 Introduction

The chapter deals with two important Rapidly Varied flow phenomenon *i.e.*, Hydraulic jump and surge.

Hydraulic jump is the best example of rapidly varied steady flow condition. The chapter describes about classical hydraulic jump, its classification, characteristics, evaluation of jump elements in rectangular and non-rectangular channels on horizontal and sloping bed; Hydraulic jump as a energy dissipator whereas surges are the best example of rapidly varied unsteady flow condition. The chapter includes the type of open channel surge, celerity of gravity wave and deep and shallow water wave.

4.2 Classical Hydraulic Jump

A hydraulic jump occurs when a supercritical (high velocity and small depth) flow meets subcritical (low velocity and large depth) flow causing a jump in flow depth.

Examples of Hydraulic Jump

(a) When bed slope of a channel changes from steep to mild, the supercritical depth on the steep channel jumps to the sub-critical depth on the mild channel.

(b) Downstream of a sluice gate, the supercritical flow under the gate meets the uniform flow depth (subcritical) in the downstream channel.

(c) At the toe of spillway, where high velocity flow over the spillway meets the larger flow depth downstream.

4.3 Application of Hydraulic Jump

Generally the formation of jump involves air entrainment, mixing and energy loss, which leads to various practical benefits such as:

(a) Dissipation of the energy of water falling over a weir or spillway.

(b) Thorough mixing of a chemical (*e.g.*, a purifying agent) in water by introducing it upstream of the jump.

(c) Improvement of water quality by increasing the dissolved oxygen content through air-entrainment.

(d) Reducing the uplift pressure on an Apron on the channel led by forming a jump on or before the apron thereby increasing the flow depth on the apron.

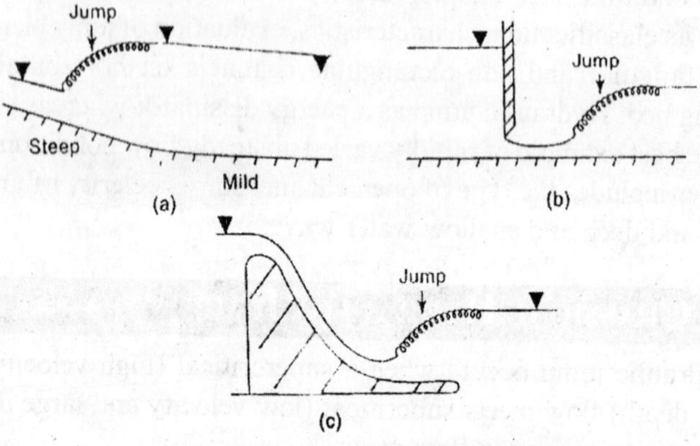

Fig. 4.1: Occurrence of hydraulic jump

Figure 4.2 shows the sketch of hydraulic jump in a horizontal channel. Sections 1 and 2 is the length of jump L_j the initial depth of the supercritical stream is y_1 and y_2 is the final depth, after the jump of the subcritical stream. The two depth y_1 and y_2 at the end of jump are called sequent depths. Due to high turbulence and shear action of the roller, there is loss of energy in the jump between sections 1 and 2.

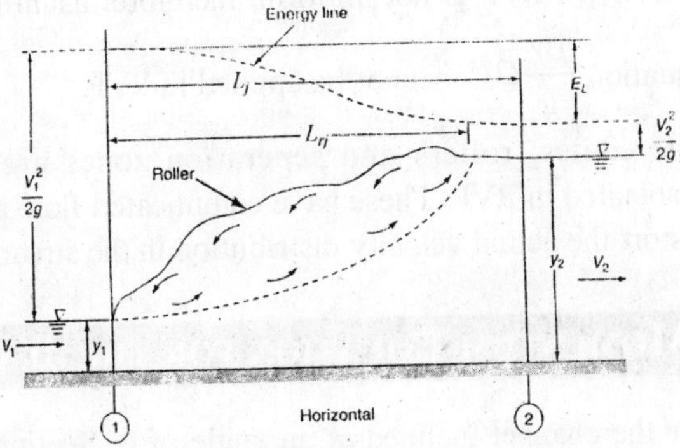

Fig. 4.2: Definition sketch of a hydraulic jump

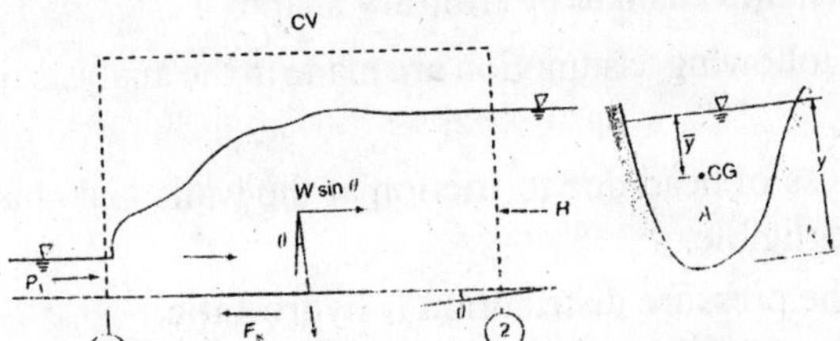

Fig. 4.3: Definition sketch for the general momentum equation

4.4 Characteristics of Rapidly Varied Flow

Following are the characteristics of rapidly varied flow:

(i) It has very large curvature of stream lines. The change in curvature is abrupt, profile will be broken, resulting very high turbulance.

(ii) Pressure distribution cannot be assumed to be hydrostatic because of very large curvature.

(iii) Kinetic energy correction factor and momentum correction coefficient are far greater than unity.

(iv) When RVF occurs in sudden transition, the physical characteristics of flow are fixed by boundry geometry of the structure as well as state of flow.

(v) In RVF, boundary friction force is quite small compared to other forces, so may be neglected.

(vi) velocity in RVF is not uniform, therefore, usual momentum equation $\frac{\beta\gamma}{g}QV$ cannot be applied in RVF.

(vii) The eddies, rollers and seperation zones are normally associated in RVF. These have complicated flow pattern and distort the actual velocity distribution in the stream.

4.5 Momentum Equation for the Jump

Consider the channel inclined at an angle of θ. Sections 1 and 2 indication the beginning and end of jump. The channel is assumed to be prismatic channel of arbitrary shape.

The following assumption are made in the analysis of hydraulic jump:

1. Loss of head due to friction at the walls and channel bed is negligible.
2. The pressure distribution is hydrostatic.
3. The channel is horizontal or it has small slope. The weight component in direction of flow is neglected.
4. The momentum correction factor is unity.

A control volume enclosing the jump is indicated by dashed lines. The flow is considered to be steady.

Applying monentum equation to the control volume

$$P_1 - P_2 - F_S + W \sin\theta = M_2 - M_1 \quad \text{...(4.1)}$$

Where, P_1 = Pressure force at section 1 = $\gamma A_1 \bar{y}_1 \cos\theta$

where, $\bar{y}_1$ depth of the centroid of the area below water surface.

Similarly P_2 = Pressure force at section 2 = $\gamma A_2 \bar{y}_2 \cos\theta$

F_s = Shear force on the control surface.

$W \sin\theta$ = longitudinal component of weight of water contained in the control volume.

M_2 = Momentun flux going out through control surface = $\beta_2 \rho Q V_2$

M_1 = Momentum flux going in through the control surface
$= \beta_1 \rho Q V_1$

Note: Pressure distribution is assumed to be hydrostatic

For small value of θ, cos θ is taken as unity.

Frictional force is usually neglected, $W \sin\theta - F_s$

Can be considered very small and hence is neglected.

For horizontal channel $\theta = 0$, $W \sin\theta = 0$.

4.6 Evaluation of Hydraulic Jump in Horizontal Rectangular Channel

Consider horizontal frictionless rectangular channel. Assuming channel width to be unity the momentum equation (4.1) can be written as

Since for horizontal channel $\theta = 0, F_s = 0, \bar{y}_1 = \frac{y_1}{2}$,

$$\bar{y}_2 = \frac{y_2}{2}, \text{ width } B = 1$$

$$\frac{1}{2}\gamma y_1^2 - \frac{1}{2}\gamma y_2^2 = \beta_2 \rho q V_2 - \beta_1 \rho q V_1 \qquad \text{...(4.2)}$$

taking $\beta_1 = \beta_2 = 1, \gamma = \rho g$

Applying continuity equation

$$A_1 V_1 = A_2 V_2$$

Since $\beta = 1,\ V_1 y_1 = V_2 y_2 = q$ (discharge intensity)

$$\rho g\left(\frac{y_1^2}{2} - \frac{y_2^2}{2}\right) = r(qV_2 - qV_1) \qquad \text{...(4.3)}$$

$$y_2^2 - y_1^2 = \frac{2q^2}{g}\left(\frac{1}{y_1} - \frac{1}{y_2}\right) \qquad \text{...(4.4)}$$

$$y_1 y_2 (y_1 + y_2) = \frac{2q^2}{g} = 2y_c^3 \qquad \text{...(4.5)}$$

Since for rectangular channel $\left(\frac{q^2}{g}=y_c^3\right)$ equation (4.5) can be written as by dividing y_1^3 both side

$$\frac{1}{2}\frac{y_2}{y_1}\left(1+\frac{y_2}{y_1}\right)=\frac{q^2}{gy_1^3}=F_1^2 \qquad ...(4.6)$$

Where F_1 = Froude number = $\frac{V_1}{\sqrt{gy_1}}$

(supercritical Froude number)

Solving equation (4.6) for $\frac{y_2}{y_1}$ we get

$$\frac{y_2}{y_1}=\frac{1}{2}\left(-1+\sqrt{1+8F_1^2}\right) \qquad ...(4.7)$$

The equation (4.7) gives the ratio of sequent depths and also relate it with initial Froude number. The equation (4.7) is popularly known as **Belanger momentum equation.**

Equation (4.5) can also be written as by dividing $y_2{}^3$ both side

$$\frac{1}{2}\frac{y_1}{y_2}\left(1+\frac{y_1}{y_2}\right)=\frac{q^2}{gy_2^3}=F_2^2 \qquad ...(4.8)$$

where F_2 = Froude number = $\frac{V_2}{\sqrt{gy_2}}$ = Subcritical Froude number.

Solving for $\frac{y_1}{y_2}$ yields

$$\frac{y_1}{y_2}=\frac{1}{2}\left(-1+\sqrt{1+8F_2^2}\right) \qquad ...(4.9)$$

4.6.1 Expression for Energy Loss

The energy loss is given by $E_L = E_I - E_2$

where E_1 and E_2 are specific energy at section (1) and section (2).

$$E_L = \left(y_1 + \frac{q^2}{2gy_1^2}\right) - \left(y_2 + \frac{q^2}{2gy_2^2}\right) \qquad ...(4.10)$$

$$= (y_1 - y_2) + \frac{1}{2}\frac{q^2}{g}\left(\frac{y_2^2 - y_1^2}{y_1^2\, y_2^2}\right) \qquad ...(4.11)$$

Substituting the value of $\frac{q^2}{g}$ from equations (4.5) to (4.11)

$$E_L = (y_1 - y_2) + \frac{1}{2}.\frac{y_1 y_2 (y_1 + y_2)}{2}\left(\frac{y_2^2 - y_1^2}{y_1^2\, y_2^2}\right)$$

On solving

$$E_L = \frac{(y_2 - y_1)^3}{4 y_1\, y_2} \qquad ...(4.12)$$

$$\frac{E_L}{y_1} = \frac{\left(\frac{y_2}{y_1} - 1\right)^3}{4\left(\frac{y_2}{y_1}\right)} \qquad ...(4.13)$$

The relative energy loss

$$\frac{E_L}{E_1} = \frac{E_L}{y_1} \Big/ \frac{E_1}{y_1} \qquad ...(4.14)$$

we know $\frac{E_1}{y_1} = 1 + \frac{F_1^2}{2}$ Put in equation (4.14)

$$\frac{E_L}{E_1} = \frac{\left(\frac{y_2}{y_1} - 1\right)^3}{4\frac{y_2}{y_1}\left(1 + \frac{F_1^2}{2}\right)} \qquad ...(4.15)$$

Put the value of $\frac{y_2}{y_1}$ from equation (4.7) to (4.15) and simplifying we get

$$\frac{E_L}{E_1} = \frac{\left(-3+\sqrt{1+8F_1^2}\right)^3}{8\left(2+F_1^2\right)\left(-1+\sqrt{1+8F_1^2}\right)} \qquad ...(4.16)$$

4.6.2 Length of Jump

Experimentally it is found that L_j is given by expression

$$L_j = 6.9\,(y_2 - y_1) \qquad ...(4.17)$$

4.7 Evaluation of Jumps in Horizontal Non-Rectangular Channel

Consider a horizontal frictionless channel of any arbitrary shape. For hydraulic jump in the channel momentum equation is applicable. Assuming $\theta = 0$ and value of $\beta_1 = \beta_2 = 1$.

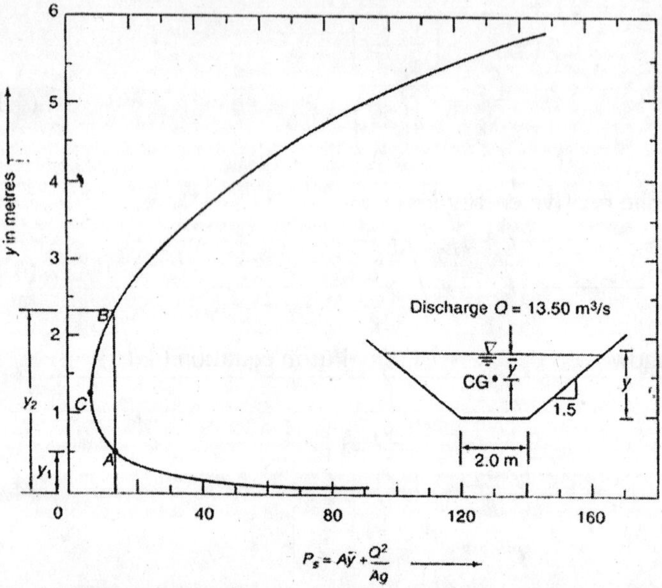

Fig. 4.4: Specific-force diagram for Example 6.7

$$P_1 - P_2 = M_2 - M_1 \quad ...(4.18)$$

$$\gamma A_1 \bar{y}_1 - \gamma A_2 \bar{y}_2 = \rho Q_2 V_2 - \rho Q_1 V_1 \quad ...(4.19)$$

$$\gamma A_1 \bar{y}_1 - \gamma A_2 \bar{y}_2 = \frac{\rho Q_2^2}{A_2} - \frac{\rho Q_1^2}{A_1} \quad ...(4.20)$$

Rearranging equation (4.18)

$$P_1 + M_1 = P_2 + M_2 \quad ...(4.21)$$

In general $$P + M = \gamma\left(A\bar{y} + \frac{Q^2}{gA}\right) = \text{Constant} \quad ...(4.22)$$

$$\frac{P+M}{\gamma} = P_s = A\bar{y} + \frac{Q^2}{gA} = \text{Constant} \quad ...(4.23)$$

The term $P_s = \dfrac{P+M}{\gamma}$ is specific force

The specific force P_s is a function of the depth of flow, channel geometry and discharge. In Fig. 4.4 a parabolic curve is obtained by plots of P_s versus y for given discharge. An ordinate drawn at a given P_s cuts the curve at two point A and B where the respective depth represents the sequent depths for a given discharge. Point C represent critical depth for given discharge.

4.7.1 Sequent Depth Ratio

We know that

$$P + M = \gamma\left(A\bar{y} + \frac{Q^2}{gA}\right) = \text{Constant}$$

Applying at two section

$$\frac{Q^2}{gA_1} + A_1\bar{y}_1 = \frac{Q^2}{gA_2} + A_2\bar{y}_2 \quad ...(4.24)$$

$$A_2\bar{y}_2 - A_1\bar{y}_1 = \frac{Q^2}{g}\left(\frac{1}{A_1} - \frac{1}{A_2}\right) \quad ...(4.25)$$

$$A_1\bar{y}_1\left(\frac{A_2\bar{y}_2}{A_1\bar{y}_1} - 1\right) = \frac{Q^2}{g}\left(\frac{A_2 - A_1}{A_1 A_2}\right) \quad ...(4.26)$$

we know $F_1^2 = \dfrac{Q^2 T_1}{g A_1^3}$ Put in equation (4.26)

$$\left(\frac{A_2}{A_1}\frac{\bar{y}_2}{y_1}-1\right) = F_1^2\left(\frac{A_1/T_1}{\bar{y}_1}\right)\left(1-\frac{A_1}{A_2}\right) \quad ...(4.27)$$

Substituting the value of A, T and $\bar{y}$ will lead to an equation relating the sequent depth ratio to the inlet Froude number.

4.7.2 Expression for Energy Loss in Non-Rectangular Channel

$$E_L = E_1 - E_2 = y_1 - y_2 + \frac{Q^2}{2g}\left(\frac{1}{A_1^2}-\frac{1}{A_2^2}\right) \quad ...(4.28)$$

4.8 Evaluation of Jumps on a Sloping Bed

Figure 4.5 shows jumps on sloping bed. In this case W sin θ is taking into account as channel is sloping and there is some value of θ. Assuming momentum correction factor to be unity $\beta_1 = \beta_2 = 1$ and channel to be frictionless. Applying momentum equation.

$$P_1 - P_2 + W\sin\theta = M_2 - M_1 \quad ...(4.29)$$

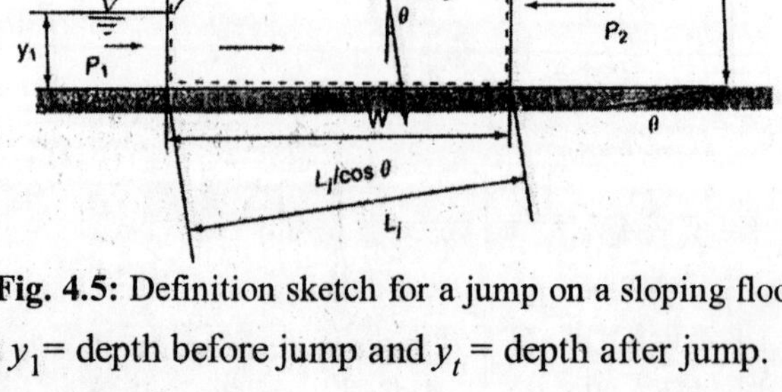

Fig. 4.5: Definition sketch for a jump on a sloping floor

Let y_1= depth before jump and y_t = depth after jump.

Assuming hydrostatic distribution at equations (1) and (2).

$$P_1 = \frac{1}{2}\gamma \bar{y}_1 \cos\theta$$

$$P_2 = \frac{1}{2}\gamma \bar{y}_t \cos\theta$$

If water surface is consider straight line joining y_1 and y_2. Then area of jump.

$$A = \frac{1}{2}(y_1 + y_t)\frac{L_j}{\cos\theta}$$

$$W = mg = \rho\, g\, v$$

$$W = \frac{1}{2}\gamma\ K L_j\,(y_1 + y_t)$$

Where K = correction factor for actual shape.

$$M_1 = \frac{\rho q^2}{y_1},\ M_2 = \frac{\rho q^2}{y_t}$$

Putting value of M_1, M_2, W, P_1 and P_2 in equation (4.29)

$$\frac{1}{2}\gamma\left[y_1^2\cos\theta - y_t^2\cos\theta + KL_j\,(y_1 + y_t)\sin\theta\right] = \rho q^2\left(\frac{1}{y_t} - \frac{1}{y_1}\right) \quad ...(4.30)$$

Arranging in the form of $\frac{y_t}{y_1}$ we get

$$\left(\frac{y_t}{y_1}\right)^2 - 1 - \frac{KL_j\tan\theta}{y_1}\left(1 + \frac{y_t}{y_1}\right) = \frac{2F_1^2}{\cos\theta}\left(\frac{y_t/y_1 - 1}{y_t/y_1}\right) \quad ...(4.31)$$

where $F_1 = \frac{V}{\sqrt{gy_1}}$

equation (4.31) can also be written as

$$\left(\frac{y_t}{y_1}\right)^3 - \frac{kL_j\tan\theta}{y_1}\left(\frac{y_t}{y_1}\right)^2 - \left(1 + \frac{kLj\tan\theta}{y_1} + \frac{2F_1^2}{\cos\theta}\right)\frac{y_t}{y_1} + \frac{2F_1^2}{\cos\theta} = 0 \quad ...(4.32)$$

This equation is used to determine sequent depth provided that KL_j is known. KL_j is the function of F_1 and θ.

4.8.1 Seqment Depth Ratio

We know that in horizontal channel $y_2 = \frac{y_1}{2}\left(-1+\sqrt{1+8F_1^2}\right)$

It was found that y_t is related to y_2 as

$$\frac{y_t}{y_2} = f(\theta)$$

experimentally it is found to be

$$\frac{y_t}{y_2} = 1.0071 \exp (3.2386 \tan q) \qquad ...(4.33)$$

4.8.2 Length of Jump (L_j)

The length of jump can be expressed by relation as

$$\frac{L_j}{y_2} = 6.1 + 4 \tan \theta \qquad ...(4.34)$$

4.8.3 Energy Loss

As we get the value of y_t, y_1 and L_j the energy loss can be calculated as

$$E_L = (E_1 + L_j \tan \theta) - E_2$$

$$E_L = y_1 \cos \theta + \frac{V_t^2}{2g} + L_j \tan \theta - y_1 \cos\theta - \frac{V_1^2}{2g} \qquad ...(4.35)$$

4.9 Classification of Jumps

(A) The hydraulic jumps in horizontal rectangular channel are classified into five categories based on Froude number F_1 of the Supercritical flow.

(a) **Undular jump of ($1.0 < F_1 \leq 1.7$)** in this type of jump, water surface is having small ripple. The sequent depth ratio is very small and $\frac{E_L}{E_1}$ is practically zero.

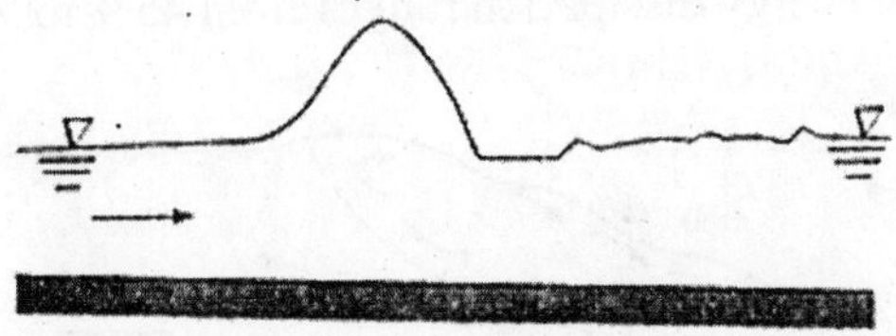

Fig. 4.6(*a*): Undular jump, $1.0 < F_1 \leq 1.7$

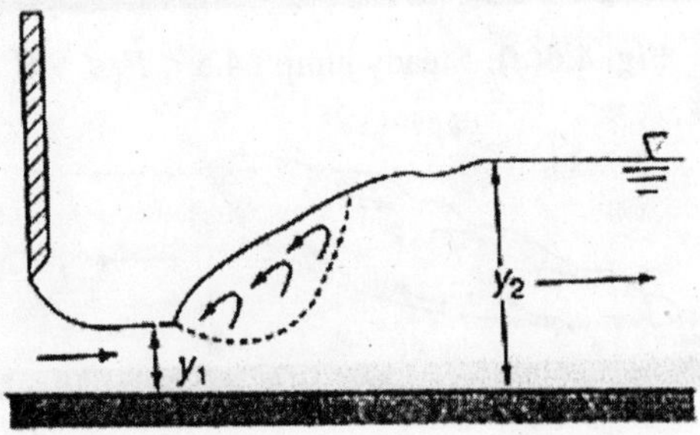

Fig. 4.6(*b*): Weak jump, $1.7 < F_1 \leq 2.5$

(*b*) **Weak jump ($1.7 < F_1 \leq 2.5$):** The surface roller gradually increases from $F_1 = 1.7$ to the end $F_1 = 2.5$. The energy dissipation is very small about 5% at $F_1 = 1.7$ and 18% at $F_1 = 2.5$.

(c) **Oscillating jump ($2.5 < F_1 \leq 4.5$):** This type of jump is having high velocity flow of water which oscillates in a random manner between the bed and the surface. The energy dissipation is moderate, about 45% at $F_1 = 4.5$.

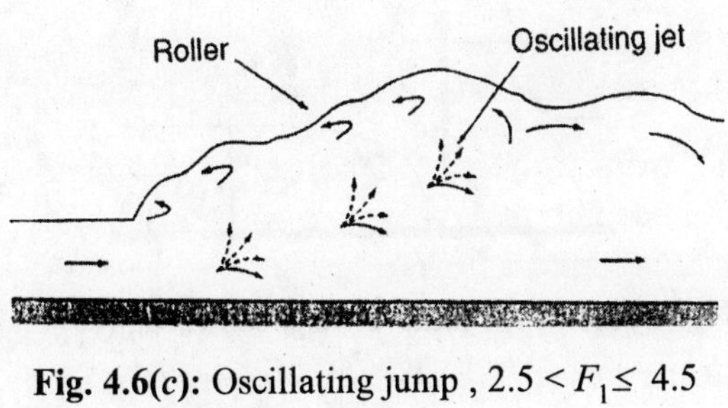

Fig. 4.6(*c*): Oscillating jump, $2.5 < F_1 \leq 4.5$

(d) **Steady jump: (4.5 < F_1 ≤ 9.0):** In this case jump is well established, the roller and jump action is fully developed. The energy dissipation ranges from 45% to 70%.

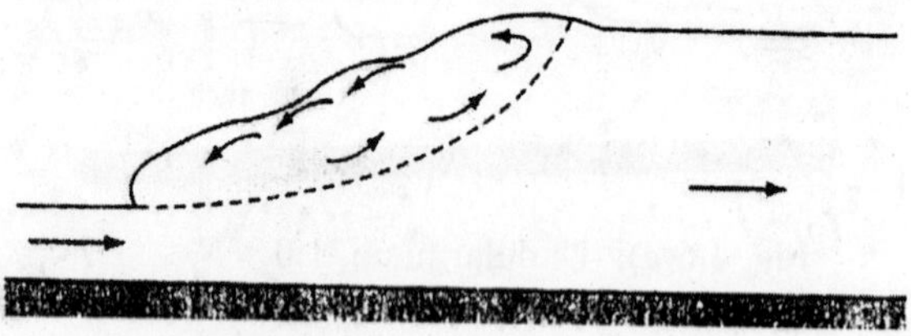

Fig. 4.6(*d*): Steady jump , 4.5 < F_1 ≤ 9.0

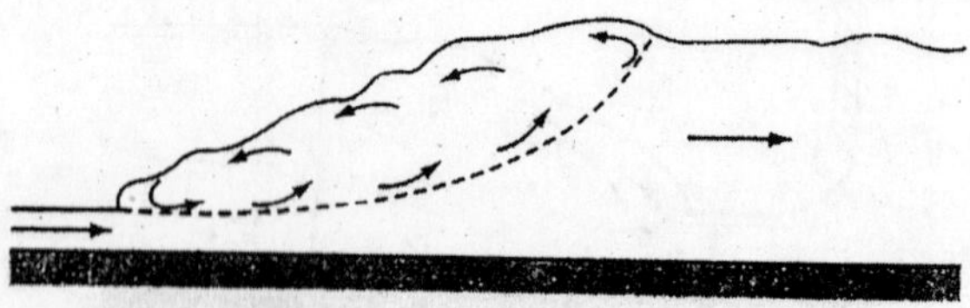

Fig. 4.6(*e*): Strong or Choppy jump, 4.5 < F_1 ≤ 9.0

(*e*) **Strong or Choppy Jump:** (4.5 < F_1 ≤ 9.0): Strong or choppy jump (F_1> 9.0) In this type of jump water surface is very rough and choppy. The sequent depth ratio is large and energy dissipation is highly pronounced which is greater than 70%.

(B) Classfication on the Basis of Location of Jump

(*a*) **Free Jump at Vena Contractar ($y_t = y_2$):** When $y_t = y_2$, a hydraulic jump will form at vena contracta. Which is called as free jump at vena contracta.

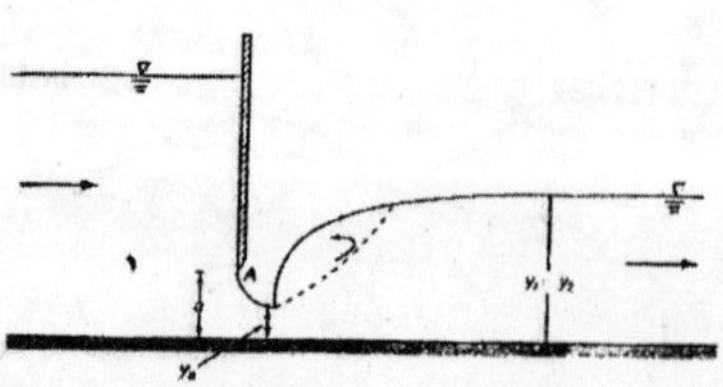

Fig. 4.7(*a*): Free jump at vena contracta, ($y_1 = y_2$)

(*b*) **Free repelled jump ($y_t < y_2$):** When $y_t < y_2$ the jump is repelled down stream of the vena contracta. The depth y_1 will be larger than Vena contracta depth y_a. Such a jump is called repelled jump with sequent depth equal or less than y_2 are called free jump.

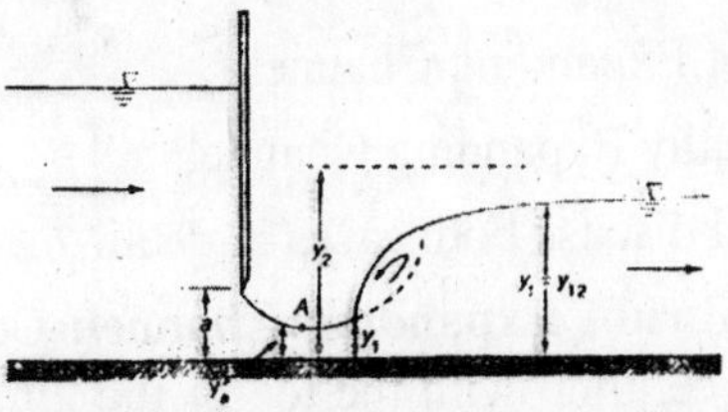

Fig. 4.8(*b*): Free repelled jump, ($y_1 < y_2$)

(*c*) **Submerged jump or drowned jump ($y_t < y_2$):** If tailwater is larger than y_2 the supercritical stream is submerged and resulting jump is called submerged or drowned jump.

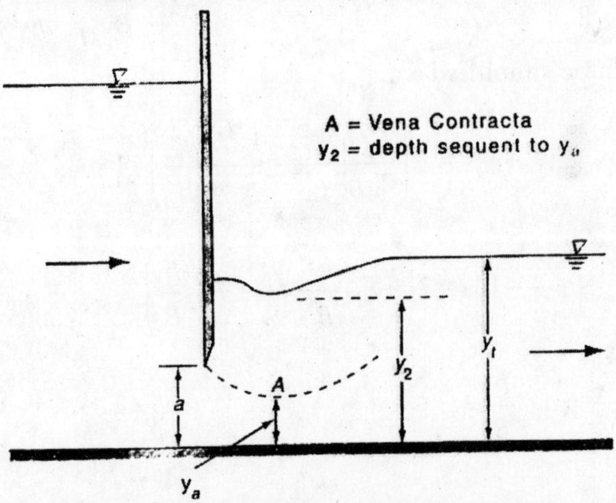

Fig. 4.7(*c*): Submerged jump, ($y_1 < y_2$)

4.10 Hydraulic Jump in Gradually & Suddenly Expanding Channels

When the tail water depth is inadequate to give a good jump in a channel of constant width and to give a good jump in a channel of constant width and it is not possible to depress the basin floor because of difficulties in excavation etc, a lateral expansion, remains the only possibility through which energy can be dissipated which is often called as spatial hydraulic jump. There are two cases in spatial hydraulic jump

a) Aprupt Expanding Channel
b) Gradually Expanding Channel.

Now, we will discuss both cases in detail

(a) **Jump at Abrubt Expanding Channel:** Consider the given figure. A free jump with the toe of the jump along XX will occur at a particular tailwater depth y_2. This depth can be calculated from momentum and continuity equation as follow:

$$P_1 + P_3 - P_2 = \rho Q(V_2 - V_1)$$

$$\frac{\rho g y_1^2 b}{2} + \frac{\rho g y_1^2}{2}(B-b) - \frac{\rho g h_2^2 B}{2} = \rho Q\left(\frac{Q}{By_2} - \frac{Q}{by_1}\right)$$

It can be simplified as:

$$y_1^2 - y_2^2 = 2\left(\frac{V_1^2 b^2 y_1^2}{gB^2 y_2} - \frac{V_1^2 b^2 h_1^2}{gby_1 B}\right)$$

or
$$\left(\frac{y_2^2}{y_1^2} - 1\right) = 2\left(-F_1^2 \frac{b^2}{B^2}\frac{y_1}{y_2} + F_1^2 \frac{b}{B}\right)$$

take
$$\frac{y_2}{y_1} = a, \; \frac{b}{B} = t$$

$$a^2 - 1 = 2F_1^2\left(t - \frac{t^2}{a}\right)$$

or
$$a^3 - a\left(1 + 2F_1^2 t\right) + 2F_1^2 t^2 = 0$$

The jump is repelled down from section XX if the tailwater depth is smaller than conjugate depth. if the tailwater depth is larger

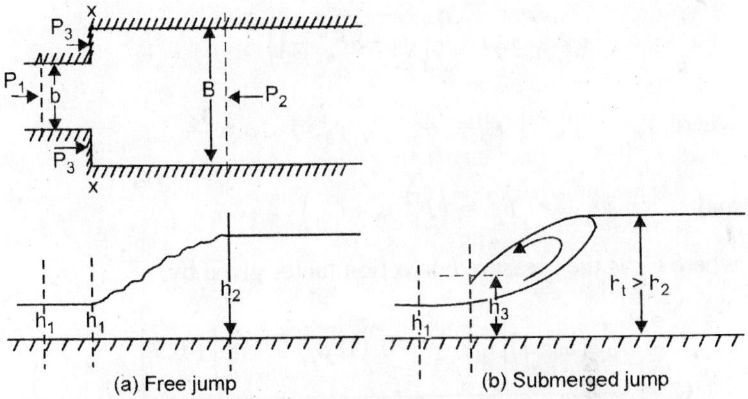

Fig. 4.8: Hydraulic jump at an abrupt expansion

than above value, the toe of the jump will move up the channel of smaller width. Such movement is prevented in case of sluice gate located at XX. A submerged jump will result in such a case. The depth y_3 can then obtained from momentum equation as follows:

$$\frac{\rho g y_3^2}{2}B - \frac{\rho g y_t^2}{2}B = \rho Q\left(\frac{Q}{By_t} - \frac{Q}{by_1}\right)$$

$$y_3^2 = y_t^2 - \frac{2Q^2}{gB}\left(\frac{By_t - by_1}{Bbyy_t}\right)$$

$$\frac{y_3}{y_t} = \sqrt{1 - 2F_t^2\left(\frac{1}{t}\left(\frac{y_t}{y_1}\right) - 1\right)}$$

in which F_t is the froude number in the downstream Channel.

Jump in Gradually Expanding Channel

The hydraulic jump in a gradually expanding channel with straight walls may be treated as a part of circular hydraulic jump. The streamlines are assumed to be radial. The water surface within the jump will be in the shape of quarter ellipse, applying momentum and continuity equations, Scientist Arbhabhirama and Abella showed that

$$r_0 a = \frac{1}{2}\left(\sqrt{1+8F_c^2}-1\right)$$

where $$r_0 = \frac{r_2}{r_1} \qquad y_0, \qquad a = \frac{y_2}{y_1}$$

and $$F_c^2 = \left(F_1^2\, r_0 + C_p\right)$$

where C_p is the pressure correction factor given by

$$C_p = \frac{r_0 a\left(r_0-1\right)\left[r_0\left(\frac{a^2}{3}+0.118a+0.048\right)+0.5\right]}{\left(r_0 a-1\right)}$$

The length of the jump L_j is obviously equal to $(r_2 - r_1)$.

The empirical equation is $\frac{L_j}{y_t} = 3.7\, F_1^{1.35}$

From the above equations for known r_1 and y_1 we can find y_2 by trial and error method.

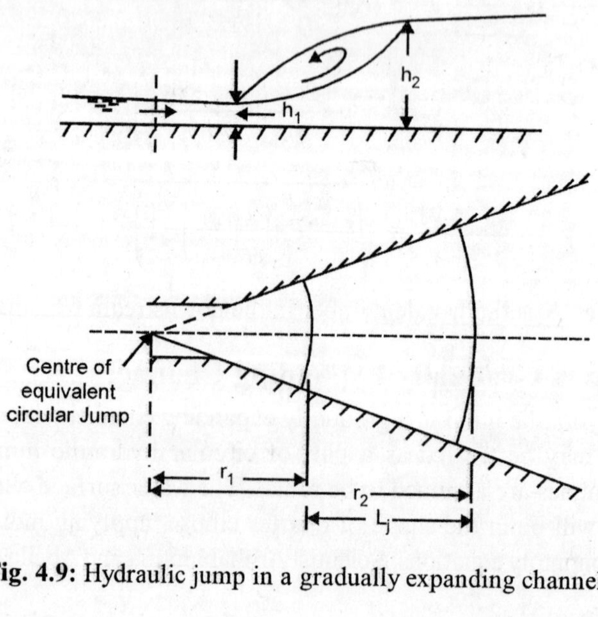

Fig. 4.9: Hydraulic jump in a gradually expanding channel

4.11 Ski Jump & Rolling Jump

In order to understand Ski jump and rolling jump lets understand by flip bucket and Roller bucket

Flip Buckets

The flip bucket energy dissipater is suitable for sites where the tail water depth is low (which would require a large amount of excavation if a hydraulic jump dissipater were used) and the rock in the downstream area is good and resistant to erosion. The flip bucket, also called ski-jump dissipater, throws the jet at a sufficient distance away from the spillway where a large scour hole may be produced. Initially, the jet impact causes the channel bottom to scour and erode. The scour hole is then enlarged by a ball-mill motion of the eroded rock pieces in the scour hole. A small amount of the energy of the jet is dissipated by the internal turbulence and the shearing action of the surrounding air as it travels in the air. However, most of the energy of the jet is dissipated in the plunge pool.

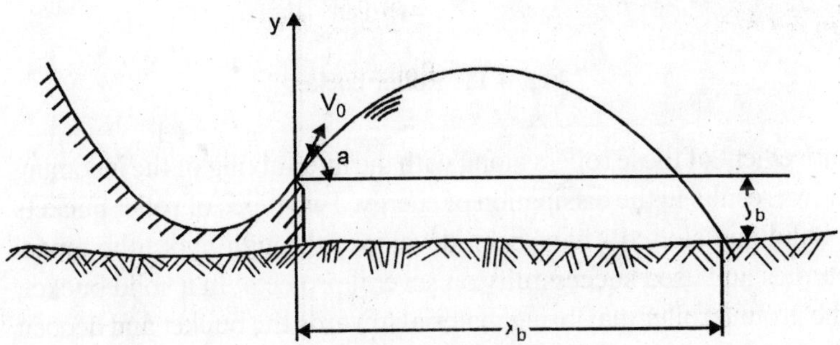

Fig. 4.10: Flip bucket

Roller bucket

A roller bucket may be used for energy dissipation if the downstream depth is significantly greater than that required for the formation of a hydraulic jump. The dissipation is caused mainly by two rollers: counter clockwise roller near the water surface above the bucket and a roller on the channel bottom downstream of the bucket. The

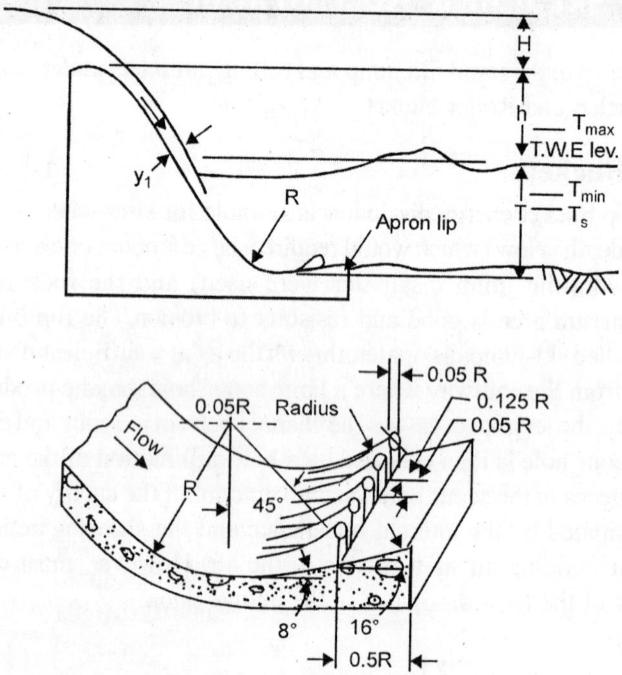

Fig. 4.11: Roller bucket

movement of these rollers along with the intermixing of the incoming flows results in the dissipation of energy. Two types of roller buckets - solid and slotted - have been developed through hydraulic model studies and used successfully on several projects. In a solid bucket, the ground roller may bring material towards the bucket and deposit it in the bucket during periods of unsymmetrical operation. In a slotted bucket, part of the flow passes through the slots, spreads laterally, and is distributed over a greater area. Therefore, the flow concentraion is less than that in a solid bucket.

4.12 Use of Jump at Energy Dissipater

As we known a lot of energy loss occurs during hydraulic jump. Because of this characteristics hydraulic jump it is used as a energy dissipator below a hydraulic structure. The simplest type of stilling basin utilizes the hydraulic jump to dissipate the energy.

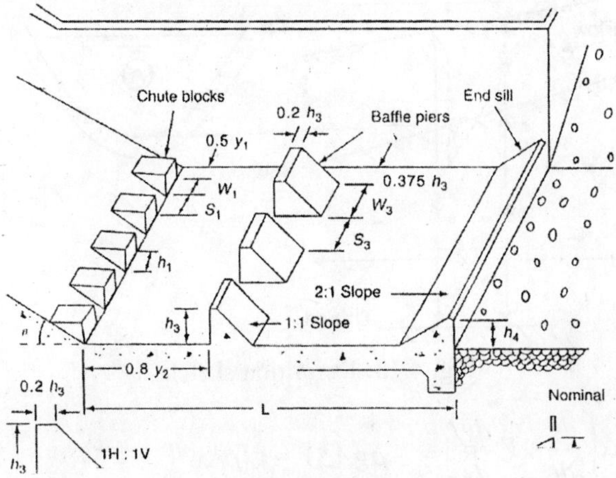

Fig. 4.12

The minimum length of stilling basin should be equal to length of jump. However, it is very expensive to reduce the length of basin various additional devices such as chute blocks, baffle blocks, and sill are provided these devices enhances the dissipation quality. The chute blocks lift a portion of the incoming high velicity jet above the basin floor, there by increase the dissipation action of jump and shorter the jump. Baffle blocks provide most of the energy disspation by the impact action and end sill prevents scouring near the basin and by lifting the out going streams. The dimensions of various devices should be shown carefully since these affect the basin efficiency.

4.13 EQUATION OF MOTION FOR UNSTEADY FLOW

Consider a prismatic channel carrying unsteady flow. The accelaration in the flow direction in such a flow is given by

$$a_x = \frac{dv}{dt} + V\frac{dv}{dx}$$

The equation of motion for such a flow under consideration would be

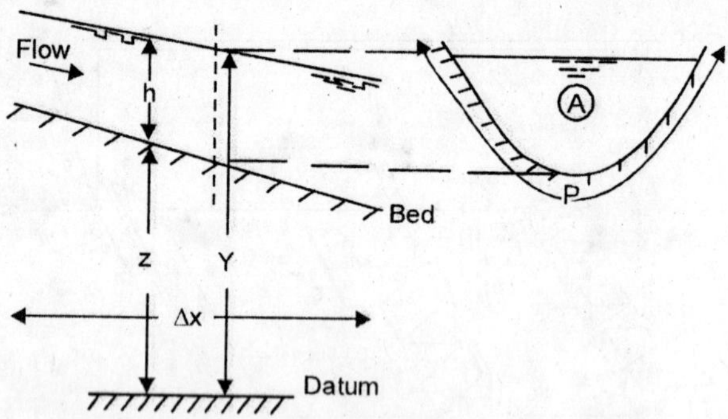

Fig. 4.13: Definition sketch

$$\rho A\Delta x\left(\frac{dv}{dt}+V\frac{dv}{dx}\right)=-\rho gA\Delta Y-\tau_0\rho\Delta x$$

or $$\tau_0=-\rho gR\left[\frac{dY}{dx}+\frac{1}{g}\frac{dv}{dt}+\frac{v}{g}\frac{dv}{dx}\right] \qquad \ldots(a)$$

But $$Y=Z+y$$

$$\frac{dY}{dx}=\frac{dz}{dx}+\frac{dy}{dx}$$

$$=-S_0+\frac{dy}{dx}$$

Also we know that $\tau_0=\rho\, g\, R\, S_f$

Substitute the value of $\frac{dy}{dx}$ and τ_0 in equation (*a*) we get

$$S_f=\left(S_0-\frac{dy}{dx}-\frac{V}{g}\frac{dv}{dx}-\frac{1}{g}\frac{dv}{dt}\right) \qquad \ldots(b)$$

Equation (*b*) is the equation of motion for unsteady flow and is called as St. Venant's equation.

The last term on right hand side vanishes in case of steady non uniform flow and equation (*b*), then reduces to the differential equation for gradually varied flow. Further $\frac{dy}{dx}$ and $\frac{v}{g}\frac{dv}{dx}$ are both zero in case of uniform flow and equation (b) reduces to $S_f = S_0$ for for steady, uniform flow.

The other governing equation is the continuity equation

$$\frac{dQ}{dx} + \frac{dA}{dt} = 0 \qquad \text{...(c)}$$

For rectangular channel equation (c) reduces to

$$\frac{dy}{dt} + y\frac{dv}{dx} + V\frac{dy}{dx} = 0 \qquad \text{..(d)}$$

Hence for solving unsteady flow problems the two differential equation *b* and *c* or *d* is taken into account. Hence for solution for unsteady flow problem equation *b* and *c* or *d* is used.

4.14 Open Channel Surge

The term open channel surge is actually rapidly varied unsteady flow which occurs when there is a sudden change in the discharge or depth or both. A surge producing increase in depth is called positive surge and the one which causes a decrease in depth is known as negative surge.

A surge can travel either in the upstream or downstream direction. Depending upon movement it can be classified into four basic types:

(a) Positive surge moving downstream.

(b) Positive surge moving upstream.

(c) Negative surge moving downstream.

(d) Negative surge moving upstream.

Positive surge are basically stable in nature while negative surges are unstable in nature and their form changes with the advances of surge.

Fig. 4.14 shows formation of surges when sluice gate is suddenly raised or lowered. When the gate is raised it generates, upstream of the gate, a negative surge that moves up stream and down stream of the gate, a positive surge that travel downstream on the other hand, lowering of the gate leads to the formation of positive surge moving upstream and negative surge moving down stream of the gate.

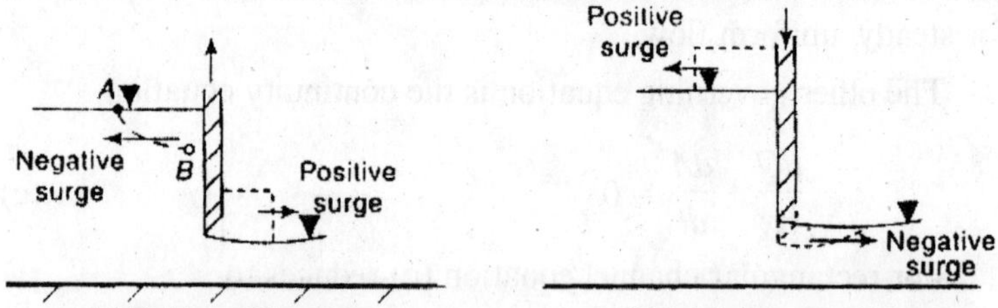

(a) Raising of the gate (b) Lowering of the gate

Fig. 4.14: Positive and negative surges at a sluice gate

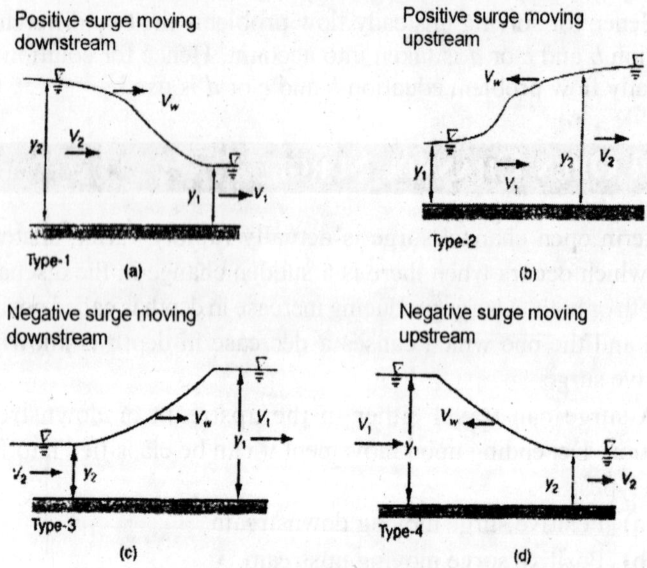

Fig. 4.15: Types of surges

4.14.1 Positive Surge Moving Downstream

Consider sluice gate is suddenly raised to cause positive surge moving down stream. Sections 1 and 2 refer to condition before and after passage of surge. The absolute velocity V_w is assumed to be constant the unsteady state is converted to steady state by applying relative velocity concept and hence applying velocity

($-V_w$) at all sections. The equivalent state is shown by Fig. 4.16(b). Remember the channel is considered to be horizontal and friction less.

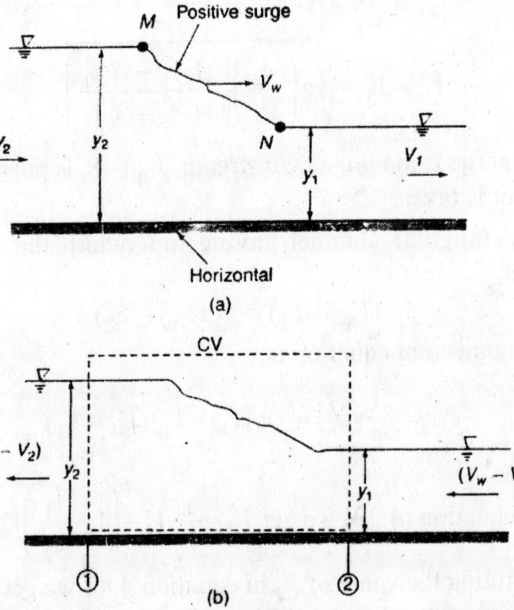

Fig. 4.16(a): Positive surge moving downstream (b) Equivalent steady flow

Now applying continuity equation to control volume CV.

$$A_2\,(V_w - V_2) = A_1\,(V_w - V_1) \qquad \text{...(4.36)}$$

Applying momentum equation to control volume at section 1 and 2

$$P_1 - P_2 = M_2 - M_1$$

$$\gamma A_1 \bar{y}_1 - \gamma A_2 \bar{y}_2 = \rho Q V_2 - \rho Q V_1$$

$$\beta_1 = \beta_2 = 1$$

$$\gamma A_1 \bar{y}_1 - \gamma A_2 \bar{y}_2 = \frac{\gamma}{g} A_1 (V_w - V_1)\left[(V_w - V_2) - (V_w - V_1)\right] \qquad \text{...(4.37)}$$

From equation (4.36) we get

$$V_2 = \frac{A_1}{A_2} V_1 + \left(1 - {A_1}/{A_2}\right) V_w$$

Substituting this in equation (4.37) we get

$$(V_w - V_1)^2 = g\frac{A_2}{A_1}.\frac{1}{A_2 - A_1}(A_2\bar{y}_2 - A_1\bar{y}_1)$$

$$V_w = V_1 + \sqrt{g\left(\frac{A_2}{A_1}\right)\left(\frac{A_2\bar{y}_2 - A_1\bar{y}_1}{(A_2 - A_1)}\right)} \qquad ...(4.38)$$

Since surge is moving down stream, $V_W - V_1$ is positive, hence positve root is taken.

For rectangular channel having unit width the continuity equation is

$$y_1\,(V_W - V_1) = y_2\,(V_W - V_2) \qquad ...(4.39)$$

Also momentum equation is

$$\frac{1}{2}\gamma y_1^2 - \frac{1}{2}\gamma y_2^2 = \frac{\gamma}{g}y_1(V_W - V_1)(V_W - V_2) \qquad ...(4.40)$$

From equation (4.39) we get $V_2 = \frac{y_1}{y_2}V_1 + \left(1 - \frac{y_1}{y_2}\right)V_W$

Substituting the value of V_2 in equation 4.40 we get

$$\frac{(V_W - V_1)^2}{gy_1} = \frac{1}{2}\frac{y_2}{y_1}\left(\frac{y_2}{y_1} + 1\right) \qquad ...(4.41)$$

4.14.2 Positive Surge Moving Upstream

Consider sluice gate is suddenly closed to cause the positive surge moving up stream. Sections 1 and 2 refer to condition before and after passage of surge. The unsteady flow is converted to equivalent steady flow by using relative velocity concept with absolute velocity of surge V_W.

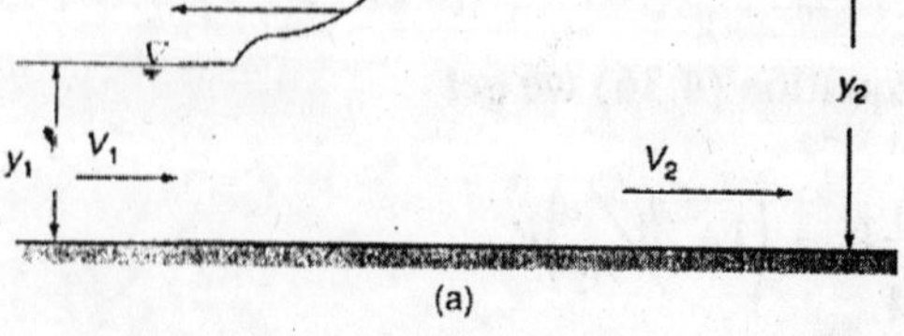

(a)

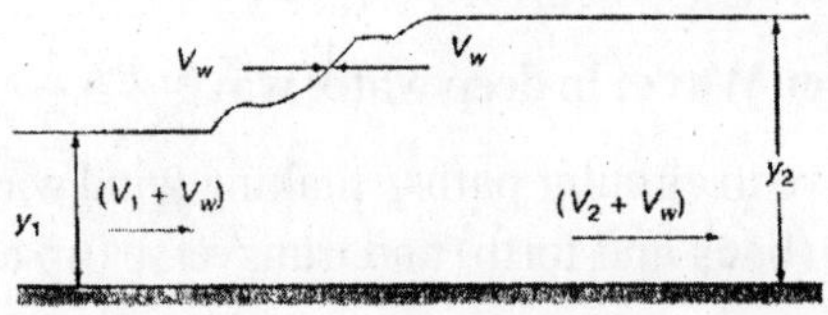

Fig. 4.17: (*a*) Positive surge moving upstream
(*b*) Simulated steady flow

Consider unit width of horizontal frictionless rectangular channel.

Applying continuity equation we get

$$y_1 (V_W + V_1) = y_2 \ (V_W + V_2) \qquad ...(4.42)$$

Applying momentum equation

$$\frac{1}{2}\gamma y_1^2 - \frac{1}{2}\gamma y_2^2 = \frac{\gamma}{g} y_1 (V_W + V_1)\left[(V_W + V_2) - (V_W + V)_1\right]$$

$$\frac{1}{2}\gamma y_1^2 - \frac{1}{2}\gamma y_2^2 = \frac{\gamma}{g} y_1 (V_w + V_1)(V_2 - V_1) \qquad ...(4.43)$$

From equation (4.42) we get

$$V_2 = \frac{y_1}{y_2} V_1 - \left(1 - \frac{y_1}{y_2}\right) V_w$$

Substitution the value of V_2 in equation 4.42

We get $$\frac{(V_w + V_1)^2}{g y_1} = \frac{1}{2}\frac{y_2}{y_1}\left(\frac{y_2}{y_1} + 1\right) \qquad ...(4.44)$$

Positive surges moving downstream and upstream are called as moving hydraulic jump.

4:15 Deep and Shallow Water Wave

Wind waves that propogate along the interface between water and air are called surface gravity wave. As the wind blows pressure and wind forces perturb the equilibrium of water surface. These forces transfer energy from air to water, forming waves.

Deep Water Wave: In deep water wave $d \geq \frac{1}{2}\lambda$, particles near the surface move in circular paths, making wind wave combination of longitudinal (back and forth) and transverse (up and down) wave motions.

Shallow Water Wave: In shallow water wave $d < \frac{\lambda}{2}$ particles near the surface compressed and move in the form of ellipse.

where d = Water depth

λ = Wavelength

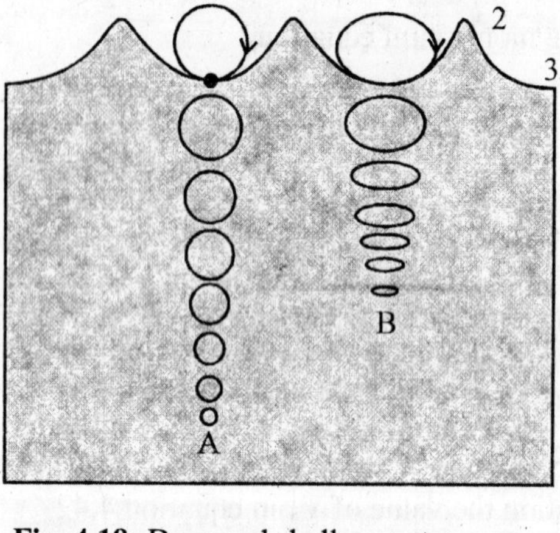

Fig. 4.18: Deep and shallow water wave.

A = At deep water the orbital motion of fluid particles decreases rapidly with increasing depth below the surface.

B = At shallow water wave, movement is elliptical.

Consider the sinusoidal wave as shown in Fig. 4.14.

$$\text{Amplitude} = a.$$

$$\text{wavelength} = \lambda.$$

$$\text{Period} = T.$$

$$\text{Frequency } w = 2\pi/T$$

$$\text{Wavenumber } k = 2\pi/\lambda$$

$$\text{Phase speed } C = \frac{\lambda}{T} = \frac{w}{k}$$

$$\text{Group speed} = C_g = \frac{\partial w}{\partial k}$$

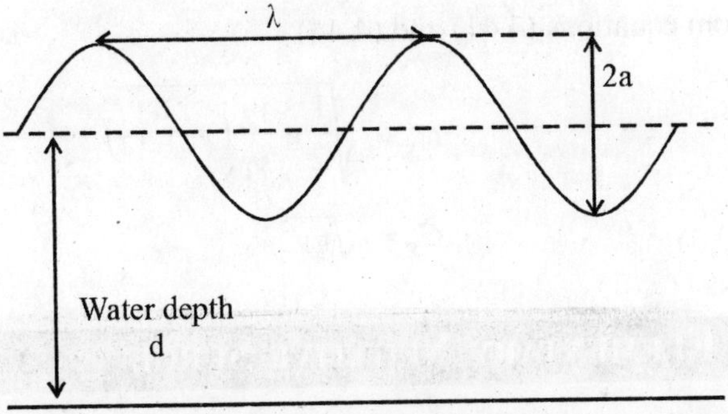

Fig. 4.19: Sinusoidal wave

The phase speed of surface wave is well approximated by

$$C = \sqrt{\frac{g\lambda}{2\pi} \tan h\left(\frac{2\pi d}{\lambda}\right)}$$

$g =$ accelaration due to gravity

In shallow water wave

when $d << \lambda$ $\quad C = \sqrt{gd}$ phase speed independent of wavelength

$C_g = \sqrt{gd} = C$ group speed same as phase speed.

In deep water wave

when $d > \frac{\lambda}{4}$, $\quad C = \sqrt{\frac{g\lambda}{2\pi}} = \sqrt{\frac{g}{k}}$, Longer wave move faster,

$\text{Cg} = \frac{1}{2}\sqrt{\frac{g}{k}} = \frac{1}{2}C$, group speed is half phase speed.

4.16 Celerity of the Surge

The velocity of the surge relative to the initial velocity of surge in channel is known as celerity of surge. It is represented by C_S.

For surge moving downstreams = $C_S = V_w - V_1$

For surge moving upstreams = $C_S = V_W + V_1$

From equations (4.41) and (4.44)

$$C_S = \sqrt{\frac{1}{2} g \frac{y_2}{y_1}(y_1 + y_2)} \quad \text{...(4.45)}$$

if $y_2 \rightarrow y_1$

$$C_S = \sqrt{gy}$$

4.17 Elementary Negative Surge

In negative surge a point M on the top of the surge moves faster than point on lower water surface. This results in the stretching of the wave profile. Since the shape of negative surge varies with time due to stretching activity by varying values of V_w along its height. The negative surge is composed of a series of elementary negative wave of celerity $\sqrt{gy}$ superimposed on existing flow.

Consider one elementary negative wave of height δy the motion is converted to an equivalent steady state flow by the superimposition of velocity $(-V_w)$ on the system.

Applying continuity equation

$$(V_w - V_2) y_2 = (V_w - V_1) y_1 \quad \text{...(4.46)}$$

Put $V_1 = V$, $y_1 = y$, $V_2 = V - \delta V$ and $y_2 = y - \delta y$

and simplifying we get

$$\delta V y = (V_w - V)\delta y \quad \text{...(4.47)}$$

$$\frac{\delta V}{\delta y} = \frac{V_w - V}{y} \quad \text{...(4.48)}$$

Applying momentum equation to control volume

$$\frac{\gamma}{2}\left(y_1^2 - y_2^2\right) = \frac{\gamma}{2}(V_w - V_1) y_1 \left[(V_w - V_2) - (V_w - V_1)\right]$$

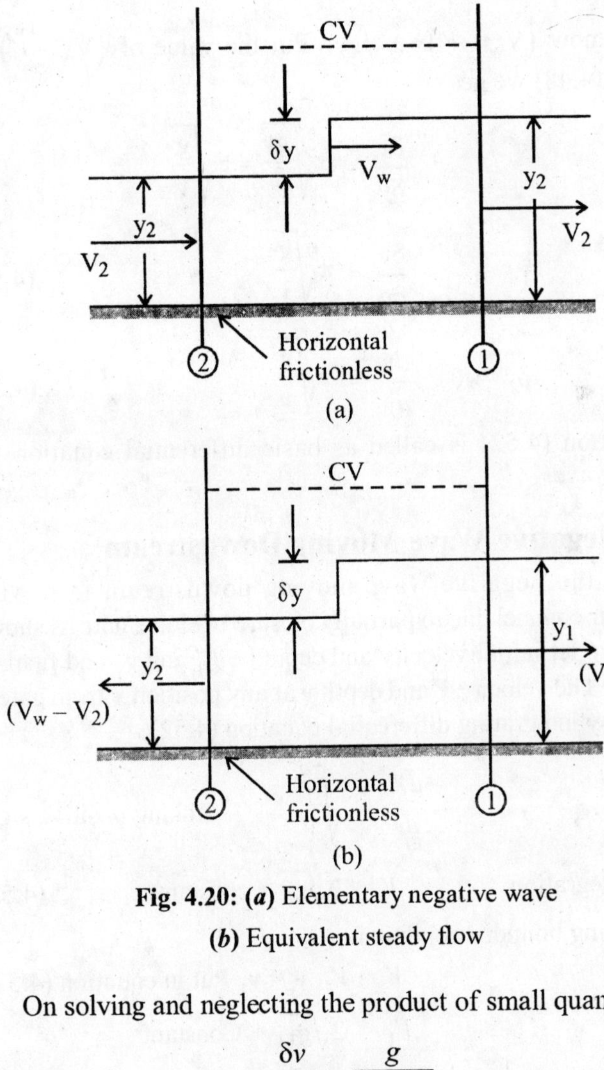

Fig. 4.20: (*a*) Elementary negative wave

(*b*) Equivalent steady flow

On solving and neglecting the product of small quantity

$$\frac{\delta v}{\delta y} = \frac{g}{V_w - V} \qquad ...(4.49)$$

From equations (4.48) and (4.49) we get

$$(V_w - V)^2 = C^2 = gy$$

$$C = \pm\sqrt{gy} \qquad ...(4.50)$$

$C =$ celerity of negative wave

We know $\left(V_w - V\right) = \pm\sqrt{gy}$. Put the value of $\left(V_w - V\right)$ in equation (4.48) we get

$$\frac{\delta v}{\delta y} = \pm\sqrt{\frac{gy}{y}} = \pm\sqrt{\frac{g}{y}}$$

$$\frac{\delta v}{\delta y} = \pm\sqrt{\frac{g}{y}} \qquad \text{...(4.51)}$$

as $\delta y \to 0$

$$\frac{dv}{dy} = \pm\sqrt{\frac{g}{y}} \qquad \text{...(4.52)}$$

Equation (4.52) is called as basic differential equation for negative wave.

4.17.1 Negative Wave Moving Downstream

Consider the negative wave moving downstream in a wide rectangular channel due to partially closure of sluice gate as shown in Fig. 4.16 let initial velocity and depth be V_1 and y_1 and final be V_2 and y_2. The velocity V and depth y at any position x from gate is obtained by integrating differential equation (4.52).

$$\frac{dV}{dy} = \sqrt{\frac{g}{y}} \qquad \text{(taking positive sign)}$$

On Integration

$$V = 2\sqrt{gy} + \text{constant} \qquad \text{...(4.53)}$$

Applying boundry condition

At $V = V_1, y = y_1$ Put in equation (4.53)

$$V_1 = 2\sqrt{gy_1} + \text{Constant}$$

$$\text{Constant} = V_1 - 2\sqrt{gy_1} \qquad \text{...(4.54)}$$

Put in equation (4.53), we get

$$V = V_1 + 2\sqrt{gy} - 2\sqrt{gy_1} \qquad \text{...(4.55)}$$

As we know wave travel downstream

$$C = V_w - V = \sqrt{gy}$$

$$V_w = V + \sqrt{gy} \qquad ...(4.56)$$

Put the value of V from equation (4.55) in (4.56)

$$V_w = V_1 + 3\sqrt{gy} - 2\sqrt{gy_1} \qquad ...(4.57)$$

If surge V_w move x distance in time 't' then

$$x = V_w t \qquad ...(4.58)$$

Put equation (4.57) in equation (4.58) we get

$$x = \left(V_1 + 3\sqrt{gy} - 2\sqrt{gy_1}\right)t \qquad ...(4.59)$$

Equation (4.59) represent profile of negative wave in terms of x, y and t.

From equation (4.55) we get value $\sqrt{gy}$ of and put in equation (4.59) we get

$$x = \left(-\frac{V_1}{2} + \frac{3}{2}V + \sqrt{gy_1}\right)t$$

or

$$V = \left(\frac{2}{3}\frac{x}{t} + \frac{V_1}{3} - \frac{2}{3}\sqrt{gy_1}\right) \qquad ...(4.60)$$

Equation (4.60) gives the value to velocity in terms of x and t.

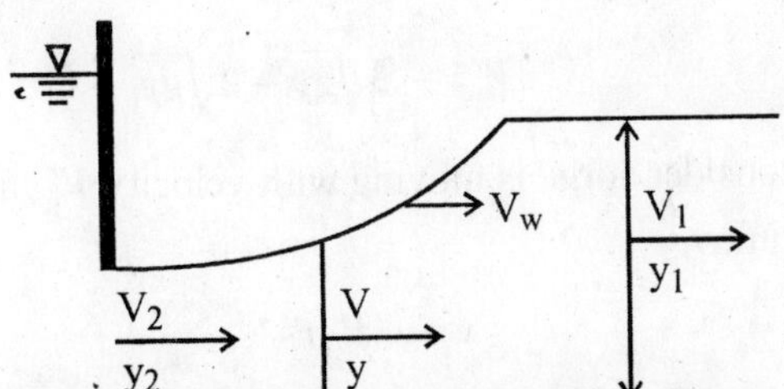

Fig. 4.21: Type-3 negative surge moving downstream

4.17.2 Negative Surge Moving Upstream

Consider negative wave produced by instaneously raising a sluice gate located at downstream end of channel as shown in Fig. 4.22.

Considering the basic differential equation of negative wave from equation (4.52).

$$\frac{dV}{dY} = -\sqrt{\frac{g}{y}}$$

On integrating $\qquad V = -2\sqrt{gy} + \text{constant} \qquad \text{...(4.61)}$

Let V_1 and y_1 be velocity and depth before passage of wave V_2 and y_2 after the passage of wave. Applying boundry condition to equation (4.61).

At $V = V_1, y = y_1$

$$V_1 = -2\sqrt{gy_1} + \text{constant}$$

or $\qquad \text{constant} = V_1 + 2\sqrt{gy_1}$ put in equation (4.61)

$$V = V_1 + 2\sqrt{gy_1} - 2\sqrt{gy} \qquad \text{...(4.62)}$$

As negative wave move upstream

$$C = V_w + V = \sqrt{gy}$$

$$V_w = \sqrt{gy} - V \qquad \text{...(4.63)}$$

Put the value of V from equation (4.62) in equation (4.63), we get

$$V_w = 3\sqrt{gy} - 2\sqrt{gy_1} - V_1 \qquad \text{...(4.64)}$$

Let us consider surge is moving with velocity V_w and moves $\dot{x}$ distance in time t.

then $\qquad x = -V_w t \qquad \text{...(4.65)}$

(Considering negative sign for upstream direction)

Put value of equation (4.64) in (4.65) we get

$$(-x) = (V_w t) = \left(3\sqrt{gy} - 2\sqrt{gy_1} - V_1\right)t \qquad \text{...(4.66)}$$

Remember Dam break problem is the case of negative wave moving upstream.

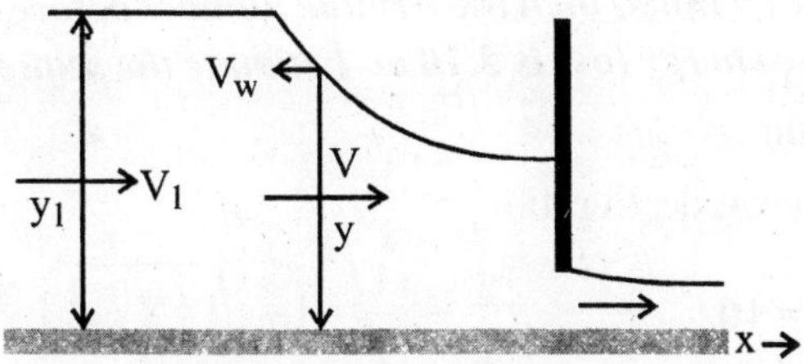

Fig. 4.22: Negative surge moving upstream

Solved Problems

Problem 1: ***A stationary hydraulic jump occur in a rectangular channel with the initial and sequent depth being equal to 0.30 m and 1.20 m respectively. Estimate (i) Discharge per unit width (ii) Energy loss.***

Solution:

$$\frac{y_2}{y_1} = \frac{1}{2}\left[-1+\sqrt{1+8F_1^2}\right]$$

$$\frac{1.20}{0.30} = \frac{1}{2}\left[-1+\sqrt{1+8F_1^2}\right]$$

$$F_1 = 3.162$$

$$F_1 = \frac{V_1}{\sqrt{gy_1}} \Rightarrow \frac{V_1}{\sqrt{9.81\times 0.30}} = 3.162$$

$$V_1 = 5.424 \text{ ms}$$

(*i*) Discharge per unit width $q = V_1 y_1 = 5.424 \times 0.3$

$$= 1.6272 \text{ m}^3\text{/s /m}$$

(*ii*) Energy loss $\quad E_L = \dfrac{(y_2 - y_1)^3}{4y_1y_2} = \dfrac{(1.20-0.30)^3}{4\times 0.30\times 1.20}$

$$= 0.506 \text{ m}$$

Problem 2: ***If, in a hydraulic jump occuring in a horizontal***

rectangular channel, then the Froude number before the jump is 10.0 and the energy loss is 3.10 m. Estimate the sequents depths.

Solution:

The sequent depth ratio

For $F_1 = 10$
$$\frac{y_2}{y_1} = \frac{1}{2}\left(-1+\sqrt{1+8\,F_1^2}\right)$$
$$= 13.651$$

Energy loss

$$E_L = \frac{(y_2 - y_1)^3}{4y_1y_2}$$

$$\frac{E_L}{y_1} = \frac{(y_2/y_1 - 1)^3}{4\left(y_2/y_1\right)}$$

$$\frac{3.10}{y_1} = \frac{(13.651-1)^3}{4(13.651)}$$

y_1 = depth before the jump = 0.0836 m

y_2 = depth after the jump = 1.141 m. Since $y_2/y_1 = 13.651$.

Problem 3: ***In a hydraulic jump occuring in a rectangular channel of 3.0 m width, the discharge is 7.8 m³/s and the depth before the jump is 0.30 m. Estimate the (i) sequent depths and (ii) the energy loss in the jump.***

Solution:

$$v_1 = \frac{Q}{A} = \frac{7.8}{3\times 0.30} = 8.67 \text{ m/s}$$

$$F_1 = \frac{V_1}{\sqrt{gy_1}} = \frac{8.67}{\sqrt{9.81\times 0.30}} = 5.054$$

the sequent depth is given by

For $F_1 = 5.054$
$$\frac{y_2}{y_1} = \frac{1}{2}\left(-1+\sqrt{1+8F_1^2}\right)$$

$$\frac{y_2}{y_1} = 6.664$$

$$\therefore \quad y_2 = 6.664 \times 0.03 = 1.99 \sim 2 \text{ m}$$

$$y_1 = 0.30 \text{ m}$$

(*ii*) Energy loss $\quad E_1 = \dfrac{(y_2 - y_1)^3}{4y_1 y_2}$

$$= \frac{(2-.3)^3}{4 \times 2 \times 3}$$

$$= 2.047 \text{ m}$$

Problem 4: ***A rectangular channel carrying a supercritical stream is to be provided with a hydraulic jump type of energy dissipator. It is desired to have an energy loss of 4.0 m in the hydraulic jump when the inlet Froude number is 8.5. What are the sequent depths in this jump?***

Solution:

$$\frac{y_2}{y_1} = \frac{1}{2}\left[-1 + \sqrt{1 + 8F_1^2}\right]$$

$$= \frac{1}{2}\left[-1 + \sqrt{1 + 8 \times 8.5^2}\right]$$

$$= 11.53$$

$$\frac{E_L}{y_1} = \frac{(y_2 / y_1 - 1)^3}{4\left(\dfrac{y_2}{y_1}\right)}$$

$$\frac{4}{y_1} = \frac{(11.53-1)^3}{4 \times 11.53} \Rightarrow y_1 = 0.158 \text{ m}$$

Since $\quad \dfrac{y_2}{y_1} = 11.53$

$$y_2 = 11.53 \times 0.158$$

$$= 1.822 \text{ m}$$

Problems 5: ***An overflow spillway is 40.0 m high. At the design energy head of 2.5 m over the spillway find the sequent depths. Neglect energy loss due to flow over the spillway face. (Assume $C_d = 0.68$)***

Solution:

$$q = \frac{2}{3} C_d \sqrt{2g} \times H_d^{3/2}$$

$$= \frac{2}{3} \times 0.68 \times \sqrt{2 \times 9.8} \times (2.5)^{3/2}$$

$$q = 7.93 \text{ m}^3/\text{s/m}$$

By using the equation

$$P + Hd = y_1 + \frac{V_1^2}{2g}$$

(Energy loss over the spillway is neglected)

$$y_1 + \frac{(7.93)^2}{2gy_1^2} = 42.5$$

$$y_1 = 0.28 \text{ m}$$

$$V_1 = \frac{q_1}{y_1} = \frac{7.93}{0.28} = 28.32 \text{ m/s}$$

$$F_1 = V_1 / \sqrt{gy_1} = 28.32 / \sqrt{9.81 \times 2.81}$$

$$= 17.088$$

$$\frac{y_2}{y_1} = \frac{1}{2}\left[-1 + \sqrt{1 + 8F_1^2}\right]$$

$$\frac{y_2}{0.28} = \frac{1}{2}\left[-1 + \sqrt{1 + 8 \times (17.088)^2}\right]$$

$$y_2 = 6.627 \text{ m}$$

Problems 6: ***For a hydraulic jump in a rectangular channel the velocity and depth after the jump are known to be 0.80 m/s and 1.75 m. Calculate the depth before the jump and the energy loss and the power dissipated per meter width.***

Solution:

$$F_2 = \frac{V_2}{\sqrt{gy_2}} = \frac{0.80}{\sqrt{9.81 \times 1.75}} = 0.1931$$

The sequent depth ratio is given in terms of F_2 as

$$\frac{y_1}{y_2} = \frac{1}{2}\left(-1+\sqrt{1+8F_2^2}\right)$$

$$= \frac{1}{2}\left(-1+\sqrt{1+8\times 0.1931^2}\right)$$

$$\frac{y_1}{y_2} = 0.0697$$

$$y_1 = 0.0697 \times 1.75 = 0.122 \text{ m}$$

Energy loss $$E_L = \frac{(y_2 - y_1)^3}{4 y_1 y_2}$$

$$= \frac{(1.750 - 0.122)^3}{4 \times 0.122 \times 1.750}$$

$$= 5.054 \text{ m}$$

Power dissipated $$P = \gamma Q E_L$$

$$= 9790 \times (0.80 \times 1.75) \times 5.054$$

$$= 69270 \text{ watt/m width}$$

Problems 7: ***A hydraulic jump occurs in a rectangular channel and the depths of flow before and after the jumps are 0.5 and 0.20 m. Calculate the critical depth.***

Solution:

$$y_1 = 0.5 \text{ m}; \qquad y_2 = 0.20 \text{ m}$$

The sequent depths are related to critical depth are as follow:

$$y_1\, y_2\, (y_1 + y_2) = 2yc^3$$

$$= 0.5 \times 0.2\,(0.5 + 0.2) = 2yc^3$$

$$= y_c = 0.327 \text{ m}$$

Problems 8: Show that the Froude number F_1 and F_2 before

and after the jump respectively, in a hydraulic jump taking place in a rectangular channel are related as:

$$F_2^{\,2} = \frac{8F_1^2}{\left(-1+\sqrt{1+8F_1^2}\right)^3}$$

Solution:

$$F_2^{\,2} = \frac{V_2^2}{gy_2} = \frac{q^2}{gy_2^{\,3}}$$

$$= \frac{q^2}{gy_1^3\,(y_2/y_1)^3} = \frac{F_1^2}{(y_2/y_1)^3}$$

But the sequent depth ratio y_2/y_1 is related to F_1 as

$$\frac{y_2}{y_1} = \frac{1}{2}\left(-1+\sqrt{1+8F_1^2}\right)$$

Hence
$$F_2^2 = \frac{8F_1^{\,2}}{\left(-1+\sqrt{1+8F_1^2}\right)^3}$$

Problem 9: ***At the toe of a hydraulic jump the Froude number and depth of flow are 9.0 and 0.4 m respectively. Estimate the specific energy head at the heel of the jump.***

Solution:

Let y_2 = sequent depth to y_1 then

$$\frac{y_2}{y_1} = \frac{1}{2}\left(-1+\sqrt{1+8F_1^2}\right)$$

$$\frac{y_2}{0.4} = \frac{1}{2}\left(-1+\sqrt{1+8\times 81}\right)$$

$$\frac{y_2}{0.4} = 12.24$$

$$y_2 = 4.90 \text{ m}$$

Energy loss $$E_L = \frac{(y_2 - y_1)^3}{4y_1y_2}$$

$$= \frac{(4.90 - 0.4)^3}{4 \times 0.4 \times 4.90}$$

$$= 11.623 \text{ m}$$

Initial specific energy $$E_L = y_1 + \frac{V_1^2}{2g} = y_1\left(1 + \frac{F_1^2}{2}\right)$$

$$= 0.4\left(1 + \frac{9^2}{2}\right) = 16.6 \text{ m}$$

Specific energy at the heel

$$E_2 = E_1 - E_L$$

$$= 16.6 - 11.623 = 4.977 \text{ m}$$

Problem 10: ***A 2.0 m wide rectangular channel has a flow of 3.60 m³/s with a velocity of 0.8 m/s. If a sudden release of additional flow at the upstream and of the channel causes the depth rise by 50 percent, determine the absolute velocity of the resulting surge and the net flow rate.***

Solution:

The surge move in downstream direction and the absolute velocity of the wave v_w is positive. By superimposing $(-v_w)$ on the system the equivalent steady flow is obtained.

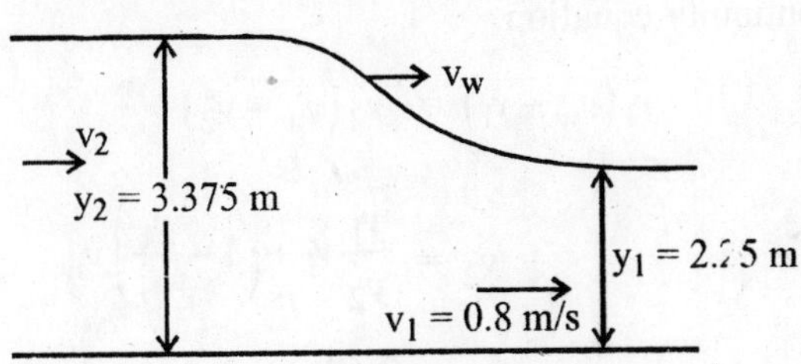

(a)

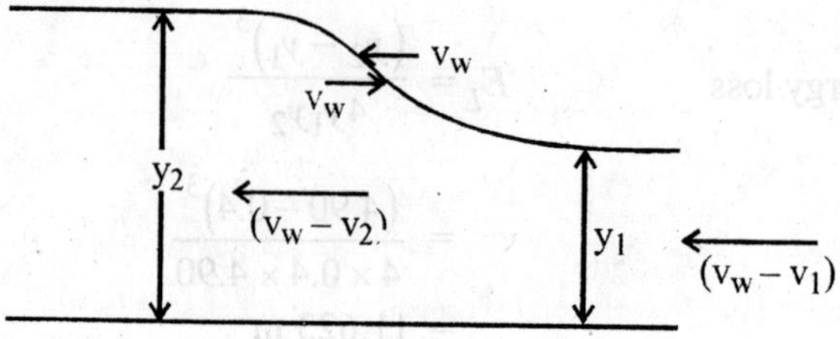

(*b*)

(*a*) Positive surge moving downstream.
(*b*) Simulated steady flow.

Here, $v_1 = 0.8$ m/s, $y_1 = \dfrac{3.6}{0.8 \times 2} = 2.25$ m

$$\frac{y_2}{y_1} = 1.5$$

$$y_2 = 1.5 \times 2.25 = 3.375 \text{ m}$$

For a positive surge moving downstream in a rectangular channel,

$$\frac{(v_w - v_1)^2}{gy_1} = \frac{1}{2}\frac{y_2}{y_1}\left(\frac{y_2}{y_1}+1\right)$$

$$\frac{(v_w - v_1)^2}{gy_1} = \frac{1}{2}\times 1.5 \times (1.5+1)$$

$$v_w = 7.233 \text{ m/s}$$

By continuity equation

$$y_1(v_w - v_1) = y_2(v_w - v_2)$$

$$v_2 = \frac{y_1}{y_2}v_1 + \left(1 - \frac{y_1}{y_2}\right)v_w$$

$$= 0.67 \times 0.8 + (1 - 0.67)\, 7.233$$

$$= 2.922 \text{ m/s}$$

Net flow rate $Q_2 = By_2v_2$

$= 2 \times 3.375 \times 2.922$

$= 19.72\ m^3/s$

Problem 11: ***Show that in a positive surge moving down a rectangular channel with absolute velocity V_w, the depth before the passage of the surge y_1 and after the passage of the surge y_2 are related as***

$$\frac{y_1}{y_2} = \frac{1}{2}\left(1+\sqrt{1+8Fa^2}\right)$$

where
$$F_a^{\ 2} = \frac{\left(v_w - v_1\right)^2}{gy_1}$$

Solution:

The unsteady flow due to the passage of the surge is shown in the Fig. (*a*). By superimposing a velocity v_w steady flow is simulated.

Here by continuity

$$y_1\,(v_w - v_1) = y_2\,(v_w - v_2)$$

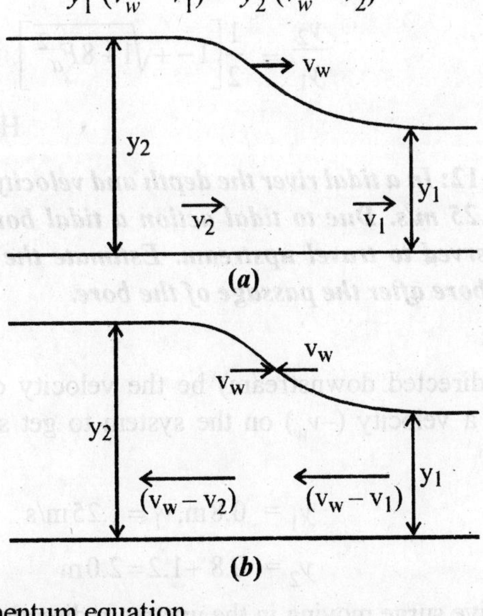

By momentum equation

$$\frac{\gamma y_1^2}{2} - \frac{\gamma y_2^2}{2} = \rho(v_w - v_1)y_1$$

$$[(v_w - v_2) - (v_w - v_1)]$$

$$= y_1^2 - y_2^2 = \frac{2}{g}(v_w - v_1)y_1\left\{\frac{y_1}{y_2}(v_w - v_1) - (v_w - v_w)\right\}$$

$$(y_2 + y_1)(y_2 - y_1) = \frac{2}{g}(v_w - v_1)^2 \frac{y_1}{y_2}(y_2 - y_1)$$

$$\left(\frac{y_2}{y_1} + 1\right)\frac{y_2}{y_1} = \frac{2}{g}\frac{(v_w - v_1)^2}{y_1} = 2F_a^2$$

$$\left(\frac{y_2}{y_1}\right)^2 + \left(\frac{y_2}{y_1}\right) - 2F_a^2 = 0$$

Considering positive root

$$\frac{y_2}{y_1} = \frac{1}{2}\left[1 - +\sqrt{1 + 8F_a^2}\right]$$

Hence Proved

Problem 12: ***In a tidal river the depth and velocity of flow are 0.8 m and 1.25 m/s. Due to tidal action a tidal bore of height 1.2 m is observed to travel upstream. Estimate the height and speed of the bore after the passage of the bore.***

Solution:

Let V_w (directed downstream) be the velocity of the bore. Superimpose a velocity $(-v_w)$ on the system to get simulated as shown:

Here $y_1 = 0.8\,\text{m}, v_1 = 1.25\,\text{m/s}$

$$y_2 = 0.8 + 1.2 = 2.0\,\text{m}$$

For positive surge moving in the upstream direction,

$$\frac{(v_w + v_1)^2}{gy_1} = \frac{1}{2}\left(\frac{y_2}{y_1}\right)\left(\frac{y_2}{y_1}+1\right)$$

$$= (v_w + 1.25)^2 = \frac{9.8 \times 0.8}{2} \times \frac{2}{0.8} \times \left(\frac{2}{0.8}+1\right)$$

$$= v_w = 4.61 \text{ m/s}$$

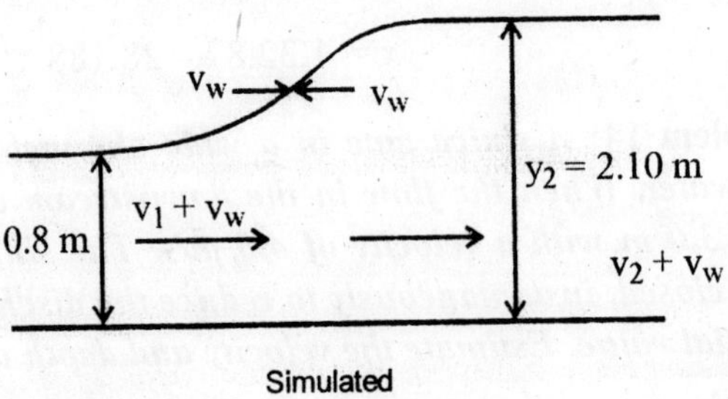

By continuity equation

$$y_2 (v_w + v_2) = y_1 (v_w + v_1)$$
$$2 (4.61 + v_2) = 0.8 (4.61 + 1.25)$$
$$v_2 = -2.26 \text{ m/s}$$

The river has a velocity of 2.26 m/s directed upstream after the passage of the bore.

Problem 13: ***A reservoir having a depth of 50 m undergoes an instantaneous, ideal, dam break. Estimate the depth and discharge intensity at the dam site and the water surface profile of the negative wave 3 second after dam break.***

Solution:

The water surface profile with positive x in the downstream of the gate axis.

$$-x = \left(3\sqrt{gy} - 2\sqrt{gy_1}\right)t$$

$$-x = \left(3\sqrt{9.81 \times y} - 2\sqrt{9.81 \times 50}\right)t$$

$$x = 44.29t - 9.393\sqrt{y}.t$$

At $\quad x = 0, y = y_0 = 44.29 / 9.393$

$$= 4.713$$

velocity at $x = 0, v = v_0 = \frac{2}{3}\sqrt{gy_1}$

$$= \frac{2}{3}\sqrt{9.81 \times 50}$$

$$= 14.76 \text{ m/s}$$

Profile after 3 sec: $x = 44.29 \times 3 - 9.396 \times 3\sqrt{y}$

$$x = 1.32.87 - 28.188\sqrt{y} \quad \textit{Ans.}$$

Problem 14: ***A sluice gate in a wide channel controls the flow of water. When the flow in the downstream channel at a depth of 3.0 m with a velocity of 4.0 m/s. The sluice gate was partially closed, instantaneously to reduce the discharge to 25% of its initial value. Estimate the velocity and depth at the gate.***

Solution:

Let suffix 1 refers to flow condition before the gate closure and suffix 2 conditionafter tha passage of the negative wave.

prior velocity $V_1 = 4.0$ m/s

New discharge instently $q = \frac{4 \times 3}{4} = 3.0 \text{ m/s} = V_1 y_1$

from equation (4.55)

$$V = V_1 + 2\sqrt{gy} - 2\sqrt{gy_1}$$

$$V_2 = 4.0 + 2\sqrt{9.81 y_2} - 2\sqrt{9.81 \times 3}$$

$$V_2 = 6.2642\sqrt{y_2} - 6.849$$

$$V_2 y_2 = 3.0$$

Solving by trial and error.

$$y_2 = 1.834 \text{ m}$$

$$V_2 = 1.635 \text{ m/s.}$$

Problem 15: ***Show that in a hyudraulic jump formed in horizontal, frictionless, rectangular channel the energy loss E_L relative to critical depth y_c can be expressed as:***

$$\left(\frac{E_L}{y_c}\right)^3 = \frac{(a-1)^9}{32)(a+1)a^4}$$

where a = sequent – depth ration = y_2/y_1

Solution: We know that

$$E_L = \frac{(y_2-y_1)^3}{4y_1y_2}$$

$$E_L = \frac{y_1^3\left(\frac{y_2}{y_1}-1\right)^3}{4y_1y_2}$$

or $$E_L = \frac{\left(\frac{y_2}{y_1}-1\right)^3}{4\frac{y_2}{y_1}}$$

We know $$E_L = \frac{(a-1)^3}{4\frac{y_2}{y_1}}$$ (Take cube both side)

Now $$E_L{}^3 = \frac{(a-1)^9}{64\frac{y_2^3}{y_1^6}} \quad ...(a)$$

$$\frac{y_2^3}{y_1^6} = \frac{y_2^4}{y_1^4}\times\frac{1}{y_1y_2}\times\frac{1}{y_1} \quad ...(b)$$

We know that $$\frac{2q^2}{g} = y_1\,y_2\,(y_1+y_2)$$

or $$2y_c^3 = y_1\,y_2\,(y_1+y_2) \quad ...(c)$$

from (b) and (c) we get

$$\frac{y_2^3}{y_1^6} = \frac{a^4}{2y_c^3}(1+a) \qquad \ldots(d)$$

Put (*d*) in (*a*) we get

$$E_L{}^3 = \frac{(a-1)^9 y_c^3}{32a^4(1+a)}$$

or

$$\left(\frac{E_L}{y_c}\right)^3 = \frac{(a-1)^9}{32a^4(1+a)} \quad \textbf{Proved.}$$

Problem 16: ***For hydraulic jump taking place in a horizontal frictionless, triangular channel show that the sequent depth y_1 and y_2 are related to pre jump froude number F_1 as***

$$F_1^2 = \frac{2a^2(a^3-1)}{3(a^2-1)}, \text{ where } a = \frac{y_2}{y_1}$$

Solution: Specific force in given by

$$\frac{Q^2}{gA_1} + A_1\bar{z}_1 = \frac{Q^2}{gA_2} + A_2\bar{z}_2 \qquad \ldots(a)$$

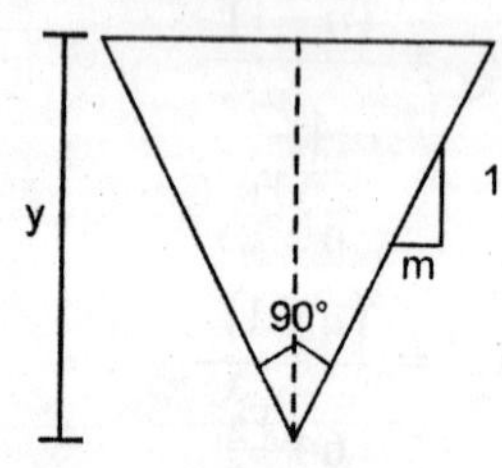

$$A = \frac{1}{2} \times 2my \times y$$

$$2\theta = 90°$$

as $\theta = 45°$ so $m = 1$

$\therefore \qquad A = y^2$

Now from equation (*a*) we get

$$\frac{Q^2}{g}\left(\frac{1}{A_1^2} - \frac{1}{A_2^2}\right) = A_2\bar{z}_2 - A_1\bar{z}_1$$

or $$\frac{Q^2}{g}\left(\frac{1}{y_1^2}-\frac{1}{y_2^2}\right)=\frac{y_2^3}{3}-\frac{y_1^3}{3}$$

$$\frac{Q^2}{g}\left(1-\frac{y_1^2}{y_2^2}\right)\frac{1}{y_1^2}=\frac{y_1^3}{3}\left(\frac{y_2^3}{y_1^3}-1\right)$$

or $$\frac{3Q^2}{gy_1^5}=\frac{\left[\left(\frac{y_2}{y_1}\right)^3-1\right]}{\left[1-\left(\frac{y_1}{y_2}\right)^2\right]}$$

as we know $$a=\frac{y_2}{y_1}$$

$$\frac{3Q^2}{gy_1^5}=\frac{(a^3-1)}{\left(1-\frac{1}{a^2}\right)}$$

$$\frac{3Q^2}{gy_1^5}=\frac{(a^3-1)}{(a^2-1)}a^2$$

$$\frac{3Q^2}{gy_1^4\,y_1}=\frac{(a^3-1)}{(a^2-1)}$$

$$\frac{3\times 2V_1^2}{gy_1}=\frac{2a^2(a^3-1)}{(a^2-1)}$$

$$\left(\because F=\frac{\sqrt{2}\cdot V}{\sqrt{gy}}\text{ for triangular channel}\right)$$

or $$3F_1^2=\frac{2a^2(a^3-1)}{(a^2-1)}$$

or $$F_1^2 = \frac{2a^2(a^3-1)}{3(a^2-1)}$$

Problem 17: ***The tidal bore in an estuary is observed to be travelling upstream at a velocity of 8 m/s. The depth any velocity of flow under the steady, uniform condition prevailing prior to the arrival of bore are 3.2 m and 1 m/s respectively. Estimate the height of bore?***

Solution: This is a case of positive surge moving upstream, so

$$\frac{(V_w+V_1)^2}{gy_1} = \frac{1}{2}\left(\frac{y_2}{y_1}\right)\left(\frac{y_2}{y_1}+1\right)$$

$V_w = 8$ m/s, $V_1 = 1$ m/s $y_1 = 3.2$ m

$C = V_w + v_1 = 8 + 1 = 9$ m/s

or $$C^2 = gy_1\left[\frac{1}{2}\left(\frac{y_2}{y_1}\right)\left(\frac{y_2}{y_1}+1\right)\right]$$

$$9^2 = \frac{9.81\times 3.2}{2}\left[\frac{y_2}{y_1}+\left(\frac{y_2}{y_1}\right)^2\right]$$

$$5.166 = \frac{y_2}{y_1}+\left(\frac{y_2}{y_1}\right)^2$$

Put $$\frac{y_2}{y_1} = x$$

We get $$x^2 + x - 5.166 = 0$$

or $$x = 1.827$$

or $$\frac{y_2}{y_1} = 1.827$$

$\therefore$ $$y_2 = 1.827 \times 3.2 = 5.847 \text{ m}$$

The height of bore $= y_2 - y_1$

$= 5.847 - 3.2$

$= 2.647$

Problem 18: Show that in an horizontal rectangular channel to create hydraulic jump

$$y_c^3 = y_1 y_2\left(\frac{y_1 + y_2}{2}\right)$$

where y_1, y_2 and y_c are depth before the jump, after the jump and critical depth.

Solution: Accordance to specific force equation we have

$$\frac{Q^2}{gA_1} + A_1\bar{z}_1 = \frac{Q^2}{gA_2} + A_2\bar{z}_2$$

or

$$\frac{Q^2}{g}\left[\frac{1}{A_1} - \frac{1}{A_2}\right] = A_2\bar{z}_2 - A_1\bar{z}_1$$

We know for rectangular channel

$$\bar{z}_1 = \frac{y_1}{2},\ \bar{z}_2 = \frac{y_2}{2}$$

Also

$$A_1 = \text{B}y_1,\ A_2 = By_2$$

Now

$$\frac{Q^2}{g}\left[\frac{1}{y_1} - \frac{1}{y_2}\right] = \frac{y_2^2}{2} - \frac{y_1^2}{2}$$

$$\frac{2q^2}{g}\left[\frac{1}{y_1} - \frac{1}{y_2}\right] = \left(y_2^2 - y_1^2\right)$$

We know

$$y_c^3 = \frac{q^2}{g}$$

Therefore

$$2y_c^3\frac{(y_2 - y_1)}{y_1 y_2} = (y_2 - y_1)(y_2 + y_1)$$

$$2y_c^3 = y_1 y_2\left(\frac{y_1 + y_2}{2}\right) \quad \textbf{proved.}$$

Problem 19: Show that for a submerged hydraulic jump just upstgream of sluice gate in a horizontal rectangular channel

$$\frac{y_s}{y_1} = \sqrt{2F_1^2\left(\frac{y_1}{y_2}-1\right)+\left(\frac{y_2}{y_1}\right)^2}$$

where y_1 is the depth of opening of sluice gatge, y_2 is the depth of flow downstream, y_s is the water depth on upstream side of the sluice gate and F_1 is the froude number of flow through sluice opening.

Soluton: Momentum equation

$$M_2 - M_1 = P_1 - P_2$$

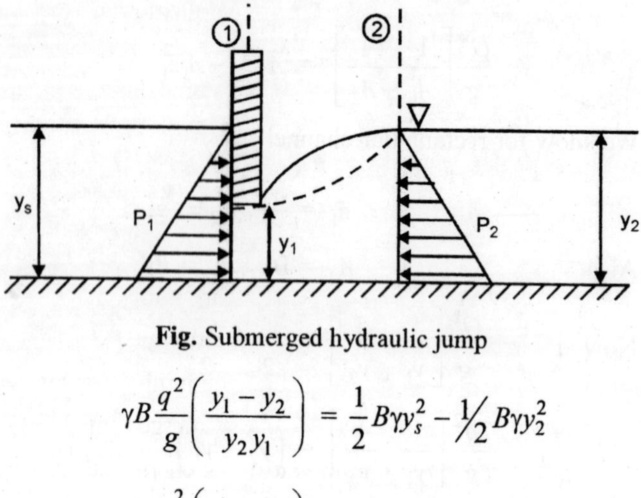

Fig. Submerged hydraulic jump

$$\gamma B\frac{q^2}{g}\left(\frac{y_1-y_2}{y_2 y_1}\right) = \frac{1}{2}B\gamma y_s^2 - \frac{1}{2}B\gamma y_2^2$$

$$\frac{q^2}{g}\left(\frac{y_1-y_2}{y_2 y_1}\right) = y_s^2 - y_2^2$$

Dividing throughout by y_1^2

$$\frac{q^2}{gy_1^3}\left(\frac{y_1}{y_2}-1\right) = \left(\frac{y_5}{y_1}\right)^2 - \left(\frac{y_2}{y_1}\right)^2$$

Substitute $F_1^2 = \dfrac{q^2}{gy_1^3}$

$$2F_1^2\left(\frac{y_1}{y_2}-1\right)=\left(\frac{y_s}{y_1}\right)^2-\left(\frac{y_2}{y_1}\right)^2$$

or
$$\left(\frac{y_s}{y_1}\right)=\sqrt{2F_1^2\left(\frac{y_1}{y_1}-1\right)+\left(\frac{y_2}{y_1}\right)^2}$$

Proved

Review Questions

1. Define hydraulic jump? Discuss the application of hydraulic jump.
2. What are the assumption made in the analysis of hydraulic jump?
3. Derive Belanger momentum equation.
4. Discuss the classification of jump.
5. Explain the term:

   (*a*) Undular jump (*b*) Weak jump

   (*c*) Oscillating jump (*d*) Steady jump

   (*e*) Strong jump (*f*) Free jump at contracta

   (*g*) Free repelled jump (*h*) Submerged jump
6. Explain how hydraulic jump is used as energy dissipator.
7. Derive expression for energy loss in rectagular channel.
8. What do you mean by surge? Discuss its classification.
9. Derive expression for the following:

   (*a*) Positive surge moving downstream.

   (*b*) Positive surge moving upstream.

   (*c*) Negative surge moving upstream.

   (*d*) Negative surge moving downstream.
10. Explain the term:

    (*a*) Celerity of wave.

    (*b*) Deep and Shallow water wave.

11. A hydraulic jump occurs in a horizontal rectangular channel with sequent depths of 0.50 m and 6.2 m. Calculate the rate of flow per unit width, energy loss and the initial Froude number.
12. A hydraulic jump occurs in a horizontal rectangular channel at an initial Froude number 11.0. What percent of initial energy is lost in this jump.
13. The Froude number of sub-critical flow after the hydraulic jump in a horizontal rectangular channel is 0.4. Estimate the Froude number of super critical approach flow.
14. At what initial Froude number F_1 would the energy loss in a hydraulic jump in a horizontal rectangular channel be 50% of the initial specific head?
15. A hydrualic jump in a 2.5 m wide horizontal rectangular channel was used as flow measuring device. If the flow depth before and after the jump are 0.2 m and 2.0 m/s, estimate the discharge.
16. Water is being discharged from a reservior under a sluice gate at a rate of 18 m^3/s in a horizontal rectangular channel of width 3.0 m. At a point where the depth is 0.4 m a hydraulic jump is known to occur. Determine the
    (*i*) Velocity before and after the jump.
    (*ii*) Energy loss in the jump.
    (*iii*) The power dissipated.
17. In a hydraulic jump on a horizontal rectangular channel the depth and Froude number before the jump are 0.10 m and 8.0 respectively. Estimate the energy loss and specific head at the end of the jump.
18. Show that the Froude numbers F_1 and F_2 in a hydraulic jump occuring in a rectangular channel are related by.

    (*a*) $$F_2^2 = \frac{8F_1^2}{\left(-1+\sqrt{1+8F_1^2}\right)^3}$$

(*b*) $F_1^2 = \dfrac{8F_2^2}{\left(-1+\sqrt{1+8F_2^2}\right)^3}$

19. A hydraulic jump in a rectangular channel has Froude number at the beginning of the jump $F_1 = 4$. Find F_2 at the end of the jump.
20. The depth and velocity in a rectangular channel are 0.8 m and 1.6 m/s respectively. If a gate at the downstream end of the channel is closed completely (abruptly). What will be the height and absolute velocity of the resulting surge?
21. A rectangular channel 5.0 m wide is conveying 20 m^3/s of discharge at a depth of 2.0 m. If a downstream sluice gate is abruptly lowered to reduce the discharge by 50%. Estimate the characteristics of resulting surge.
22. A tidal bore moves up a wide river with a velocity of 8 m/s. If the river had a depth of 2.0 m and velocity of 0.8 m/s before the passing of the bore, estimate the height of the bore and velocity of flow in the river after the passage of bare estimate the height of the surge.
23. In a rectangular channel a positive surge of velocity 5.0 m/s was seen moving down the stream. If the depth and velocity after the passage of the surge are 2.8 m and 4.0 m/s. Estimate the height of the surge.

Unit 5

Flow Measurement in Open Channel Flow

5.1 Introduction

The flow in small channel is generally measured by means of weirs and in large channel by using the current meter. Sometimes the gates which care installed to regulate the flow in channels are also used for flow measurement. We shall discuss various types of weirs which are generally used *for* flow measurement in open channels, and also the flow under a sluice gate.

5.2 Weir

The weir has been a standard device for the measurement of flow in open channels. It is an obstruction in the channel that causes the liquid to rise behind the weir and then flows over it. By measuring the height of upstream liquid surface, the rate of flow is determined. Weirs may be constructed from a sheet of metal, masonry or from other material. If the jet of liquid, known as the *nappe,* springs free as it leaves the upstream face, the weir is known as the sharp-crested weir. The broad crested weirs are those which support the falling nappe over its crest, in the longitudinal direction.

The weirs may be of different shapes such as the rectangular, triangular, trapezoidal, parabolic etc. They may discharge under free conditions or may function as submerged weirs under appropriate downstream conditions. When the liquid level downstream of weir is above its crest level, the weir is said to be a *submerged weir*.

5.2.1 Sharp-Crested Rectangular Weir

It may be considered as large orifice of the rectangular shape placed in the channel in such a way that the head on its upper edge is zero. Consequently the upper edge is eliminated leaving only the lower edge known as the crest. The rate of flow is determined by measuring the head H over the weir crest, at a distance upstream at least four times the maximum head to be used. As the liquid approaches the weir, its surface assumes a curvature with convexity upwards. The drop in surface in the plane of the weir is known as the *drawdown.* The drawdown at the weir crest is about 0.15 H. Consider an elementary strip of area Bdh at a depth h below the free surface.

Theoretical flow through the strip

$$dQ = Bdh\sqrt{2gh}$$

The theoretical flow passing over the weir is

$$Q = \int_0^H B\sqrt{2gh}\, dh = \frac{2}{3} B\sqrt{2g}\, H^{3/2} \qquad \text{...(5.1)}$$

This ideal flow will be decreased slightly by fluid friction and decreased much more due to following reasons;

(1) The flow area in the plane of the crest is not-BH, but less on account of the drawdown.

(2) The velocity at $h = 0$, is not zero

(3) The streamlines in the plane of the crest are curved and not normal to the area of the flow. To correct for these discrepancies, it is necessary to introduce an experiment coefficient of discharge. This then gives the basic weir formulas:

$$Q = C_d\left(\frac{2}{3}B\sqrt{2g}H^{3/2}\right) = \frac{2}{3}C_d B\sqrt{2g}\, H^{3/2} \qquad \text{...(5.2)}$$

If the velocity of approach V_0 at the sectin H was measured is appreciable, the limits of integration should be $(H + V_0^2/2g)$ and $(V_0^2/2g)$ and the resulting discharge equation will be

$$Q = \frac{2}{3} C_d B \sqrt{2g} \left[\left(H + \frac{V_0^2}{2g} \right)^{3/2} - \left(\frac{V_0^2}{2g} \right)^{3/2} \right] \quad \ldots(5.3)$$

The following empirical formula may be use to determine the discharge coefficient

$$C_d = 0.611 + 0.075 \text{ H/Z}$$

In which Z is the height of the weir crest above the channel floor.

5.2.2 Suppressed Rectangular Weir

When the length of crest of the weir is the same as the width of the channel, the weir is said to be a *suppressed weir*. The width of the nappe for suppressed weir is equal to the length of the crest. Thus in this case, the effect of sides or ends of the weir on contraction of the nappe is eliminated or suppressed.

Thus for suppressed weirs,

Length of weir crest = Width of the channel.

The discharge equations developed above are applicable to suppressed weirs.

5.2.3 Contracted Weirs

When the crest length B of a rectangular weir is less than the width of the channel, there will be a lateral contraction of the nappe so that its length is less than B. Such type of weirs have end contractions and are known as contracted weirs or weirs with end contractions. The effective length of the nappe is thus reduced by the amount of contraction at each end.

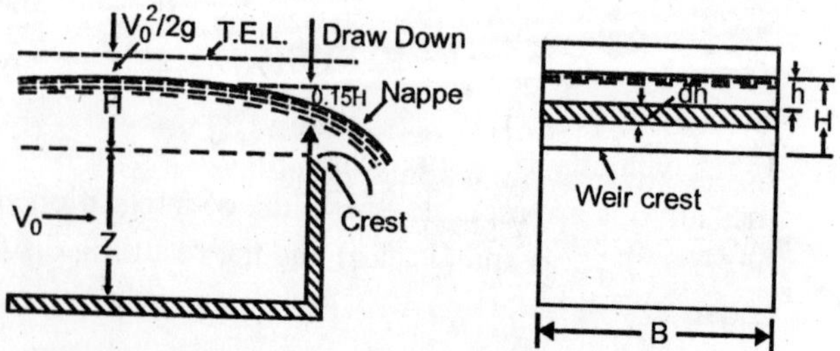

Fig. 5.1: Flow over a sharp-crested rectangular weir

If *W* is the width of the channel and *B* is the length of the weir crest and if $B < W$, the weir will be a contracted weir. Francis concluded that the effect of each side contraction is to reduce the length of nappe by 0.1 *H*: Thus if there are *n* end contractions, the length of the nappe will be reduced by 0.1 *nH* and will be equal to (*B* – 0.1 nH) and the discharge equation (5.2), will then be written as

$$Q = \frac{2}{3} C_d (B - 0.1nH) \sqrt{2g} H^{3/2} \qquad ..(5.4)$$

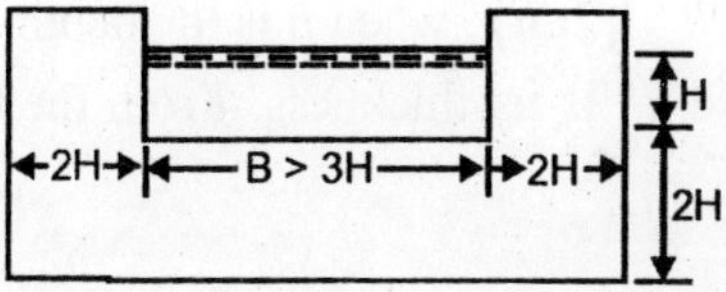

Fig. 5.2: Standard Contracted Weir

From Eq. (5.4), it would appear that at higher heads a condition may be reached when this formula would indicate zero flow, and at still higher heads would compute negative flow. If Eq. (5.4) is to be used the minimum proportions as indicated in Fig. 5.2 are desirable.

5.2.4 Francis Weir Formula

On the basis of a large number of experiments on sharp-crested rectangular weirs, J.B. Francis found that the flow varied as $H^{1.47}$, but for greater convenience he adopted $H^{1.5}$. He then selected a constant average value of $C_d = 0.622$, so that or suppressed weir with a negligible velocity of approach.

$$Q = \frac{2}{3} C_d B \sqrt{2g} H^{3/2}$$

$$= \frac{2}{3} \times 0.622 \sqrt{2' \times 9.81} BH^{3/2}$$

$$= 1.837\ BH^{3/2} \text{ (MKS and SI units)} \qquad ...(5.5)$$

In F.P.S. units, substituting g = 32.2 ft/s2,

The discharge equation becomes

$$Q = 3.33\ BH^{3/2} \text{ (FPS units)} \qquad ...(5.6)$$

5.2.5. Triangular Weir

A triangular weir is particularly useful where the discharge is to be vary over a large range and the same accuracy is desired for both small and large discharges. For measuring small discharges accurately and to avoid surface tension effects which are associated with flow at low heads and to obtain higher measurable heads, a triangular weir is preferable over the rectangular weir is also known as a V-notched weir. Fig. 5.3 shows a triangular weir with a vertex angle 2θ. The rate of flow through an elementary strip of area $dA = 2xdh$ is $dQ = 2xdh\sqrt{(2dh)}$, when h is the depth of strip below the liquid surface and dh is its thickness. From the triangle OAB, we have

$$\frac{x}{H-h} = \tan\theta$$

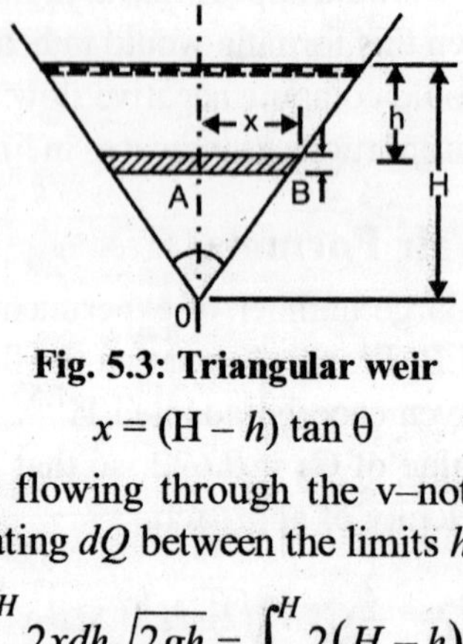

Fig. 5.3: Triangular weir

$$x = (H - h)\tan\theta$$

The discharge flowing through the v–notched weir may be obtained by integrating dQ between the limits $h = 0$ and $h = H$

$$Q = \int_0^H 2xdh\sqrt{2gh} = \int_0^H 2(H-h)\tan\theta\sqrt{2gh}dh$$

$$= \frac{8}{15}\sqrt{2gh}\tan\theta H^{5/2} \qquad \text{...(5.7)}$$

Eq. (5.7) gives the theoretical discharge. The actual discharge is less than the theoretical one due factors already discussed while dealing with the rectangular weirs. The actual discharge is obtained by multiplying Eq. (5.7) with the coefficient of discharge C_d thus

$$Q = \left(\frac{8}{15} C_d \sqrt{2g} \tan\theta\right) H^{5/2} \quad \text{...(5.8)}$$

For a given notch, the discharge varies as $H^{2.5}$, and the discharge head relation ship may be expressed as $Q = KH^{5/2}$ in which K is almost a constant for the weir and s expressed by the bracket quantity in Eq. (5.8).

5.2.6 Trapezoidal Weir

A trapezoidal weir has an opening of the trapezoidal shape and may be regarded as a combination of a rectangular and a triangular weir. Fig. 5.4 shows a trapezoidal weir.

The trapezoidal weir section can be broken into rectangular portion of width B and a triangular portion of vertex angle 2θ, for the purpose of computing discharge. Discharge through the trapezoidal weir is equal to the sum of the discharge through the rectangular and triangular portions, thus

$$Q = \frac{2}{3} C_d B \sqrt{2g} H^{3/2} + \frac{8}{15} C_d \sqrt{2g} \tan\theta H^{5/2} \quad \text{...(5.9)}$$

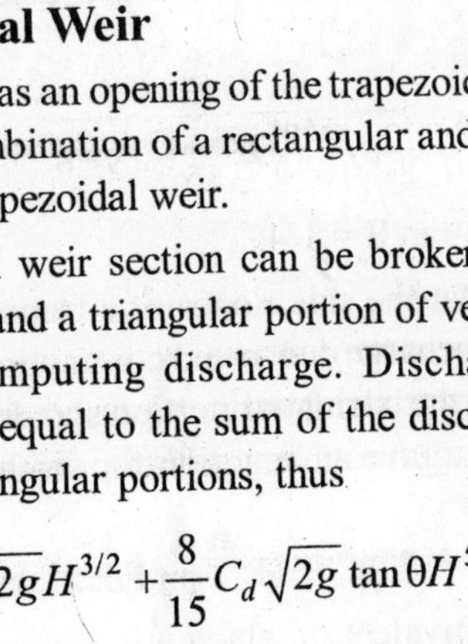

Fig. 5.4 Trapezoidal weir

5.2.7. Cipolletti Weir

This is a type of trapezoidal weir in which the sides have a slope of 1 horizontal to 4 vertical. This weir is claimed to have an advantage over a rectangular weir and that is the decrease in discharge due to end contraction is balanced by the discharge through the triangular portion and that a rectangular weir formula can then be used to compute the discharge. These conclusions were based on the experiments by Cipolletti on such weirs and, therefore, such a weir known after him as Cipolletti weir.

The reduction in the discharge of a rectangular weir due to two end contractions

$$= 2/3\ C_d\ \sqrt{2g}\ (0.2H)\ H^{3/2}$$

The discharge through the triangular weir of vertex angle 2θ

$$= \frac{8}{15} C_d \sqrt{2g}\ \tan\theta H^{5/2}$$

Since in a Cipolletti weir these two discharge must balance, thus,

$$\frac{2}{15} Cd\sqrt{2g} H^{5/12} = \frac{8}{15} C_d \sqrt{2g}\ \tan\theta H^{5/2}$$

Solving for θ, tan θ = 1/4.

This shows that the side slopes of a trapezoidal weir in which the reduction in discharge due to end contractions is balanced by the discharge through the triangular portion should be 1 horizontal to 4 vertical. Cipollelti gave an equation for discharge through such a weir as

$$Q = 3.367\ BH^{2/2}\ \text{(in F.P.S. units)} \qquad ...(5.10)$$

The metric equivalent of which is

$$Q = 1.86\ BH^{3/2} \qquad ...(5.11)$$

In which B and H are measured in mature and Q is in m3/s. For determining discharge in irrigation canals, where this type of weir may be found in common use, we need only the observation of head H over the weir crest, as is evident from Eq. (5.10) or (5.11).

5.2.8 Submerged Weir

When the water on the downstream side of a weir rises above the level of the crest, the weir is said to be a submerged weir. .

The flow over the submerged weir may be considered by dividing the flow into two portions.

(1) flow over the upper part of the section (lying above the line AB) of depth $(H_1 - H_s)$ may be considered as a free discharge into the air, it is a weir flow and

(2) flow through the remaining lower part of depth H_1 (lying below the line *AS*) may be considered as a discharge through a submerged orifice.

Thus for the upper portion, the weir formula gives the flow rate

$$Q_1 = \frac{2}{3}C_{d1}\sqrt{2g}B(H_1 - H_2)^{3/2}$$

and for the lower portion, the orifice formula yields,

$$Q_2 = C_{d2}BH_2\sqrt{2g(H_1 - H_2)}$$

If we assume $C_{d1} = C_{d2} = C_{d3}$, then the total discharge

$$Q = Q_1 + Q_2$$

$$= C_d B\sqrt{2g(H_1 - H_2)}\left[\frac{2}{3}H_1 - \frac{1}{3}H_2\right] \qquad ...(5.12)$$

The discharge coefficient C_d varies with the submergence ratio, H_2/H_1, the variation of which for a given weir may be obtained from experiments.

J.R. Villemonto studied the discharge characteristics of a number of sharp-crested submerged weirs. He conducted tests on rectangular, triangular, parabolic and proportional weirs and showed that the result for all types could be represented by a single equation as given below:

$$Q = Q_1\left[1 - \left(\frac{H_2}{H_1}\right)^n\right]^{0.385} \qquad ...(5.13)$$

where Q = discharge for submerged conditions

Q_1 = free discharge given by $Q_1 = KH_{1^n}$

n = exponents in the free discharge equation.

Eq. (5.13) shows that the discharge of the weir under submerged conditions depends upon the submergence ratio H_2/H_1 and that the weir will yield zero discharge when H_2/H_1, that is when the upstream and the downstream water surfaces are at the same level above the weir crest.

5.3 Special Sharp Crested Weir

5.3.1 Linear Proportional Weir

The linear proportional weir, with its linear head-discharge characteristic is used as a control for float-regulated dosing devices, as a flow meter and as an outlet for grit chambers (sedimentation tanks). The linear proportional weir was invented by Stout (1897). This weir is only of theoritical interest as its width at base is infinite This was improved by Sutro (1908) to develop a practical linear P-weir and is well known as the *Sutro weir.* The Sutro weir has a rectangular base over which a designed shape is fitted. It is found that for flows above the base weir the discharges are proportional to the heads measured above of reference plane located at one-third the depth of the base weir from figure blow.

$$Q = b\left(h + \frac{2}{3}a\right) \quad (5.14)$$

where b = the proportionality constant.

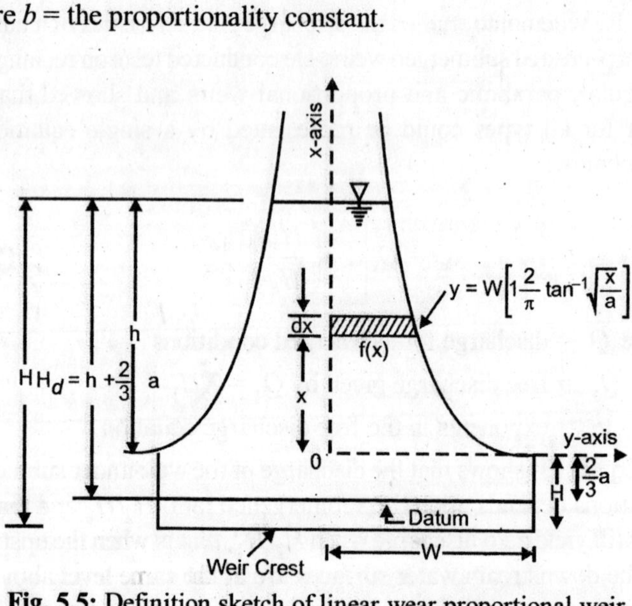

Fig. 5.5: Definition sketch of linear wear proportional weir

5.3.2 Shape of the Sutro Weirs

Let the weir have the base on a rectangular weir of width 2W and depth a. For convenience, the horizontal and vertical axes at the origin O are chosen as y and x-axes respectively. The weir is assumed to be symmetrical about the x-axis.

The discharge through the rectangular weir when the depth of flow is h above the origin is

$$q_1 = \frac{4}{3} W C_d \sqrt{2g}\left[(h+a)^{3/2} - h^{3/2}\right] \tag{5.15}$$

where C_d = coefficient of discharge.

The discharge through the upper portion above the origin called the *complimentary weir* is

$$q_2 = 2C_d\sqrt{2g}\int_0^h \sqrt{h-x}\, f(x)dx \tag{5.15a}$$

The total discharge through the weir is

$$Q = q_1 + q_2$$

We wish this discharge to be proportional to the head measured above the reference plane situated $\frac{a}{3}$ above the crest of the weir. This reference plane is chosen arbitrarily by Sutro for mathematical convenience. Thus

$$\begin{aligned} Q &= q_1 + q_2 \\ &= \frac{4}{3} W C_d \sqrt{2g}\left[(h+a)^{3/2} - h^{3/2}\right] \\ &\qquad + 2C_d\sqrt{2g}\int_0^h \sqrt{h-x}\, f(x)dx \\ &= b\left(h + \frac{2a}{3}\right) \quad \text{for } h \ge 0 \end{aligned} \tag{5.16}$$

where b is the proportionality constant.

As there is no flow above the base weir when h =0, we have by substituting $h = 0$ in Eq. 5.16.

$$b = WKa^{1/2} \qquad \text{...(5.17)}$$

where $K = 2C_d\sqrt{2g}$

Substituting this value of b in Eq. 5.16

$$\frac{2}{3}W\left[(h+a)^{1/2} - h^{3/2}\right] + \int_0^h \sqrt{h-x}\,f(x)dx = Wa^{1/2}\left(h + \frac{2a}{3}\right) \qquad \text{...(5.18)}$$

Re-arranging,

$$\int_0^h \sqrt{h-x}\,f(x)dx = Wa^{1/2}\left(h+\frac{2a}{3}\right) - \frac{2}{3}W\left[(h+a)^{3/2} - h^{3/2}\right]$$

$$= W\left[\frac{2}{3}a^{3/2} + a^{1/2}h + \frac{2}{3}h^{3/2} - \frac{2}{3}(a+h)^{3/2}\right]$$

$$= \frac{2}{3}W\left[h^{3/2} - \frac{3}{8}a^{-1/2}h^2 + \frac{1}{16}a^{-3/2}h^3 - \frac{3}{128}a^{-5/2}h^4 + \ldots\right] \qquad \text{...(5.18a)}$$

It is required to find the function $f(x)$ such that Eq. 5.18a is satisfied for all positive values of h. This is achieved by expressing $f(x)$ in a series of powers of x and determining their coefficients. A general term x^m in $f(x)$ results in a term

$$\int_0^h \sqrt{h-x}\,x^m dx = \int_0^h x^m\left(h^{1/2} - \frac{1}{2}h^{-1/2}x + \frac{1}{8}h^{-3/2}x^2 - \ldots\right)dx$$

$$= \text{Const } (h)^{m+(3/2)} \qquad \text{...(5.19)}$$

so that the first term in Eq. 5.18a can be obtained by a constant term in the series for $f(x)$ and the other terms by taking m half an odd integer. Consequently we assume,

$$f(x) = y = A_1 + A_2 x^{1/2} + A_3 x^{3/2} + A^4 x^{5/2} + \ldots \qquad \text{...(5.20)}$$

Substituting this in Eq. 5.18a

$$\int_0^h \left(A_1\sqrt{h-x} + A_2\sqrt{hx-x^2} + \ldots\right)dx$$

$$= \frac{2A_1}{3}h^{3/2} + \frac{\pi A_2 h^2}{8} + \frac{\pi A_3 h^3}{16} + \frac{5\pi A_4 h^4}{128} + ..$$

$$= \frac{2}{3}W\left[h^{3/2} - \frac{3}{8}a^{-1/2}h^2 + ...\right]$$

This leads to

$$A_1 = W$$

$$A_2 = -\frac{2}{\pi}a^{-1/2}W$$

$$A_3 = \frac{2}{3\pi}a^{-3/2}W$$

and so on

Substituting these coefficients in Eq. 5.20

$$f(x) = W\left[1 - \frac{2}{\pi}\left\{\frac{x^{1/2}}{a^{1/2}} - \frac{x^{3/2}}{a^{3/2}} + \frac{x^{5/2}}{a^{5/2}}\right\}\right]$$

$$= W\left[1 - \frac{2}{\pi}\tan^{-1}\sqrt{\frac{x}{a}}\right] \quad (5.21)$$

The discharge equation for the Sutro weir can now be summarised as

$$Q = b\left(h + \frac{2}{3}a\right) = b\left(H - \frac{a}{3}\right) = bH_4 \quad (5.22)$$

where $b = WKa^{1/2}$ and $K = 2C_d\sqrt{2g}$

H = depth of flow in the channel. (It is usual practice to make the crest of the base coincide with the bed of the channel)

h = head measured from the top of the rectangular base weir

H_d = depth of water over the datum.

The sharp edged Sutro weir is found to have an average coefficient of discharge of 0.62.

A simple weir geometry, called *quadrant plate weir*, which has the linear head-discharge relationship. This weir has the advantage of easy fabrication and installation under field conditions.

5.3.3 General Equation for the Weir

Cowgill[8] and Banks[9] have shown that the curve describing the weir producing a discharge $Q = bh^m$ for $m \geq 1/2$ is given by

$$y = y(x) = \frac{b}{2C_d\sqrt{2g\pi}} \cdot \frac{\Gamma(m+1)}{\Gamma(m-1/2)} \cdot (x)^{(m=3/2)} \quad \ldots(5.33)$$

when $$Q = \int_0^h 2C_d f(x)\sqrt{2g(h-x)}dx$$

where
- h = head measured above the crest of the weir
- x, y = coordinates along vertical and horizontal axis respectively
- b = a coefficient of proportionality
- C_d = coefficient of discharge of the weir
- Γ = gamma function

The relationship between the exponent m and the profile of the weir is shown in Fig. 5.6 it is clear from that an attempt to design a

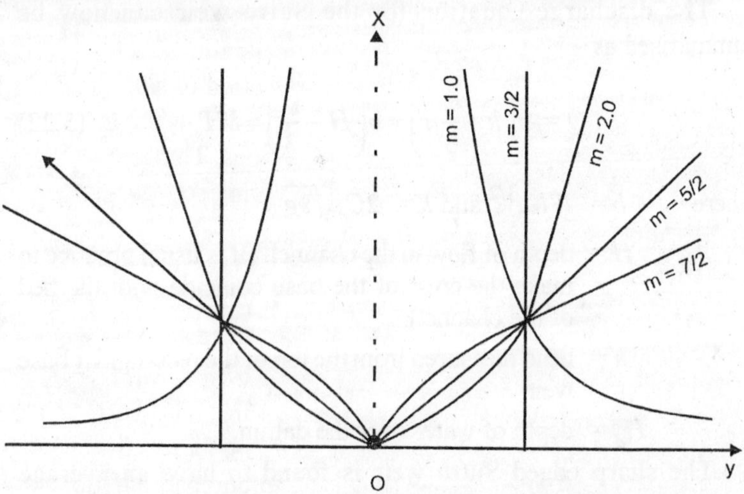

Fig. 5.6 Weir profiles for Different values of exponent m

weir producing a discharge proportional to h^m for $m < 3/2$ inevitably leads to a curve which will be asymptotic at the base giving rise to infinite width, which is physically unrealizable. The linear proportional weir ($m = 1$) is one such case. Sutro overcame this defect by the ingenious method of proving a rectangular base. A rational explanation for the selection of the datum was provided by Keshava Murthy which is enunciated in the theorem of slope discharge continuity.

The slope-discharge-continuity theorem states: 'In any physically realizable were having a finite number of finite discontinuities in its geometry, the rate of change of discharge is continuous at all points of discontinuity.' Physically this means that the curve describing the discharge versus the head for any compound weir cannot have more than one slope at any point. This is clear as otherwise it would mean, theoretically, there could be more than one value for the discharge in the infinitesimal strip in the neighbourhood of the discontinuity which is physically meaningless. The proof of this theorem which is based on the use of the theorem of Laplace transforms is beyond the scope of this book.

The P-weir consists of a known base over which a designed complimentary weir is fixed. Each weir is associated with a reference plane or datum which is determined by evaluating a new parameter λ called the *datum constant* by the application of slope discharge-continuity theorem. The problem of design of P-weirs is solved by the technique of solution of integral equations. This is explained by the design of *quadratic weirs* in the following section.

5.3.4 Quadratic Weir-notch Orifice

A quadratic weir is a proportional weir in which the discharge Q is proportional to the square root of the head h. This weir has applications in bypass flow measurement.

It is evident from the general Eq. 5.23 that an attempt to design a weir to pass a discharge proprietorial to h^m, for $m < 3/2$ inevitably leads to a curve having infinite width at the bottom which is physically unrealizable. In order to obviate this, the quadratic weir is provided with a base in the form of a rectangular weir of width $2W$ and depth

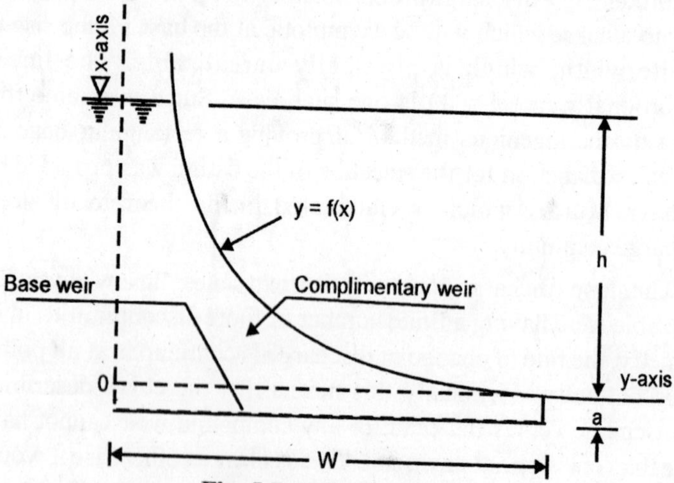

Fig. 5.7 Definition sketch

a, over which a designed curve is fitted.

Referring to the weir is assumed to be sharp-edged and symmetrical over the *x*-axis. When the flow is *h* above the base, the discharge through the rectangular weir, below the *y*-axis, is

$$q_1 = \frac{2}{3}WK\left[(h+a)^{3/2} - h^{3/2}\right] \qquad \ldots(5.24)$$

where $K = 2C_d\sqrt{2g}, C_d$ = the coefficient of discharge. The discharge from the complementary weir above the origin is

$$q_2 = K\int_0^h \sqrt{h-x}f(s)dx \qquad \ldots(5.24a)$$

Total discharge $Q = q_1 + q_2$

$$= \frac{2}{3}WK\left[(h+a)^{3/2} - h^{3/2}\right] + K\int_0^h \sqrt{h-x}f(x)dx \qquad \ldots(5.25)$$

We wish to have the discharge $Q = b\sqrt{h+\lambda a}$, where *b* is the proportionality constant and λ is the datum constant. In other words, the discharge is proportional to the square root of the head measured above a reference plane. The constant b and λ are to be evaluated. They are determined by the two conditions of continuity of discharge and the requirement of the slope discharge continuity theorem. Rewritting Eq. 5.25

$$Q = \frac{2}{3}WK\left[(h+a)^{3/2} - h^{3/2}\right] + K\int_0^h \sqrt{h-x}\,f(x)dx$$

$$= h\sqrt{h+\lambda a} \text{ for } h \geq 0 \qquad \text{...(5.26)}$$

When $h = 0$, there is no flow above the base weir. Hence, substituting $h = 0$ in Eq. 7.30

$$\frac{2}{3}WKa^{3/2} = b\sqrt{\lambda a}, \qquad \text{...(5.27)}$$

Differentiating Eq. 5.26 on both sides by using Leibnitz's rule for differentiating under the integral sign, and re-arranging

$$\int_0^h \frac{f(x)}{\sqrt{h-x}}dh = \frac{b}{K\sqrt{h+\lambda a}} - 2W\left[(h+a)^{1/2} - h^{1/2}\right] = \phi(h)$$

$$\text{...(5.28)}$$

Applying the slope-discharge-continuity theorem, i.e., putting $h = 0$ in Eq. 5.28, we have

$$\frac{b}{K\sqrt{\lambda a}} = 2Wa^{1/3} \qquad \text{...(5.29)}$$

Solving Equations 7.31 and 7.33

$$\lambda = \frac{1}{3} \text{ and } b = \frac{2}{\sqrt{3}}WKa \qquad \text{...(5.30)}$$

The reference plane for this weir is situated at $\frac{2a}{3}$ above the crest of the weir. Equation 5.28 is in the Abel's form of integral equation whose solution is:

$$y = f(x) = \frac{1}{\pi}\int_0^b \frac{\phi'(h)}{\sqrt{x-h}}dh$$

Substituting for $\phi'(h)$

$$y = \left[1 - \frac{2}{\pi}\tan^{-1}\sqrt{\frac{x}{a}} - \frac{6}{\pi}\frac{\sqrt{x/a}}{\left(1+\frac{3x}{a}\right)}\right] \quad ..(5.31)$$

The weir profile drawn to scale is shown in Fig. 5.25. Beyond $\frac{x}{a} = 2.0$, the weir as an orifice for all practical purpose. As this weir gives discharges proportional to the square root of the head (measured above the reference plane), both while acting as a notch as well as an orifice, this device is also called a *notch-orifice.*

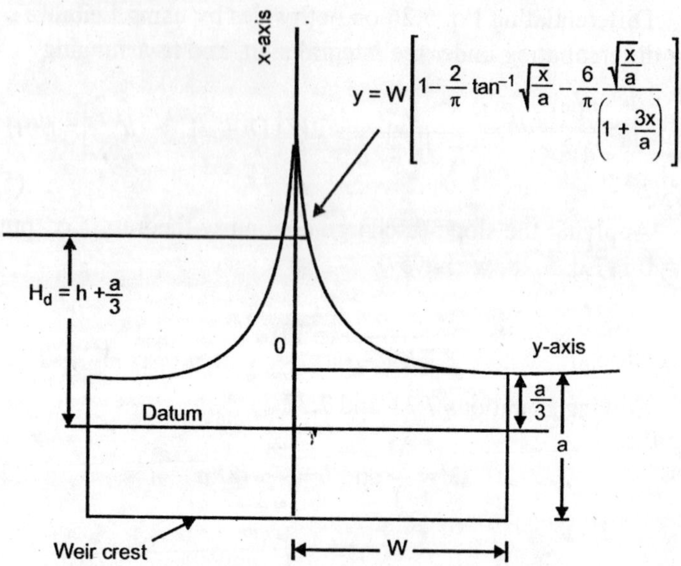

Fig. 5.8 Quadratic weir

The discharge equation for the quadratic weir can now be written as

$$Q = b\sqrt{h + \frac{a}{3}} = b\sqrt{\left[H - \frac{2}{3}a\right]} = b\sqrt{H_d} \quad ...(5.32)$$

where $b = \frac{2}{\sqrt{3}} WKa$ and $K = 2C_d\sqrt{2g}$

and h = depth of flow above the rectangular base

H_d = head above the reference plane

H = depth of flow

Quadratic weirs having non-rectangular lower portions (base weirs) .

The quadratic weir has an average coefficient of discharge of 0.62. In a quadratic weir the error involved in the discharge calculation for a unit per cent error in head is only 0.5 per cent as against 1.5% in a rectangular weir and 2.5% in a V-notch. Hence this is more sensitive tan the rectangular weir and V-notch.

5.4. Broad-Crested Weir

Weirs in which the sheet of flowing liquid is supported by the surface of the crest are called broad crested weirs. If the length of weir crest is less than about 2/3 H, the nappe will ordinarily spring clear and the weir is in fact a sharp crested one. As the length of crest increases the form of the nappe changes, and the sharp-crested weir coefficients no longer apply.

The discharge over a broad-crested weir may be computed by the following formula:

$$Q = C\,BH^{3/2} \qquad \text{...(5.33)}$$

If the velocity of approach is significant, the above formula would change to

$$Q = CB\left[\left(H + \frac{V_0^1}{2g}\right)^{3/2} - \left(\frac{V_0^2}{2g}\right)^{3/2}\right] \qquad \text{...(5.34)}$$

For flat crested weirs with crest length ranging from 0.3 to 3 m (1 to 10 ft) and square upstream corner, for L/H values of the coefficient C of about 2.7 in FPS units (in metric units, $C = 1.49$). As L/H decrease below 2, C increases reaching a value of 3.3 (its metric equivalent being 1.825) which corresponds to the sharp crested condition and the nappe clears the crest.

The separation of nappe from the crest which occurs at a sharp corned entrance can be avoided by adequate rounding of the entrance. The rounding reduces the contraction and increases the coefficient of discharge.

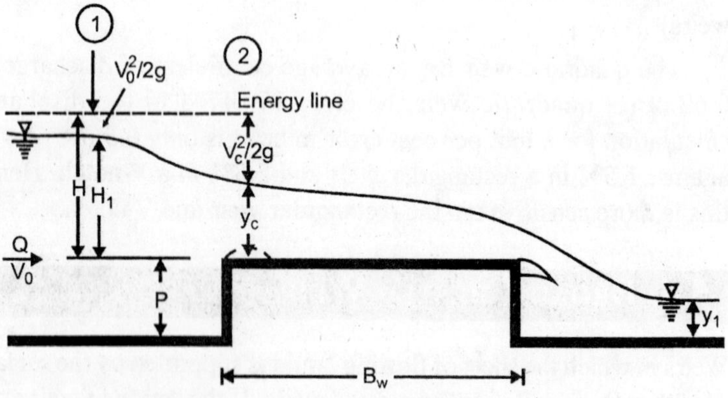

Fig. 5.9: Definition sketch of broad-crested weir

5.4.1 Classification

Based on the value of H_1/B_w the flow over a broad-crested weir with an upstream sharp corner is classified as follows.

1. H/Bw $\lesssim$ 0.1: In this range the critical flow control section is at the downstream end of the weir and the resistance of the weir surface plays as important role in determining the value of C_d (Fig. 5.10a).
2. 0.1 $\lesssim H_1/B_w \lesssim$ 0.35: The critical depth control occurs near the upstream end of the weir and the discharge coefficient varies slowly with H_1/B_w in this range (Fig. 5.10b). This kind can be called a *true broad-crested weir.*
3. 0.35 $\lesssim H_1/B_w \lesssim$ about 1.5: The water-surface profile will be curvilinear all over the weir. The control section will be at the upstream end (Fig. 5.10c). The weirs of this kind can be termed

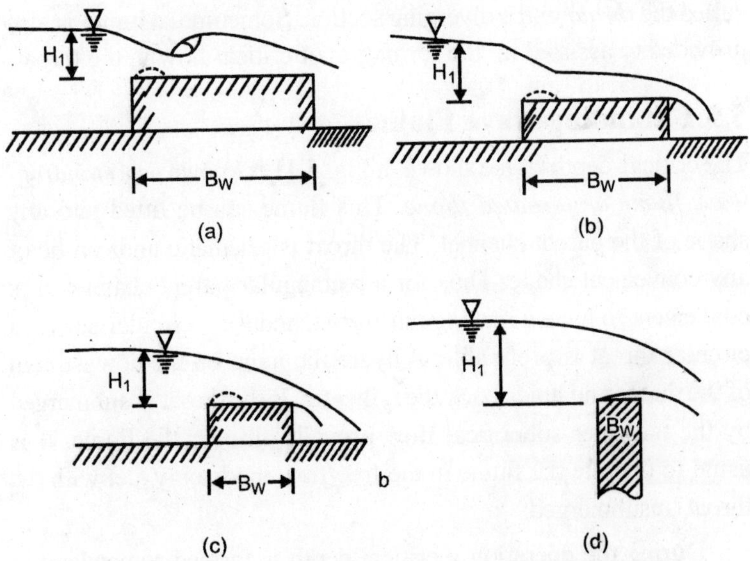

Fig. 5.10: (a) Long-crested weir ($H_1/B_w \lesssim 0.1$)
(b) Broad-crested weir ($0.1 \lesssim H/B_w \lesssim 0.35$)
(c) Narrow-crested weir ($0.35 \lesssim H_1/B_w \lesssim 1.5$)
(d) Sharp-crested weir ($H_1/B_2 \geq 1.5$)

as *narrow-created* weirs. The upper limit of this range depends upon the value of H_1/P.

4. $H_1/B_w >$ about 1.5: The flow separates at the upstream at the upstream corner and jumps clear across the weir crest. The flow surface is highly curved (Fig. 5.10d), and the weir can be classified as sharp-crested.

5.5 Critical-Depth Flumes

Critical-depth flumes are flow-measuring devices in which a control section is achieved through the creation of a critical-flow section by a predominate width constriction. In practice, these are like broad-created weirs but with a major change that these are essentially flow-measuring devices and cannot be used for flow-regulation purposes. A typical-depth flume consists of a constricted portion called the *throat* and a diverging section. Sometimes a hump is also provided to assisted in the formation of critical flow in the throat.

5.5.1 Standing-wave Flume

The critical-depth flume shown in Fig. 5.11 is known as a *standing-wave flume or throated flume.* This flume can be fitted into any shape of the parent channel. The throat is prismatic and can be of any convenient shape. Thus for a rectangular parent channel, it is convenient to have a rectangular throat, and for a circular sewer, a circular throat is preferable. A hydraulic jump on the downstream of the throat and holds back the tailwater. If the throat is submerged by the tailwater subcritical flow prevails all over the flume. It is usual to operate the flume in the free-flow mode only, i.e. with the throat unsubmerged.

During the operation a critical depth is formed somewhere in the throat and as such its discharge equation is similar to that of a broad-crested weir.

However, it is usual to relate the discharge to the upstream depth H_1 which can be typically recorded by an automated float equipment.

Thus for a rectangular throat section, the discharge is given by

$$Q = C_f B_t H_1^{3/2} \qquad ...(5.35)$$

where C_f= overall discharge coefficient of the flume = $f(H_1/L)$. For a well-designed flume, C_f is of the order of 1.62. It may be noted that in standing-wave flumes, H_1 is the difference in the water-surface elevation upstream of the inlet and the elevation of the crest at the throat. Constriction flumes operating in the subcritical flow range are called *venturi flumes*.

The modular limit (H_2/H_1) of standing wave flumes is high, being of the order of 0.90. It is usual to take it as 0.75 to incorporate a small safety factor and to avoid the region of transition from the free to submerged-flow mode.

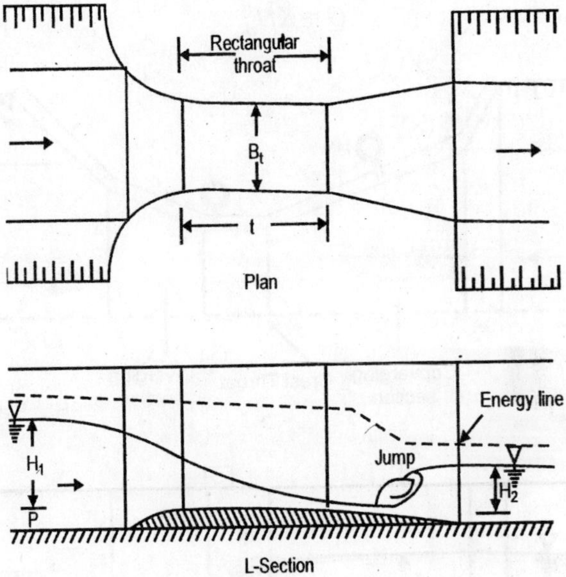

Fig. 5.11: Standing-wave flume

Large varieties of standing-wave flumes with different types of modifications of the basic type described above, resulting in different geometric shapes and corresponding flow characteristics are in use. However, the basic favourable features of all these throated flumes can be summerised as (i) low energy loss, (ii) rugged construction, (iii) easy passage for floating and suspended material load, and (iv) high modular limit. These features are responsible for extensive use of throated flumes as flow measuring devices in water-treated plants and in irrigation practice.

5.5.2 Parshall Flume

The Parshall flume is a type of critical-depth flume popular in the USA. This flume consists of a converging section with a level floor, a throat with a downstream sloping floor and a diverging section with an adverse slope bed (Fig. 5.12). Unlike in the standing-wave flume, the head (H_a) is measured at a specified location in the converging section. The discharge in the flume in the free flow mode is given by

$$Q = KH_a^n \qquad \ldots(5.36)$$

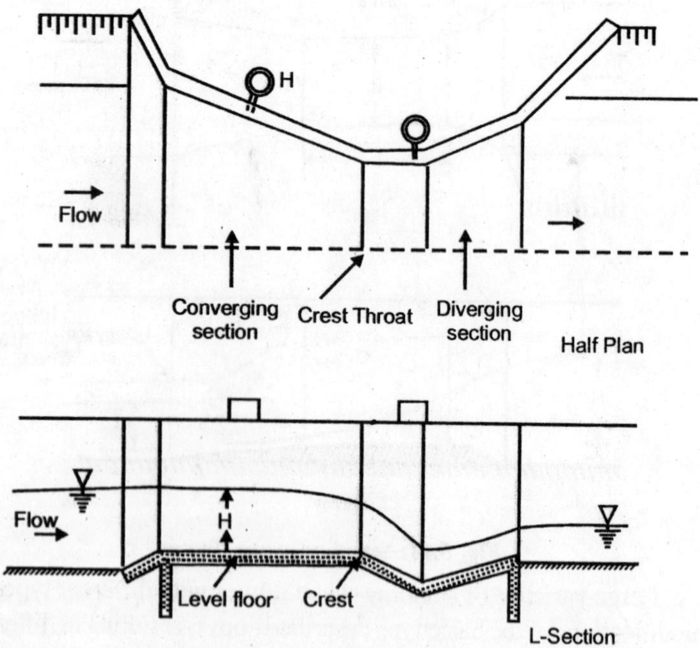

Fig. 5.12: Parshall flume

Where K and n are constants for a given flume.

5.6 End Depth in a Free Overfall

A free overfall is a situation in which there is a sudden drop in the bed causing the flow to separate from the stream bed and move down the step with a free nappe. The situation is analogous to the flow over a sharp-crested weir of zero height. A free overfall causes not only a GVF profile in the subcritical flow, but also offers the possibility of being used as a flow measuring device in all flow regimes.

The flow in the nappe emerging out of the overfall is obviously affected by gravity. With the atmospheric pressure existing above and below the nape, the water-surface profile is a parabola. Due to the need for continuity of the water-surface profile, the gravity effect extends a short distance on the waters-surface behind the edge, causing an acceleration of the flow.

5.7 Measurement of Discharge in Rivers and Canals

The most widely used method for estimating the river and canal flow is the current meter methods. Other methods which are rather crude may be the surface float, rod float and surface slope methods. Each of these are described briefly.

1. Current meter method: The mean velocity of flow in open channels is known to occur at 0.6 times the depth below the water surface. For a more accurate value, the velocity measurements are made at 0.2 depth and 0.8 depth from the water surface and their average gives the mean velocity. For determining discharge either in a river or in a big canal, a straight uniform reach which has a fair regular bottom should be selected. The site should be such as to be accessible from either bank as observations are to be made from one bank to the other. The river cross-section is divided into a number of sub sections of known width and depth. The mean velocity for each sub-section is determined using a current meter. The exact location where velocity measurements are to be taken may be reached in a boat. Abridge location may perhaps be the best site for discharge measurements as it offers a stable cross-section and the ease with which velocity observations can be taken. Having found the mean velocity for each sub-section, the discharge passing through each sub-section can be computed, and then the total discharge, which is the sum of the discharge through each sub-section can then the total discharge, which is the sum of the discharge through each sub-section, is obtained. Fig. 5.13 shows the river cross-section and the probable location of current meter stations.

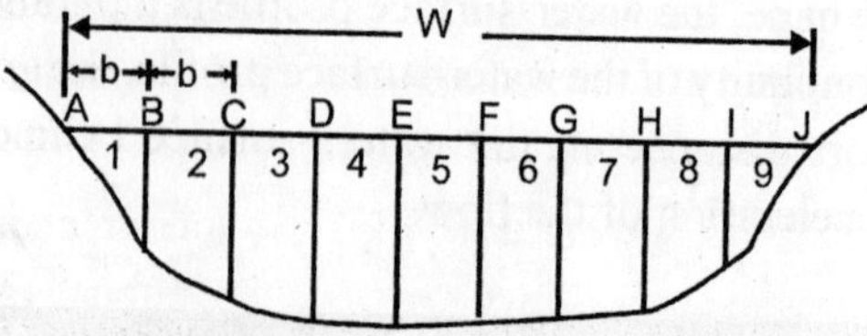

Fig. 5.13: Discharge measurement by current meter

The entire cross-section has been shown divided into 9 sub-sections of equal width b (this number depends on the width and the accuracy desired). The current meter stations where the velocities are to be observed have been marked by capital letters A, B, C and so on. If d_1, d_2 etc. represent the mean depth for each sub-section marked 1,2 respectively, then the discharge through the sub-section will be

$$Q_1 = bd_1\left(\frac{O+V_B}{2}\right)$$

Where d_1 is the mean of depths at A and B,

$$d_1 = \frac{d_A + d_B}{2}, \quad Q_2 = bd_2\left(\frac{V_B + V_C}{2}\right), \quad d_2 = \frac{d_B + d_C}{2}$$

2. Surface Float method: A surface float, which may be a corked bottle loaded inside so that its top floats just above the water surface may be used. The time taken by the float in travelling about 100 meters of straight length will permit determination of the surface velocity. The mean velocity of flow may be 0.80 to 0.95 of the surface velocity depending upon the depth and the boundary roughness. The mean velocity may thus be estimated from which the dischgarge $Q = AV$ can be determined. The cross-sectional area is to be determined from the depth observations.

3. Rod Float method: A rod float consists of a telescopic rod loaded at the bottom so that it floats vertically at about 0.6 depth from the water surface. The time taken by the rod to move in the downstream direction for 100 metres will enable determination of the mean velocity and the knowledge of the mean channel cross-section will permit computation of discharge.

4. Surface Slope method: This method which is also known *as* the slope-area method as involves the use of either Manning's equaion or Chezy's equation for determination of mean velocity

$$V = \frac{1}{n} R^{2/3} S^{1/2}$$

The slope S is determined by observing the fall h in the water surface level in a distance L, thus yielding $S = h/L$. The hydraulic radius R can be estimated from the knownchannel cross-section. The only problem is to select a proper value of roughness n, which, of course, is a matter of experience and judgement.

The method is particularly useful for estimating flood-flow, in natural streams. During the periods of high floods it may not be possible to measure discharge by a more accurate method such as the current meter method thereby necessitating use of an indirect method which might utilize the history of the flood by referring to the impressions or high water marks left by it. The water surface slope and the cross-sectional area are determined by the high water marks along each of the banks. A straight reach of channel of considerable length must be selected if the slope and the area measurements are to be truly representative. The estimation of Manning's roughness n may pose some difficulty and is a source of considerable error in such measurements.

5.8 Flow under a Sluice Gate

The flow under a sluice gate differs from that through an orifice on account of the fact that the emerging jet is not exposed to the atmospheric pressure throughout, but is guided by a horizontal floor. Consequently, the pressure in the jet is not atmospheric but is distributed hydrostatically. Fig. 5.14 shows a sluice gate installed in a channel of horizontal floor. The figure also shows distribution of pressure on the gate as well as on the channel floor-Noteworthy here is the influence of curvature of streamlines as the flow passes underneath the gate. The upward convexity of streamlines on the upstreamside of the gate is responsible for decreasing the pressure on the floor, below the hydrostatic value. With a similar reasoning, the pressure on the floor downstream of the gate are larger than hydrostatic. When the streamlines remain their original character–straight and parallel to the floor–the pressures then are distributed hydrostatically.

The continuity equation between sections (1) and (2) yields

$$q = V_1 H = V_2\, C_c d \qquad \text{...(5.37)}$$

in which q is the discharge per unit length of the gate V_1 and V_2 are the respective velocities at the sections under consideration, H is the water depth upstream of the gate, d is the gate opening, and C_c is the coefficient of contraction of the jet.

Writing the Bernoulli's equation between these sections, neglecting the energy loss in between

$$\frac{V_1^2}{2g} + H = \frac{V_2^2}{2g} + C_c d$$

From Eqs. (5.37) and Eq. (5.37a), one can write

$$V_2 = \sqrt{\frac{2gH}{1 + C_c d / H}}$$

and using (5.37), $q = \dfrac{C_c}{\sqrt{1 + C_c d / H}}\, d\sqrt{2gH}$

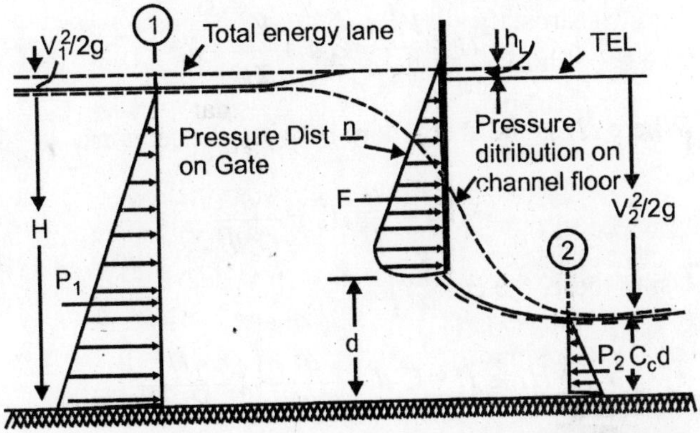

Fig. 5.14 Flow under a sluice gate

Introducing a coefficient of discharge,

$$q = dC_d\sqrt{2gH} \quad ...(5.38)$$

$$C_d = C_c / \sqrt{1+C_c d/H} \quad ...(5.39)$$

Fig. 5.14 shows the variation of C_d and C_c with the non-dimensional gate opening d/H. Also shown on the figure is the variation of C_d with $H/(H+Z2)$ for a sharp crested rectangular weir. The opposite trend exhibited by the C_d curves (for the weir and the sluice gate) is attributable to different pressure conditions within the contracted jet.

The degree of contraction for flow under a sluice get remains practically same over a considerable range of d/H, thereby yielding an almost horizontal curve for $C_d d/H$. At low values of d/H, H and $V_2^2/2g$ differ by a narrow margin.

The continuity, energy (i.e. Bernoulli's) and momentum turn equations can be applied to evaluate the energy loss and the force exerted by the liquid upon the sluice gate. Applying the energy equation between sections (1) and (2).

$$H + \frac{V_1^2}{2g} = C_c d + \frac{V_2^2}{2g} + h_L \qquad ...(5.40)$$

Where H_L is the energy loss between the sections.

$$V_1 = \frac{q}{H}, V_2 = \frac{q}{C_c^2 d^2}$$

Using these, the energy loss

$$H_L = H - C_c d + \frac{q^2}{2gH^2}\left(1 - \frac{H^2}{C_c^2 d^2}\right)$$

Expressing it in a non-dimensional form

$$\frac{H_L}{H} = 1 - C_c \frac{d}{H} + \frac{q^2}{2gH^3}\left(1 - \frac{H^2}{C_c^2 d^2}\right) \qquad ...(5.41)$$

The momentum equation, applied to the control volume enclosed by section (1) and (2), the gate and the channel floor yields.

$$P_1 - P_2 + F = \frac{\gamma Q}{g}(V_2 - V_1) \qquad ...(5.42)$$

where P_1 and P_2 are the hydrostatic forces at the respective sections, F is the force exerted by the gate upon the control volume. Here the frictional force on the channel floor has been considers negligible. Substituting for P_1 and P_2, and using the continutity relationship, one gets

$$\frac{1}{2}\gamma BH^2 - \frac{1}{2}\gamma B(C_c d)^2 + F = \frac{\gamma Bq}{g}\left(\frac{q}{C_c d} - \frac{q}{H}\right)$$

Which yields,

$$F = \frac{\gamma Bq^2}{gC_c d}\left(1 - C_c \frac{d}{H}\right) - \frac{\gamma B}{2}\left(H^2 - C_c^2 d^2\right) \qquad ...(5.43)$$

Here B is the length of the sluice gate. Writing Eq. non-dimentionally

$$\frac{F}{\gamma BH^2} = \frac{q^2}{gH^2C_c d}\left(1 - C_c\frac{d}{H}\right) - \frac{1}{2}\left(1 - C_c^2\frac{d^2}{H^2}\right) \quad ...(5.44)$$

5.8.1 Sluice Gate Operating Under Submerged Conditions

When the sluice gate operates under submerged conditions, under certain downstream depths (in relation to the upstream ones) the discharge is considerably decreased as compared to the corresponding free-discharge. This is reflected in the lower values of the coefficients of discharge, the representative values of which are given in the following table:

Table Showing effect *t* of submergence on Discharge

H/d				Discharge coeffcient, C_d		
	For free flow			*H/d*		
	2	4	6	8	10	
2	0.510	0.000	0.00	0.00	0.00	0.00
4	0.550	0.550	0.00	0.00	0.00	0.00
6	0.570	0.570	0.40	0.00	0.00	0.00
8	0.580	0.580	0.50	0.31	0.00	0.00
10	0.580	0.580	0.58	0.42	0.29	0.00
12	0.585	0.585	0.585	0.47	0.37	0.25
14	0.585	0.585	0.585	0.51	0.43	0.33

In the above table, h represent the depth of submergence, i.e. water depth in the downstream channel.

Solved Problems

Problem 1: ***Water flows over a rectangular sharp crested weir 1.2 m long, the head over the sill of the weir being 0.65 m. The approach channel is 1.4 m wide and the depth of the flow in the channel is 1.2 m. Starting from first principles, determine the rate of discharge over the weir. Consider also the velocity of approach and effect of end contraction. Take the coefficient of discharge for the weir as 0.6.***

Solution: The discharge Q for rectangular sharp crested contracted weir is given by

$$Q = \frac{2}{3} C_d (B - 0.1nH)\sqrt{2g}H^{3/2}$$

$$= \frac{2}{3} \times 0.6(1.2 - 0.1 \times 2 \times 0.65) \times \sqrt{2 \times 9.81} \times 0.65^{3/2}$$

$$= 0.99 \text{ m}^3/\text{s}$$

$$\text{Velocity of approach } V_0 = \frac{Q}{A} = \frac{0.99}{1.4 \times 1.2} = 0.589 \text{ m/s}$$

$$H + \frac{V_0^2}{2g} = 0.65 + \frac{0.589^2}{2 \times 9.81} = 0.667 \text{ m}$$

Discharge considering the velocity of approach:

$$Q = \frac{2}{3} C_d (B - 0.1nH)\sqrt{2g}\left[\left(H + \frac{V_0^2}{2g}\right)^{3/2} - \left(\frac{V_0^2}{2g}\right)^{3/2}\right]$$

$$Q = \frac{2}{3} \times 0.6(1.2 - 0.1 \times 2 \times 0.65)\sqrt{2 \times 9.81}$$

$$\left[\left(0.65 + \frac{0.589^2}{2 \times 9.81}\right)^{3/2} - \left(\frac{0.589}{2 \times 9.81}\right)^{3/2}\right]$$

$Q = 1.02$ m³/s **Ans.**

Problem 2 : ***Show that the time required to reduce the water level from H_1 to H_2 by means of rectangular weir is given by expression:***

$$t = \frac{3A}{C_d L\sqrt{2g}}\left(\frac{1}{\sqrt{H_2}} - \frac{1}{\sqrt{H_1}}\right)$$

where A is the area of reservoir, C_d is the discharge coefficient and L is the length of weir?

Solution: The discharge Q for rectangular weir is given by

$$Q = \frac{2}{3} C_d L \sqrt{2g}\, H^{3/2}$$

The rate at which reservior surface is falling

$$-\frac{dH}{dt} = \frac{Q}{A}$$

Negative sign is taken because level of water decreases

The time to empty the reservoir surface from a level H_1 to H_2 is given by:

$$t = -A\int_{H_1}^{H_2} \frac{dH}{Q} = -A\int_{H_1}^{H_2} \frac{dH}{\frac{2}{3} C_d L\sqrt{2g} H^{3/2}}$$

$$= \frac{3A}{2C_d L\sqrt{2g}} \int_{H_1}^{H_2} \frac{dH}{H^{3/2}} = \frac{3A}{C_d L\sqrt{2g}}\left(\frac{1}{\sqrt{H_2}} - \frac{1}{\sqrt{H_1}}\right)$$

Problem 3: ***A rectangular notch 60 cm long is used to measure a discharge of 25 litre/sec. An error of 1 mm, was made in measuring the head our the notch. Calculate the percentage error in the discharge. Take = 0.6.***

Solution: The discharge Q through rectangular notch is given by expression :

$$Q = \frac{2}{3} C_d \sqrt{2g} L H^{3/2} \qquad (1)$$

$$Q = 25 \text{ litre/sec} = \frac{25}{1000} = 0.025 \text{ m}^3/\text{s}$$

$$0.025 = \frac{2}{3} \times 0.6 \times \sqrt{2 \times 9.81} \times (0.6) H^{3/2}$$

$$H = 0.082 \text{ m} = 8.2 \text{ cm}$$

differentiating (1) with respect to H

$$dQ = \frac{2}{3}C_d\sqrt{2g}\cdot L\left(\frac{3}{2}H^{1/2}dH\right) \quad (2)$$

dividing (2) by (1) we get

$$\frac{dQ}{Q} = \frac{3}{2}\frac{dH}{H} = \frac{3}{2}\left(\frac{2}{82}\right) = 0.036$$

Percentage error in $Q = 3.6\%$

Problem 4: ***While conducting flow measurment using rectangular notch, an error of 2.5% in head over the notch, and an error of 5% in the length of notch occured. Determine the percentage error in discharge***

Solution: The discharge Q is given by:

$$Q = \frac{2}{3}C_d\sqrt{2g}\ LH^{3/2} = KLH^{3/2}$$

$$\frac{dQ}{dL} = KH^{3/2} \quad \text{and} \quad \frac{dQ}{dH} = \frac{3}{2}KLH^{1/2}$$

Total change in discharge:

$$dQ = \frac{dQ}{dL}dL + \frac{3}{2}\frac{dQ}{dH}dH$$

$$dQ = \frac{Q}{L}dL + \frac{3}{2}\frac{Q}{H}dH = Q\left(\frac{dL}{L} + \frac{3}{2}\frac{dH}{H}\right)$$

$$\frac{dQ}{Q} = \left(\frac{dL}{L} + \frac{3}{2}\frac{dH}{H}\right)$$

$$\frac{dH}{H} = 2.5\%, \frac{dL}{L} = 5\%$$

$$\frac{dQ}{Q} = \left(5 + \frac{3}{2}\times 2.5\right) = 8.75\%$$

The percentage error in discharge is 8.75%.

Problem 5: ***Water flows over a rectangular weir 1.5 m wide at a depth of 15 cm and afterwards passes through a triangular right angled weir. Find the depth of water through the triangular weir. Discharge coefficient for the rectangular and triangular weir are 0.62 and 0.59 respectively.***

Solution: Discharge over rectangular weir is given by

$$Q_1 = \frac{2}{3} C_d \sqrt{2g} B H_1^{3/2}$$

$$= \frac{2}{3} \times 0.62 \times \sqrt{2 \times 9.81} \times 1.5 \times (0.15)^{3/2}$$

$$= 0.159 \text{ m}^3/\text{s}$$

The same discharge will pass through triangular noth.

$$Q_2 = \frac{8}{15} C_d \sqrt{2g} H_2^{5/2}$$

$$0.159 = \frac{8}{15} \times 0.59 \times \sqrt{2 \times 9.81} \times H_2^{5/2}$$

$$H_2 = 0.419 \text{ m}$$

Problem 6: ***Find the head over a right angled V-notch if the discharge over it is 0.06 m³/s. Take $C_d = 0.6$. If 2 mm error was made in measuring this head. What is the corresponding error in discharge?***

Solution: Discharge through 90° V-notch is given by

$$Q = \frac{8}{15} C_d \sqrt{2g} H^{5/2}$$

$$H^{5/2} = \frac{15Q}{8C_d\sqrt{2g}} = \frac{15 \times 0.06}{8 \times 0.6 \times \sqrt{2 \times 9.81}}$$

$$H = 0.28 \text{ m.}$$

Now error in Q due to 2 mm error in head

$$dQ = \frac{8}{15} C_d \sqrt{2g} \left(\frac{5}{2}\right) H^{3/2} dH$$

As we know $\frac{Q}{H} = \frac{8}{15} C_d \sqrt{2g} H^{3/2}$

So we can write

$$dQ = \frac{5}{2} Q \frac{dH}{H}$$

$$\frac{dQ}{Q} = \frac{5}{2} \frac{dH}{H} \Rightarrow dQ = 0.06 \times \frac{5}{2} \times \frac{2 \text{ mm}}{280 \text{ mm}}$$

$$= 1.07 \times 10^{-3} \text{ m}^3/\text{s}$$

$$= 1.07 \text{ litre/sec.}$$

Problem 7: ***In a expriment on a 90° V notch, the flow is collected in a 1 m diameter vertical cylindrical tank. It is found that depth of water increaes by 0.7 m in 15 sec when the head over notch is 0.2 m. Determine the coefficient of discharge.***

Solution: Discharge for 90° V-notch is given by

$$Q = \frac{8}{15} C_d \sqrt{2g} H^{5/2}$$

Quantity of water collected in cylindrical tank in 15 seconds.

$$V = \frac{\pi}{4} \times 1^2 \times 0.7 = 0.549 \text{ m}^3$$

$$Q = \frac{V}{t} = \frac{0.549}{15} = 0.0366 \text{ m}^3/\text{s}$$

$$C_d = \frac{15Q}{8\sqrt{2g} H^{5/2}} = \frac{15 \times 0.0366}{8\sqrt{19.62} \times 0.2^{5/2}} = 0.866$$

Problem 8: ***If there is an error of 2% in observing the head over (i), rectangular weir (ii) triangular weir, detemine the error in the discharge resulting from it. Assume identical condition for both the weir.***

Solution: (i) Rectngular weir

$$Q = KH^{3/2} \quad \text{(i)}$$

$$\frac{dQ}{dH} = \frac{3}{2}KH^{1/2}$$

$$dQ = 1.5\sqrt{H}\ dH \qquad \text{(ii)}$$

Dividing (ii) and (i) we get

$$\frac{dQ}{Q} = 1.5\ \frac{dH}{H}$$

Since $\frac{dH}{H} = 2\%$

$$\frac{dQ}{Q} = 1.5 \times 2 = 3\%$$

(ii) Tringular weir:

$$Q = KH^{5/2} \qquad \text{(i)}$$

$$\frac{dQ}{dH} = \frac{5}{2}KH^{3/2}$$

$$dQ = \frac{5}{2}KH^{3/2}\,dH \qquad \text{(ii)}$$

Dividing (ii) and (i) we get

$$\frac{dQ}{Q} = 2.5\ \frac{dH}{H}$$

since $\frac{dH}{H} = 2\%$

$$\frac{dQ}{Q} = 2.5 \times 2 = 5\%$$

Problem 9: ***Show that for a triangular broad crested weir flowing free the discharge equation can be expressed as***

$$Q = \frac{16}{25} C_{d1} \tan\theta \sqrt{\frac{2g}{5}}\ H^{5/2}$$

Solution: $H = y_c + \frac{V_C^2}{2g} = 1.25 y_C$

or $$y_c = \frac{4}{5}H$$

$$A = my_C^2 = \frac{16}{25}\tan\theta\, H^2, \text{ where } \theta = \text{semi vertex eagle}$$

$$F = \frac{V\sqrt{2}}{\sqrt{gy_C}} = 1 \quad \text{or} \quad V = \sqrt{\frac{2g}{5}}\, H^{1/2}$$

$$Q = C_{d1} VA \text{ or } Q = \frac{16}{25} C_{d1} \tan\theta \sqrt{\frac{2g}{5}}\, H^{5/2}$$

Problem 10: ***A channel has its area given by A = ky³ where k = a constant. 1m subscribed flow in this channel estimate the ratio of the end-depth to critical depth.***

Solutions: This is an exponential channel with a = 3.0. For subcritical flow by

$$\varepsilon = 1 + \frac{1}{2a} = 1.167$$

$$f(\eta) = \frac{1}{a\eta^{2a}} = \frac{1}{3}\frac{1}{\eta^6}$$

The general equation of end depth is

$$6\varepsilon - 4\eta - 3f(\eta) = 0$$

Substitute for ε and f(η), $(6 \times 1.167) - 4\eta - \frac{3}{3}\frac{1}{\eta^6} = 0$

i.e. $$\frac{1}{\eta^6} + 4\eta - 7 = 0$$

Solving by trial and error $\eta = y_e/y_c = 0.80$

Problem 11: ***A rectangular channel carries a supercritical flow with a Floude number of 2.0. Find the end-depth ratio at a free overfall in this channel.***

Solution: In supercritical flow, for an exponential channel, $\varepsilon = f_n(a, F_0)$

Here a = 1.0

$$\varepsilon = \frac{1}{F_0^{2/(2a+1)}}\left(1+\frac{F_0^2}{2a}\right)$$

$$= \frac{1}{2^{2/(2+1)}}\left(1+\frac{2^2}{(2\times1)}\right) = 1.89$$

$$f(\eta) = \frac{1}{a\eta^{2a}} = \frac{1}{\eta^2}$$

The general equation of end-depth, is

$$6\varepsilon - 4\eta - f(\eta) = 0$$

Substituting for ε and $f(\eta)$, $6 \times 1.89 - 4\eta - \dfrac{3}{\eta^2}$

i.e. $\quad 4\eta^3 - 11.34 + 3 = 0$

solving by trial-and-error $\quad \eta = y_e/y_c = 0.577.$

Problem 12: ***A Sutro weir has a rectangular base of 30-cm width and 6-cm height. The depth of water in the channel is 12 cm. Assuming the coefficient of discharge of the weir as 0.62, determine the discharge through the weir. What would be the depth of flow in the channel when the discharge is doubled? (Assume the crest of the base weir to coincide with the bed of the channel).***

Solution: Given $a = 0.06$ m, $W = 0.30/2 = 0.15$ m, $H = 0.12$ m

$$K = 2\,C_d\,\sqrt{2g} = 2\times0.62\times\sqrt{2\times9.81} = 5.4925$$

$$b = W\,K\,a^{1/2} = 0.15\times5.4925\times(0.06)^{1/2} = 0.2018$$

$$Q = b\left(H - \frac{a}{3}\right)$$

$$= 0.2018\left(0.12 - \frac{0.06}{3}\right) = 0.02018 \text{ m}^3/\text{s}$$

$$= 20.18 \text{ lit/sec}$$

When the discharge in double

$$Q = 2 \times 0.02018 = .04036 \text{ m}^3\text{s}$$

$$0.04036 = 0.2018\left(H - \frac{0.06}{3}\right)$$

$$H = 0.2 + 0.2 = 0.22 \text{ m}$$
$$= 22 \text{ cm}$$

Problem 13: ***Design a Sutro weir for use in a 0.30 m wide rectangular channel to have linear discharge relationship in the discharge range from 0.25 m to 0.60 m3/s. The base of the weir well have to span the full width of the channel. Assume C_d = 0.62.***

Solution: Here $2W = 0.30$m. and $C_d = 0.62$.

$$K = 2C_d\sqrt{2g} = 2 \times 0.62 \times \sqrt{2 \times 9.81} = 5.49$$

$$Q_{\min} = \frac{2}{3}Wka^{3/2} = 0.25$$

$$\frac{2}{3} \times 0.15 \times 5.49 \times a^{3/2} = 0.25$$

$$a = 0.592 \text{ m}$$

$$b = WK\sqrt{a} = 0.15 \times 5.49 \times \sqrt{0.592} = 0.6337$$

$$Q = bH_d$$

for $Q = 0.60 \text{ m}^3\text{s}$, $\quad H_d = \frac{Q}{b} = \frac{0.60}{0.6337} = 0.9468 \text{ m}$

$$= H - \frac{a}{3} = h + \frac{2}{3}a$$

$H = 1.1444$ m and $h = 0.0552$ m

$$y = f(x) = 0.15\left|1 - \frac{2}{\pi}\tan^{-1}\sqrt{\frac{x}{0.592}}\right|$$

Problem 14: ***A quadratic weir is designed for installation in a rectangular channel of 30-cm width. The rectangular base of the weir occupies the full width of the channel and is 6 cm in***

height. The crest of the base weir coincides with the channel bed. (a) Determine the discharge through the weir when the depth of flow in the channel is 15 cm. (b) What would be the depth of flow upstream of the weir when the discharge in the channel is 25 litres/s?

[Assume $C_d = 0.62$]

Solution: Given $a = 0.06$ m, $W = 0.30/2 = 0.15$ m $H = 0.15$ m

$$K = 2C_d\sqrt{2g} = 2\times 0.62\times\sqrt{2\times 9.81} = 5.4925$$

$$b = \frac{2}{\sqrt{3}}WKa = \frac{2}{\sqrt{3}}\times 0.15\times 5.4925\times 0.06 = 0.05708$$

$$Q = b\sqrt{\left(H - \frac{2a}{3}\right)}$$

(a) $$Q = 0.05708\times\left[0.15 - \left(\frac{2\times 0.06}{3}\right)\right]^{1/2}$$

$$= 0.0189 \text{ m}^3/\text{s} = 18.93 \text{ litres/s}$$

(b) When the discharge $Q = 25$ litres/s $= 0.024$ m^3/s

$$0.025 = 0.05708\times\left[H - \left(\frac{2\times 0.06}{3}\right)\right]^{1/2}$$

$$H = 0.1918 + 0.04 = 0.2318 \text{ m}$$

$$= 23.18 \text{ cm}$$

Review Questions

1. Write short notes on the following:
   (a) Sharp crested rectangular weir
   (b) Suppressed rectangular weir
   (c) Contracted weirs
   (d) Francis weir
   (e) Triangular weir

(f) Trapezoidal weir

(g) Cipolletti weir

(h) Submerged weir

(i) Special shartp crested weir

2. Discuss the classificationof broad crested weir.
3. Explain the significance of

   (a) Critical depth flume

   (b) Standing wave flume

   (c) Parshell flume
4. Describe various methods for measuring discharge in river and canal.
5. What are sluice gates why it is provieded.
6. A rectangular weir notch 50 cm long is used for measuring a discharge of 30 *l*/sec. An error of 2 mm was made measuring the head over the notch. Calculate the percentage error in the discharge Take $C_d = 0.5$.
7. While conducting flow meausrement using rectangular notch an error of –3% in length of notch occured. Determine percentage error in discharge.
8. Find the head over a right angle V-notch if discharge over it is 0.05 m^3/s Take $C_d = 0.6$. If 2 mm error was made in measuring this head, what is the corresponding error in the discharge.
9. A triangular notch is used to measure flow in channel under a head of 20 cm if discharge to be measured within 3% accuracy what is the maximum velocity of approach that can be neglected.
10. Show that for triangular broad crested weir flowing free the discharge equation can be expressed as

$$Q = \frac{16}{25} C_{d_1} \tan\theta \sqrt{\frac{2g}{5}}\, H^{5/2}$$

Unit 6

Spatially Varied Flow

6.1 Basic Principles

Spatially varied flow (SVF) is a gradually varied flow with non-uniform discharge resulting from addition or withdrawal of water along the course of flow. The addition or withdrawal of water will cause disturbance in energy or momentum of the flow. As a result hydraulic behaviour of spatially varied flow is more complicated than that of flow at constant discharge. Further, analysis of spatially varied flow is quite different with two types of spatially varied flow SVF (i.e., increasing and decreasing discharge).

Flow with Increasing Discharge

In this type of SVF an appreciable portion of energy is lost due to turbulent mixing of the added water and water existing in the channel already. In most of the cases, the mixing is of relatively high magnitude and uncertainty. Because of uncertain losses, the momentum equation will found more convenient than the energy equation in solving this problem. Flows inside channel spill way, wash water troughs in filter plant, roof gutters, high gutters are some of the typical examples of flow with increasing discharge.

Flow with Decreasing Discharge

In this type of SVF, the water is diverted or withdrawn from the channel, however it does not effect the energy head. Therefore, the use of energy equation will be found more convenient in solving the problem. Flows in side weirs, bottom racks and siphon type irrigation systems are some of the typical examples of flow with decreasing discharge.

6.2 Differential SVF Equation for Increasing Discharge

As we have already discussed that momentum equation will found more convenient in solving equation for increasing discharge so for this purpose while applying momentum equation the following assumptions are made

(i) The flow is considered to be steady.

(ii) Momentum coefficient β is assumed to be unity

(iii) The flow is unidirectional. The lateral unevenness of water surface due to presence of spiral flow, turbulance, etc. are ignored.

(iv) Velocity distribution across the channel section is constant i.e., velocity distribution coefficient α (kinetic energy correction factor) is taken as unity. However proper value of α may be introduced, if needed.

(v) The channel is prismatic and is of small slope.

(vi) The effect of air entrainment is neglected.

(vii) The pressure distribution is assumed to be hydrostatic i.e., the flow is parallel. The flow at the outlet, however, may be curvilinear and deviate greatly from parallel flow assumption if hydraulic drop occurs. In such cases proper values of pressure distribution coefficients may be introduced, if necessary.

(viii) The Manning's formula is used to evaluate the friction loss due to shear developed along the channel wall.

Consider the control volume containing sections 1 and 2, Δx distance apart. Applying momentum equation, we have

$$M_2 - M_1 = P_1 - P_2 + w \sin\theta - F_f \tag{6.1}$$

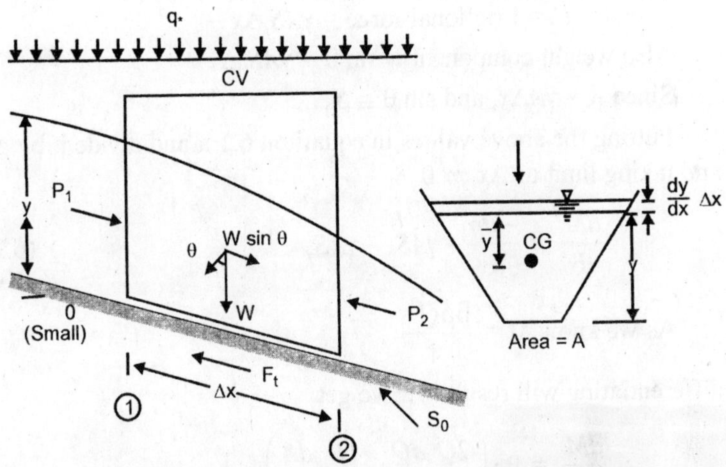

Fig. 6.1 Definition sketch of SVF with lateral inflow

where M_1 and M_2 are momentum rate of section 1 and 2 and P_1 and P_2 are the pressure fores at section 1 and 2 respectively. W is the weight of fluid in control volume and F_f is the shear force.

Equation 6.1 can also be written as

$$\Delta M = -\Delta P + w \sin\theta - F_f \tag{6.1a}$$

Momentum rate is given by $= M = \dfrac{mv}{t} = \dfrac{\rho V v}{t} = \dfrac{\rho Q^2}{A}$

Since $\left(Q = vA, \text{ also } Q = \dfrac{V}{t}\right), m = \rho V$

where v = velocity and V = volume

if Momentum correction factor is applied then

$$M = \frac{\beta \rho Q^2}{A} \qquad \beta = \text{Momentum correction factor}$$

$$\text{P = Pressure force} = \gamma\, A\bar{y}$$

where γ = eg and $\bar{y}$ = depth of the centre of gravity of the flow cross-section from the water surface.

$w \sin\theta$ = component of weight in x direction of control volume.

F_f = Frictional force = $\gamma A S_f \Delta x$

Also weight component $w \sin\theta = \gamma A S_0 \Delta x$

Since $w = \gamma A \Delta x$, and $\sin\theta \cong S_0$

Putting the above values in equatiion 6.1 a and divide it by Δx and taking limit as $\Delta x \Rightarrow 0$

$$\frac{dM}{dx} = \frac{-dp}{dx} + \gamma A S_0 - \gamma A S_f \tag{6.2}$$

As we know $M = \dfrac{\beta \rho Q^2}{A}$

differentiating will result to x,we get

$$\frac{dM}{dx} = \beta\rho\left(\frac{2Q}{A}\frac{dQ}{dx} - \frac{Q^2}{A^2}\frac{dA}{dx}\right)$$

$$= \beta\rho\left(\frac{2Q}{A}q_* - \frac{Q^2}{A^2}\frac{dA}{dy}\cdot\frac{dy}{dx}\right), \frac{dA}{dy} = T$$

$$\frac{dM}{dx} = \beta\rho\left(\frac{2Q}{A}q_* - \frac{Q^2 T}{A^2}\frac{dy}{dx}\right) \tag{6.3}$$

where $q_* = \dfrac{dQ}{dx}$ = discharge per unit length entering the channel.

Also we know that

$$P = \gamma A \bar{y}$$

On differentiating with respect to x, we get

$$\frac{dP}{dx} = \gamma\left(A\frac{d\bar{y}}{dx} + \bar{y}\frac{dA}{dx}\right) \tag{6.4}$$

By taking moment of the areas about new water surface after small change dy in depth

$$A(\bar{y} + dy) + dA\frac{dy}{2} = (A + dA)(\bar{y} + d\bar{y})$$

$$\Rightarrow \quad A\bar{y} + A\,dy + dA\frac{dy}{2} = A\bar{y} + \bar{y}\,dA + A\,d\bar{y} + dA\,d\bar{y}$$

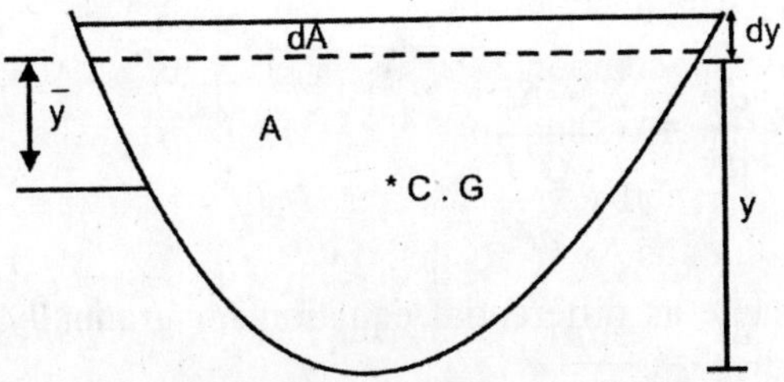

On cancelling similar term we get

$$Ady + dA\,\frac{dy}{2} = \bar{y}dA + Ad\bar{y} + dAd\bar{y}$$

Neglecting second order small quantities we have

$$Ady = \bar{y}dA + Ad\bar{y} \qquad (6.5)$$

Comparing 6.4 and 6.5 equation, we can say

$$\frac{dP}{dx} = \gamma A\,\frac{dy}{dx} \qquad (6.6)$$

Now simplifying equation 6.2 by putting values of $\frac{dM}{dx}$ and $\frac{dP}{dx}$ we get

$$\rho\beta\left(\frac{2Q}{A}q_* - \frac{Q^2T}{A^2}\frac{dA}{dx}\right) = -\gamma A\frac{dy}{dx} + \gamma AS_0 - \gamma AS_f$$

$$\frac{2Q\beta q_*}{gA^2} - \frac{\beta Q^2T}{gA^3}\frac{dy}{dx} = \frac{-dy}{dx} + S_0 - S_f$$

or

$$\frac{dy}{dx} = \frac{S_0 - S_f - 2\beta Qq_*/gA^2}{1-\beta\dfrac{Q^2T}{gA^3}} \qquad (6.7)$$

Equation 6.7 is the basic differential equation governing SVF with increasing discharge.

It should be noted that if $\beta = 1$ and $q_* = 0$ then equation 6.7

reduces to

$$\frac{dy}{dx} = \frac{S_0 - S_f}{1 - \dfrac{Q^2 T}{gA^3}} \tag{6.7a}$$

which is same as differential equation for gradually varied flow (GVF).

6.3 Computation of Profile with Increasing Discharge

In case of SVF with increasing discharge, it is required to find or locate control section or a point from which computation can start because in an SVF with increasing discharges, the critical depth line is not a straight line parallel to bed as in GVF but is a curved line. The critical depth of spatially varied flow can occur at a location somewhere between the ends of the channel, giving rise to a profile which may be subcritical during the first part and supercritical in the subsequent part of the channel. A method of calculation of critical depth and its location based on the concept of equivalent critical depth channel has been proposed by Hinds.

An alternative method suggested by Smith, having advantage of less tedious and simple as compared to Hind's method is described below.

Equation 6.7 can be written as:

$$\frac{dy}{dx} = \frac{1 - \dfrac{S_f}{S_0} - \dfrac{2\beta Q q_*}{gA^2 S_0}}{1 - \dfrac{\beta Q^2 T}{gA^3}} \tag{6.8}$$

As we know

$$Q = K\sqrt{S_f} = \text{Actual discharge} \tag{a}$$

$$Q_n = K\sqrt{S_0} = \text{Normal discharge} \tag{b}$$

$$Q_c = \sqrt{\frac{gA^3}{\beta T}} = \text{Critical discharge with } \beta \text{ as momentum correction factor} \quad \text{(c)}$$

Putting (a), (b) and (c) in equation 6.8, we get

$$\frac{dy}{dx} = S_0 \frac{\dfrac{1-K^2Q^2}{K^2Q_n^2} - \dfrac{Q^2}{Q_n^2}\left(\dfrac{2\beta K^2 q_*}{gA^2 Q}\right)}{1-\left(\dfrac{Q}{Q_c}\right)^2} \quad (6.9)$$

$$= S_0 \frac{1-\left(\dfrac{Q}{Q_n}\right)^2\left(1+\dfrac{2\beta K^2}{gA^2}\dfrac{q_*}{Q}\right)}{1-\left(\dfrac{Q}{Q_c}\right)^2}$$

Now difining Q_{n_1} = modified normal discharge

$$Q_{n_1} = \frac{Q_n}{\sqrt{1+\dfrac{2\beta K^2}{gA^2}\dfrac{q_*}{Q}}} \quad \text{(d)}$$

Putting value of equation (d) in 7.9, we get

$$\frac{dy}{dx} = S_0 \frac{1-\left(\dfrac{Q}{Q_{n_1}}\right)^2}{1-\left(\dfrac{Q}{Q_c}\right)^2} \quad (6.10)$$

The location of the transitional profile at a given x would be determined by the condition $Q_{n1} = Q_c$. The intersection of the transitional profile with the critical-depth line will satisfy the condition $Q = Q_{n1} = Q_c$ and hence would locate the control point; i.e. the section at which the actual flow would pass at critical depth.

At transitional depth, $Q_{n1} = Q_c$.

i.e.
$$\frac{Q_n}{\sqrt{1+\dfrac{2\beta K^2 q_*}{gA^2 Q}}} = \sqrt{\frac{gA^3}{\beta T}}$$

or
$$\frac{2\beta K^2 q_*}{gA^2 Q} = \frac{\beta Q_n^2 T}{A^3 g} - 1 \tag{6.11}$$

Substituting $Q_n = K\sqrt{S_0}$ and simplifying

$$\frac{q_*}{Q} = \frac{1}{2}\left(\frac{S_0 T}{A} - \frac{gA^2}{\beta K^2}\right) \tag{6.12}$$

which is the equation of the transitional profile for SVF with increasing discharge. In a general SVF with increaseing discharge,

$$Q = Q_i + \int_0^x q_* dx$$

which Q_i = channel discharge at $x = 0$. For the SVF in a lateral spillway channel $Q_i = 0$, and q_* = Constant, i.e. $Q = q_* x$ which simplifies Eq. 6.12 to

$$\frac{1}{x_t} = \frac{1}{2}\left(\frac{S_0 T}{A} - \frac{gA^2}{\beta K^2}\right) \tag{6.13}$$

the suffix t denoting the transitional profil. It is interesting to note that in an SVF due to a side-channel spillway, the transitional profile is independent of the rate of lateral inflow.

To locate the control point, the critical depth line is first calculated and plotted to scale (Fig. 6.2). Note that the critical depth line for SVF with increasing discharge is to be calculated by using the relationship $Q_c = \sqrt{\dfrac{gA^3}{\beta T}}$. The transitioal profile is then calculated by Eq. 6.12 and plotted on the same figure and the intersection of the critical-depth line with the transitional profile gives the location of the control section at while the actual flow passes as a critical depth (Fig. 6.2).

The method is of general use and can be easily incorporated into a numerical-method algorithm to compute the SVF profile using a digital computer.

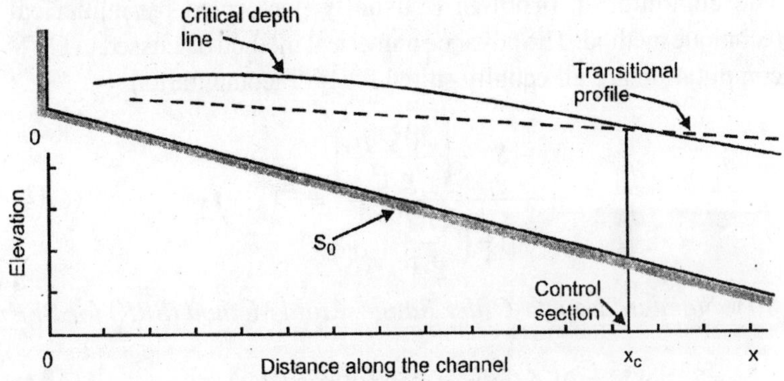

Fig. 6.2 Determine of control section through transitional profile

6.4 Classification and Solution

Li has classified SVF with lateral inflow into following categories:

- **Type A:** In this type the flow is subcritical throughout the channel and froude number goes on increasing in the downstream direction.
- **Type B:** In this type, the flow is subcritical throughout the channel but the froude number will first increase as the flow proceeds downstream, reaching a maximum value less than unity, and then decrease.
- **Type C:** It represent the condition at which there s supercritical flow in the downstream portion of the channel and hydraulic jump in the channel. The hydraulic jump will f om only if the outlet is sufficiently submerged.
- **Type D:** It represents the flow condition at which there is supercritical flow throughout the downstream portion of the

channel but where the depth of submergence at the outlet is not great enough to create hydraulic jump in the channel.

6.5 NUMERICAL METHOD

The computation problem is usually performed by numerical technique method. The advance numerical method discussed in GVV computation are all equally suited for SVF computation.

$$\frac{dy}{dx} = \frac{S_0 - S_f - \left(\frac{2\beta Q q_*}{gA^2}\right)}{1 - \beta\frac{Q^2 T}{gA^3}} = F(x, y) \tag{6.14}$$

(i) *Standard Fourth-Order Range–Kutta Method (SRK) for SVF*

$$y_{i+1} = y_i + \frac{1}{6}(K_1 + 2K_2 + 2K_3 + K_4) \tag{6.15}$$

in which $K_1 = \Delta x \cdot F(x_i, y_i)$

$$K_2 = \Delta x \cdot F\left(x_i + \frac{1}{2}\Delta x, y_i + \frac{1}{2}K_1\right)$$

$$K_3 = \Delta x \cdot F\left(x_i + \frac{1}{2}\Delta x, y_i + \frac{1}{2}K_2\right)$$

$$K_4 = \Delta x \cdot F(x_i + \Delta x, y_i + K_3)$$

(ii) *Kutta–Merson Method (KM) for SVF*

$$y_{i+1} = y_i + \frac{1}{2}(K_1 + 4K_4 + K_5) \tag{6.16}$$

in which $K_1 = \frac{1}{3}\Delta x \cdot F(x_i, y_i)$

$$K_2 = \frac{1}{3}\Delta x \cdot F\left(x_i + \frac{1}{3}\Delta x, y_i + K_1\right)$$

$$K_3 = \frac{1}{3}\Delta x \cdot F\left(x_i + \frac{1}{3}\Delta x,\, y_i + \frac{1}{2}K_1 + \frac{1}{2}K_2\right)$$

$$K_4 = \frac{1}{3}\Delta x \cdot F\left(x_i + \frac{1}{2}\Delta x,\, y_i + \frac{3}{8}K_1 + \frac{9}{8}K_3\right)$$

$$K_5 = \frac{1}{3}\Delta x \cdot F\left(x_i + \Delta x,\, y_i + \frac{3}{8}K_1 + \frac{9}{2}K_3 + 6K_4\right)$$

(iii) Trapezoidal Method (TRAP) for SVF

$$y_{i+1} = y_i + \frac{1}{2}\Delta x\left[F(x_i, y_i) + F(x_{i+1}, y_{i+1})\right] \qquad (6.17)$$

6.6 DIFFERENTIAL EQUATION FOR SVF WITH DECREASING DISCHARGE

As we have already discussed that Energy equation will found more convenient in solving equation for decreasing discharge. So for this purpose while applying energy equation the following assumption are made:

(i) The flow is steady

(ii) The channel is prismatic and is of small slope.

(iii) The withdrawal of water does not affect the energy content per unit mass of water

(iv) The pressure distribution is hydrostatic

(v) The friction losses are represented by manning's formula

(vi) The energy correction factor α is used to represent the non-uniformity of velocity distribution.

The total energy at a section is given by

$$H = z + y + \alpha\frac{V^2}{2g}$$

where z = datum head and $E = y + \alpha\dfrac{V^2}{2g}$

or $H = Z + E$

α = kinetic energy correction factor differentiating with respect to x

$$\frac{dH}{dx} = \frac{dz}{dx} + \frac{dy}{dx} + \frac{d}{dx}\left(\alpha \frac{v^2}{2g}\right) \tag{6.18}$$

But $$\frac{dH}{dx} = -S_f, \frac{dz}{dx} = -S_0$$

$$\frac{d}{dx}\left(\alpha \frac{V^2}{2g}\right) = \frac{d}{dx}\left(\alpha \frac{Q^2}{2gA^2}\right) = \frac{\alpha}{2g}\left(\frac{2Q}{A^2}\right)\frac{dQ}{dx} - \frac{2Q^2}{A^3}\frac{dA}{dy}\cdot\frac{dy}{dx}$$

$$\left(\text{Since } \frac{dA}{dy} = T, \frac{dQ}{dx} = q_*\right)$$

$$\frac{d}{dx}\left(\alpha \frac{V^2}{2g}\right) = \frac{\alpha}{2g}\left(\frac{2Q}{A^2} q_* - \frac{2Q^2 T}{A^3}\frac{dy}{dx}\right)$$

Putting the value of $\frac{dH}{dx}, \frac{dz}{dx}$ and $\frac{d}{dx}\left(\frac{\alpha V^2}{2g}\right)$

$$-S_f = -S_o + \frac{dy}{dx} + \frac{\alpha}{2g}\left[\frac{2Q}{A^2} q^* - \frac{2Q^2 T}{A^3}\frac{dy}{dx}\right]$$

or $$S_0 - S_f - \frac{\alpha Q q_*}{gA^2} = \frac{dy}{dx} - \frac{\alpha Q^2 T}{gA^3}\frac{dy}{dx}$$

or $$\frac{dy}{dx} = \frac{S_0 - S_f - \dfrac{\alpha Q q_*}{gA^2}}{1 - \dfrac{\alpha Q^2 T}{gA^3}} \tag{6.19}$$

6.7 Side Weir

A side weir is a hydraulic control structure used to divert flow from main (or parent) channel into side channel when the water level in the main channel exceeds the specified limits. As the name suggests, the structure is normally located in the side of channel and water discharges over it freely under gravity in the same way as conventional weir. The most common function of a side weir is to remove flow from channel in order to prevent the downstream flow capacity of the channel being exceeded.

Typical example of side weir include

1. A flood overflow structure in a river channel, usually designed to divert excess flow into an off stream reservoir, thereby protecting downstream infrastructure from flooding.
2. A storm water overflow in a combined sewer, designed to divert excess flow that would otherwise overload the sewer downstream.
3. A storm water overflow in the approved channel to a wastewater treatment works, to ensure that the capacity of works is not exceeded.

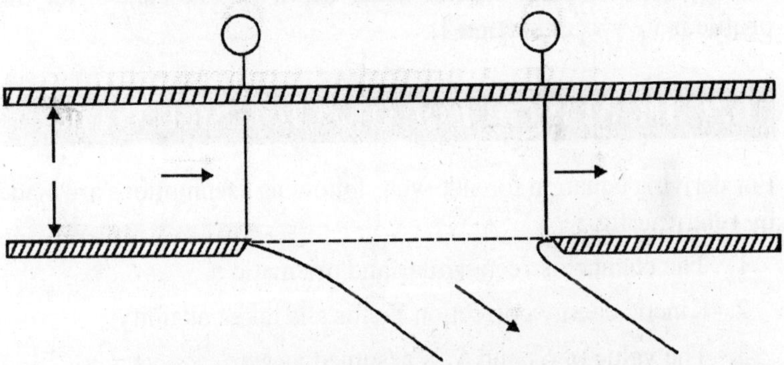

Fig. 6.3 Definition sketch of side weir flow

6.8 CLASSIFICATION OF FLOW PROFILE IN SIDE WEIR

The possible flow profiles are broadly classified into three categories: Type 1, Type 2 and Type 3.

Type 1: The channel is on a mild slope and weir heights $S > y_{c1}$ where y_{c1} is the critical depth corresponding to the incomming discharge Q_1 at Section 1 as shown in figure of Type 1. At downstream end the normal depth corresponding to discharge Q_2 will prevail. Thus $y_2 = y_t$, the tailwater depth. As section 1, the depth y_1 will such that $yc_1 < y_1 < y_0$, where y_0 = normal depth for $Q_0 = Q_1$. Along the weir the depth increase from y_1 to y_2. Upstream of Section 1 there will be an M_2 curve from y_0 to y_1. The control for SVF will be downsream depth $y_2 = y_t$.

Type 2: The channel is on mild slope ($y_0 > y_{c1}$) and $S < yc_1$. At the upstream end of weir, the depth y_1 can be considered equal to yc_1. The downstream end the depth y_2 will rise to tailwater depth y_t through jump. The control for type 2 profile is at Section 1.

Type 3: The channel is on steep slope, ($y_0 < y_{c1}$) and $S < yc_1$. The upstream depth $y_1 = y_0$ decreasing depth water profile will start from Section 1. At section 2 the depth reaches a minimum value and in the downstream channel the water surface rises through an S_3 profile to meet the tail water depth y_t. The control for the profile is $y_1 = y_0$ at section 1.

6.9 De Marchi Equation for Side Weirs

For deriving equation for side weir, following assumptions are made by referring fig.

1. The channel is rectangular and prismatic
2. Kinetic energy correction factor a is taken as unity
3. The value of S_0 and S_f is assumed as zero.
4. The side weir is assumed to be sharp crested weir with proper aseration of the nappe and to be discharging freely.
5. The specific energy is taken as constant between section 1 and 2.

As we know that differential equation for decreasing discharge is given by

$$\frac{dy}{dx} = \frac{S_0 - S_f - \dfrac{\alpha Q q_*}{gA^2}}{1 - \dfrac{\alpha Q^2 T}{gA^3}}$$

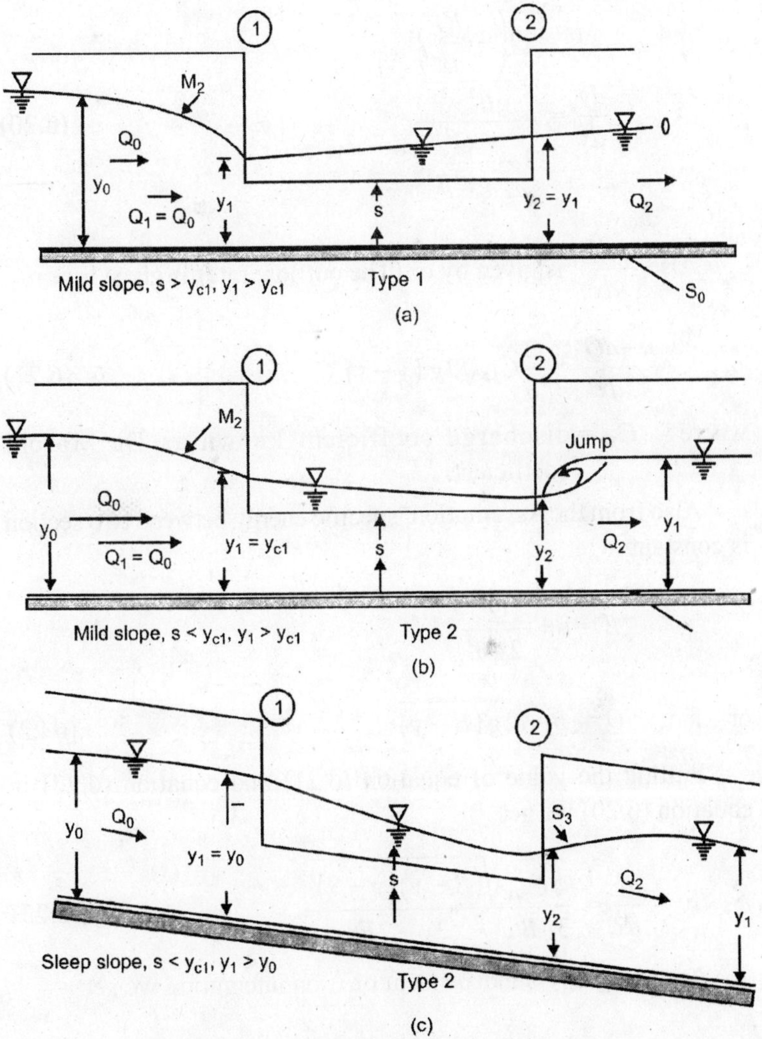

Fig. 6.4 Classification of flow over side weirs

From the above assumption, $S_0 = S_f = 0$

Also $\alpha = 1$ and $q_* = -\dfrac{dQ}{dx}, T = B,\ A = By$

Differential equation reduces to

$$\frac{dy}{dx} = \frac{\dfrac{Q\left(-\dfrac{dQ}{dx}\right)}{gB^2y^2}}{1-\dfrac{Q^2}{gy^3B^3}} \tag{6.20}$$

Also $\dfrac{-dQ}{dx}$ is given by or (The outflow rate) is given by:

$$\frac{-dQ}{dx} = \frac{2}{3}C_M\sqrt{2g}\,(y-s)^{3/2} \tag{6.21}$$

where C_M = discharge coefficient known as De Marchi coefficient.

Also from the Assumption, specific energy between two section is constant

$$E = y + \frac{Q^2}{2gA^2}$$

or
$$Q = By\sqrt{2g(E-y)} \tag{6.22}$$

Putting the value of equation (6.21) and equation (6.22) in equation (6.20) we get

$$\frac{dy}{dx} = \frac{4}{3}\frac{C_M}{B}\frac{\sqrt{(E-y)(y-S)^3}}{3y-2E} \tag{6.23}$$

Assuming C_M is independent of x, on integration we get:

$$x = \frac{3B}{2C_M}\cdot\phi_M\,(y, E, S) + \text{const} \tag{6.24}$$

where
$$\phi_M\,(y, E, s) = \frac{2E-3s}{E-s}\sqrt{\frac{E-y}{y-s}} - 3\sin^{-1}\sqrt{\frac{E-y}{E-s}}$$

Equation 6.24 is known as he *De Marchi equation* and the function

$\phi_M(y, E, s)$ is known as the *De Marchi varied flow function.* Applying Eq. 6.24 to Section 2 and 1,

$$x_2 - x_1 = L = \frac{3}{2}\frac{B}{C_M}(\phi_{M2} - \phi_{M1}) \tag{6.25}$$

knowing L, s, and (Q and y) at either 2 or 1, the discharge over he side weir Q_s can be computed by Eq. 6.25 and by the continuity equation

$$Q_s = Q_1 - Q_2 \tag{6.26}$$

De Marchi Coefficient C_M Experimental and theoretical studies by Subramanya and Awasthy have shown that in subcritical approaching flow the major flow parameter affecting the De Marhi coefficient is the Froude number is shown to be

$$\boxed{C_M = 0.611\sqrt{1 - \frac{3F_1^2}{(F_1^2 + 2)}}} \tag{6.27}$$

where $$F_1 = \frac{V_1}{\sqrt{gy_1}}$$

Equation 6.27 can be simplified as

$$C_M = 0.864\sqrt{\frac{1 - F_1^2}{2 + F_1^2}} \tag{6.28}$$

However, for supercritical approach flow the effect of the approach Froude number is insignificant and the variation of C_M for $F_1 > 2.0$ is obtained as

$$C_M = 0.36 - 0.008\, F_1$$

6.10 Uniformly Discharging Side Weirs

In many applications of side weirs, such as in irrigation systems and in the disposal of effluents, it is sometimes necessary to have side weirs in which the discharge rate $\left(-\frac{dQ}{dx}\right)$ is constant along its length. If $\frac{dQ}{dx} = q_* =$ constant, $(y - s)$ is constant along the weir. If s is kept constant, the water surface elevation will be constant for such weirs. Further, if specific energy is assumed to be constant, the water surface will be parallel to the energy line; i.e. the velocity of flow will be constant for a uniformly discharging weir. Hence,

By continuity equation, $Q_1 - q_* x = Q$

i.e. $$A_1 - \frac{q_* x}{V_1} = A \quad \text{and} \quad \frac{q_*}{V_1} = \left(\frac{A_1 - A_2}{L}\right)$$

$$A_1 - \left(\frac{A_1 - A_2}{L}\right)x = A \tag{6.29}$$

Thus, the uniformly discharge side weir can be achieved by linear reduction of area of flow. This can be acheved in two ways: (i) by contouring the channel side (Fig. 6.5a), and (ii) by contouring the channel bed (Fig. 6.5b). Further details on uniformly discharging side weirs are available in literature.

For a uniformly discharging side weir, the lateral outflow Q_s is therefore,

$$Q_s = \frac{2}{3} C_M \sqrt{2gL}\,(y - s)^{3/2} \tag{6.30}$$

where C_M = De Marchi coefficient. Further, by continuity,

$$Q_1 - Q_2 = Q_s$$

The computations for Type 1, 2 and 3 flows follow the same assumptions as indicaed in the previous section.

6.11 Flow through Bottom Racks

Similar to side weir, flow through bottom rack is another example of SVF. A bottom rack is provided at bottom of channel for diverting part of flow from the channel. Bottom rack consists of an opening in the channel bottom covered with metal rack to present transport of unwanted material through opening.

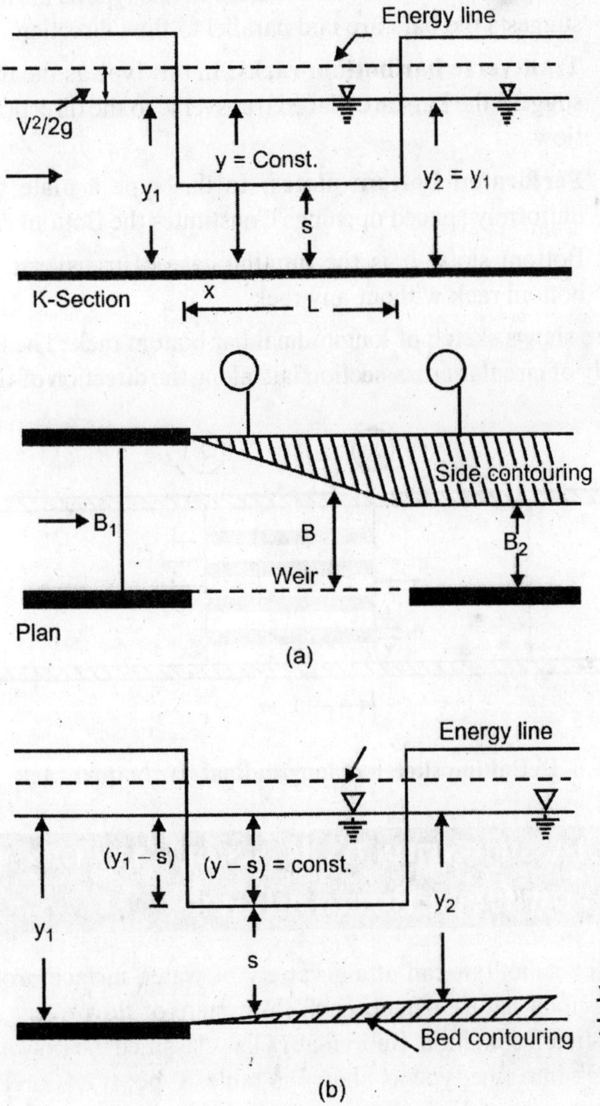

Fig. 6.5 Uniformly discharging side weirs

Bottom racks are broadly classified into four categories:

1. **Longitudinal bar bottom racks:** In this type as the name suggests the bars are laid parallel to flow direction.
2. **Transverse bar bottom racks:** In this type as the name suggess the bars are placed transverse to the direction of flow.
3. **Perforated bottom plates:** In this type a plate with uniformly spaced openings Constitutes the Bottom rack.
4. Bottom slots: It is the limiting case of transverse bar bottom rack without any rack.

Figure shows sketch of longitudinal bar bottom rack. The bars are usually of circular cross-section laid along the direction of flow.

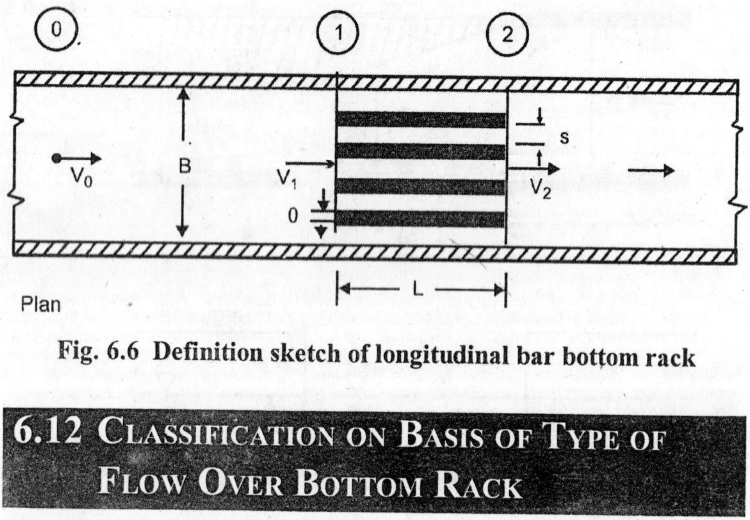

Fig. 6.6 Definition sketch of longitudinal bar bottom rack

6.12 Classification on Basis of Type of Flow Over Bottom Rack

The flow over bottom can attain variety of water surface profile depending upon nature of approach flow, state of flow over rack and downstream condition. Subramanya has classified the flow over bottom rack into fine type as shown in table. Type A_1, A_3 and B_1 occurs commonly and also have significance from design point of view.

Table 6.1 Types of Flow Over Bottom Racks

Type	Approach	Flow over the rack	Downstrea state
A_1	Subcritical	Supercritical	May be a jump
A_2	Subcritical	Partially supercritical	Subcritical
A_3	Subcritical	Subcritical	Subcritical
B_1	Supercritical	Supercritical	May be a jump
B_2	Supercritical	Partially supercritical	Subcritical

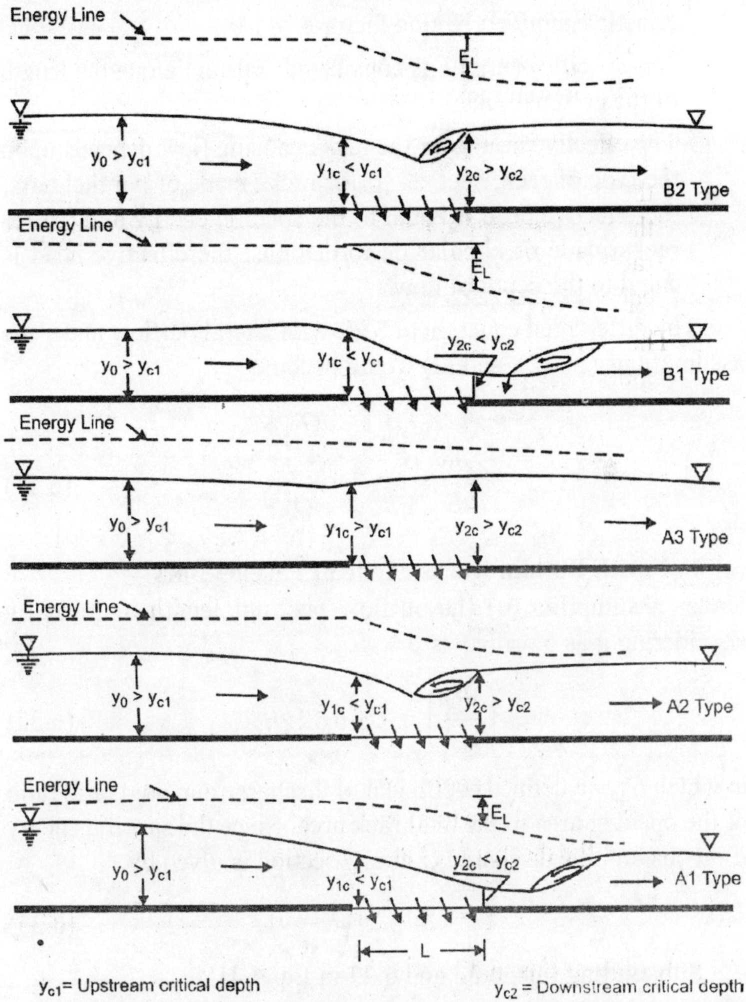

Fig. 6.7 Classification of different types of flows over bottom racks

6.13 Mostkow Equations for Bottom Racks

Mostkow derived expressions for the water-surface profile for spatially varied flow over bottom racks by making the following assumptions:

1. The channel is rectangular and prismatic.
2. Kinetic energy correction factor $\alpha = 1.0$.
3. The specific energy E is considered constant along the length of the bottom racks.
4. The effective head over the racks causing flow depends upon the type of rack, such as (i) for racks made of parallel bars, the effective head is equal to the specific energy and (ii) for racks made of circular performations, the effective head is equal to the depth of flow.

The differential equation of SVF with lateral outflow under the assumptions (1), (2), and (3) would become

$$\frac{dy}{dx} = \frac{Qy\left(-\frac{dQ}{dx}\right)}{gB^2y^2 - Q^2} \tag{6.31}$$

(a) SVF with Bottom Racks Made of Parellel Bars

Under assumption (4) the outflow per unit length of rack, by considering it as a orifice, is

$$\left(\frac{dQ}{ds}\right) = C_1\varepsilon B\sqrt{2gB} \tag{6.32}$$

in which C_1,= a defined coefficient of discharge and void ratio ratio of the opening area to the total rack area. Since the specific energy E – constant, the discharge Q at any section is given by:

$$Q = By\sqrt{2g(E-y)} \tag{6.33}$$

Substituting Eqs. 6.32 and 6.33 in Eq. 6.31

$$\frac{dy}{dx} = \frac{2\varepsilon C_1\sqrt{E(E-y)}}{3y-2E} \tag{6.34}$$

On integration,

$$x = -\frac{E}{\varepsilon C_1}\frac{y}{E}\sqrt{1-\frac{y}{E}} + \text{Const.} \tag{6.35}$$

Putting $y/E = \eta$ and using the boundary condition $y = y_1$ and $x = 0$, gives

$$x = \frac{E}{\varepsilon C_1}\left(\eta_1\sqrt{1-\eta_1} - \eta\sqrt{1-\eta}\right) \tag{8.36}$$

which is the equation of the SVF profile. As in the case of the De Marche equation the control depths for use in Eq. 6.36 which are compatible with the assumptions are:

(i) $y_1 = y_0$ for A3 and B1 type flows and

(ii) $y_1 = y_{c1}$ for A1 type flows.

Note that from Eq. 6.36, $\left(\dfrac{dQ}{dx}\right)$ is constant along the rack. Hence the total discharge Q_s diverted out is,

$$Q_s = C_1\varepsilon BL\sqrt{2gE} \tag{6.37}$$

(b) SVF Equation for Bottom Racks Made of Performated Plates

From perforated plate bottom racks the outflow discharge per unit length, under assumption (4), is

$$\left(-\frac{dQ}{dx}\right) = C_2\varepsilon B\sqrt{2gy} \tag{6.38}$$

where C_2 = discharge coefficient for performated plate flow. Substituting Eqs. 6.38 and 6.33 in Eq. 6.31, and simplifying

$$\frac{dy}{dx} = \frac{2\varepsilon C_2\sqrt{y(E-y)}}{3y-2E} \tag{6.39}$$

Integrating and using the boundary condition $y = y_1$ at $x = 0$, yields tye SVF profile for perforated bottom plates as

$$x = \frac{E}{\varepsilon C_2}\left[\frac{1}{2}\left(\cos^{-1}\sqrt{\eta} - \cos^{-1}\sqrt{\eta_1}\right) + \frac{3}{2}\sqrt{\eta_1(1-\eta_1)} - \sqrt{\eta(1-\eta)}\right] \tag{6.40}$$

in which $\eta = y/E$ and suffix 1 refers to Section 1.

Solved Problems

Problem 1: ***A horizontal frictionless rectangular lateral spillway channel of Length L has a free overfall. Show that***

$$\frac{dx^2}{dy} - \frac{x^2}{y} = \frac{-gB^2y^2}{q_*^2}$$

Also show that the solution of above equation for the condition of critical depth y_c occuring at the outlet is

$$\left(\frac{x}{L}\right)^2 = \frac{3}{2}\left(\frac{y}{y_c}\right) - \frac{1}{2}\left(\frac{y}{y_c}\right)^3$$

Solution: Considering basic differential equation for increasing discharge

$$\frac{dy}{dx} = \frac{S_0 - S_f - \dfrac{2Qq_*}{gA^2}}{1 - \left(\dfrac{Q^2T}{gA^3}\right)} \qquad (\beta = 1)$$

here $s_0 = 0$, $s_f = 0$ $A = By$, $T = B$, $q_* x = Q$

Therefore Now,

$$\frac{dy}{dx} = \frac{-2q_*^2 \dfrac{x}{gB^2y^2}}{1 - \left(q_*^2 \dfrac{x^2}{gB^2y^3}\right)}$$

or
$$\left(1-q_*^2\frac{x^2}{gB^2y^3}\right)dy=-\frac{2q_*^2x}{gB^2y^2}dx$$

Let us suppose $t = x^2$ so, we can write

$$dt = 2xdx$$

Further suppose $\frac{q_*^2}{gB^2y^2}=\frac{1}{a}$

then

$$\left(1-\frac{t}{y}\cdot\frac{1}{a}\right)dy=-\frac{1}{a}dt$$

$$\frac{dt}{dy}-\frac{t}{y}=-a \quad \Rightarrow \frac{dx^2}{dy}-\frac{x^2}{y}=\frac{-gB^2y^2}{q_*^2} \qquad (1)$$

Equation (1) is a linear differential equation of first order. The general solution of this equation is

$$x^2=\frac{-gB^2y^3}{q_*^2}+C_1y \qquad (2)$$

At the outlet, $x = L$ and $y = y_0$, put in equation we get

$$C_1=\frac{1}{y_c}\left(L^2-\frac{gB^2y_0^3}{2q_*^2}\right)$$

Put the value of C_1 in equation 2 and after rearranging, we get

$$\left(\frac{x}{L}\right)^2=\left(1+\frac{1}{2F_0^2}\right)\frac{y}{y_0}-\frac{1}{2F_0^2}\left(\frac{y}{y_0}\right)^3$$

where
$$F_0^2=\frac{q_*^2L^2}{gB^2y_0^3}$$

As we know that F_0 is the froude number of the flow at outlet. When free overfall occurs at the outlet, the flow is critical therfore $y_0 = y_c$ and $F_0 = 1$. Now equation of flow profile becomes

$$\left(\frac{x}{L}\right)^2 = \frac{3}{2}\left(\frac{y}{y_c}\right) - \frac{1}{2}\left(\frac{y}{y_c}\right)^3 \quad \text{Proved.}$$

Problem 2: ***Find out the expression to determine the critical depth and its location for a lateral spillway channel of rectangular channel use the Chezy formula.***

Solution: In case of rectangular channel

$$T = B$$

using chezy formula conveyance, $K = AC\sqrt{R}$

or $$K^2 = A^2C^2R$$

As we know that transitional profile in lateral spillway is given by:

$$\frac{1}{x_t} = \frac{1}{2}\left(\frac{S_0 T}{A} - \frac{gA^2}{\beta K^2}\right)$$

As the channel is rectangular, we have

$$T = B,\ A = By,\ K^2 = A^2C^2R$$

Put in above equation we get

$$\frac{1}{x_t} = \frac{1}{2}\left(\frac{S_0}{y} - \frac{g}{\beta C^2 R}\right)$$

or $$x_t = \frac{2\beta C^2 R y}{(\beta C^2 R S_0 - gy)}$$

Also equation of critical depth line is given by

$$Q_c = \sqrt{\frac{gA^3}{\beta T}} \quad \text{and} \quad x_c = \frac{Q_c}{q_*}$$

Since channel is rectangular

$$Q_c = \sqrt{\frac{gB^2 y_c^3}{\beta}}$$

and $$x_c^2 = \frac{gB^2 y_c^3}{\beta q_*^2}$$

At critical flow section, the critical depth line and transitional depth line intersect

Therefore,

$$x_t = x_c,\ y = y_c$$

or $$x_t^2 = x_c^2$$

$$\Rightarrow \qquad \frac{gB^2 y_c^3}{\beta q_*^2} = \frac{4\beta C^4 y_c^2 R_c^2}{\left(\beta C^2 R_c S_0 - g y_c\right)^2} \qquad \text{eq}$$

Equation ___ is the desired equation for the determination of critical depth.

Problem 3: ***Find out the expression to determine the critical depth and its location for lateral spillway channel of rectangular channel use Manning's equation.***

Solution: Using Meaning's equation, coonveyance K is given by

$$K = \frac{AR^{2/3}}{\eta} G$$

As channel is rectangular, $T = B$, $A = B_y$ $K^2 = \dfrac{A^2 R^{4/3}}{\eta^2}$

transitional profile is given by

$$\frac{1}{x_t} = \frac{1}{2}\left(\frac{S_0 T}{A} - \frac{gA^2}{\beta K^2}\right)$$

Now putting all above value, we get

$$\frac{1}{x_t} = \frac{1}{2}\left(\frac{S_0}{y} - \frac{g\eta^2}{\beta R^{4/3}}\right)$$

or $$x_t = \frac{2\beta y R^{4/3}}{\left(\beta S_0 R^{4/3} - \eta^2 g y\right)}$$

Equation of critical depth line is given by

$$Q_c = q_* x_c = \sqrt{\frac{gA^3}{\beta T}}$$

or
$$Q_c = q_* x_c = \sqrt{\frac{gB^2 y_c^3}{\beta}}$$

$$x_c^2 = \frac{gB^2 y_c^3}{\beta q_*^2}$$

At the critical flow section, the critical depth line and transitional depth line intersect hence, we have

$$x_t = x_c,\ y = y_c$$
$$x_t^2 = x_c^2$$

$$\frac{g\beta^2 y_c^3}{\beta q_*^2} = \frac{4\beta^2 y_c^2 R_c^{8/3}}{\left(\beta S_0 R_c^{4/3} - g\eta^2 y_c\right)^2}$$

Which is a required expression.

Problem 4: ***A lateral spillway channel is trapezoidal in section with B = 5.0 m, m = 1 and n = 0.015. The bed slope is 0.10. Find the location of the control point and the critical depth for a lateral discharge rate of (a) q_* = 2.0 m³/s/m, (b) q_* = 3.0 m³/s/m, Assume β = 1.25.***

Solution: The computations necessary to plot the critical-depth line and the transitional profile are done in a tabular formas shown in Table 8.1. Various depth values are assumed and x_t and x_c, the longitudinar co-ordinates of the transtional profile and critical depth line respectively are calculated x_t is calculate using equation 8.8 and $x_c = \dfrac{Q_c}{q_*}$ where $Q_c = \sqrt{\dfrac{gA^3}{\beta T}}$. The transitional profile and critical depth line for a given q_* are plotted and control point is determined by intersection of these two lines as

$q_* = 3\ m^3/s/m$ $\quad y_c = 4$ m $\quad x_c = 56$ m

$q_* = 2\ m^3/s/m$ $\quad y_c = 2.65$ m and $x_c = 40$ m

Table 8.1 Calculations of Control Point in a Lateral Spillway Channel – Example 8.1

Y	A	T	Q_c	P	$R = A/P$	K	$1/x_t$	x_t	$X_c = Q_c/q^*$	
(m)	(m²)	(m)	$\sqrt{\frac{gA^3}{\beta T}}$	(m)	(m)		Eq. (8.8)	(m)	$q^* = 3.0$	$q^* = 2.0$
1.0	6.00	7.0	15.56	7.828	0.7665	335	0.05707	17.54	5.18	7.78
2.0	14.00	9.0	48.92	10.657	1.3137	1120	0.03153	31.72	16.31	24.46
2.5	18.75	10.0	71.92	12.071	1.5533	1677	0.02618	38.20	23.97	35.96
3.0	24.00	11.0	9.31	13.485	1.7798	2349	0.02251	44.43	33.10	40.66
3.5	29.75	12.0	131.20	14.899	1.9968	3145	0.01982	50.46	43.73	65.60
4.0	36.00	13.0	167.80	16.314	2.2067	4068	0.01775	56.34	55.94	83.90
5.0	50.00	15.0	255.70	19.142	2.6120	6322	0.01475	67.78	85.32	127.85
6.0	66.00	17.0	364.30	21.971	3.0000	9160	0.01268	78.89	121.43	182.15

Problem 5: ***A 2 m wide rectangular channel conveys a discharge of 1.8 m²/s at a depth of 0.58 m. A uniformly discharging side weir with crest 0.45 m abve the bed at the commencement of the side weir is proposed to diverst a flow of 0.3 m³/s laterally. Design the length of side weir and other geometry of the channel at the weir.***

Solution: $V_1 = \dfrac{Q_1}{A_1} = \dfrac{1.8}{2\times 0.58} = 1.55$ m/s

Froude number $F_1 = \dfrac{V_1}{\sqrt{gy_1}} = \dfrac{1.55}{\sqrt{9.8\times 0.58}} = 0.65$

hence the flow is subcritical

Now $C_M = 0.611\sqrt{1-\dfrac{3F_1^2}{(F_1^2+2)}}$

$$= 0.611\sqrt{1-\dfrac{3\times 0.65^2}{(0.65^2+2)}} = 0.42$$

Critical depth at section one $= y_{c_1} = \left(\dfrac{q^2}{g}\right)^{1/3}$

$$y_{c1} = \left(\dfrac{\left(\dfrac{1.8}{2}\right)^2}{9.8}\right)^{1/3} = 0.435 \text{ m}$$

Height of weir crest = s = 0.45 m

Since $y_{c1} < S$, Type 1 flow will preocil

$y_1 = y_0 = 0.58$ m

Depth of water over the weir = $(y - s) = 0.58 - 0.45$

$= 0.13$ m

Diverted discharge $Q_s = \frac{2}{3} C_M \sqrt{2g}\, L(y-s)^{3/2}$

$$0.3 = \frac{2}{3} \times 0.42 \sqrt{2 \times 9.81} \times L \times (0.13)^{3/2}$$

$$0.3 = 0.058\, L$$

$$L = 5.17 \text{ M}$$

Q_2 = downstrem discharge = 1.8 – 0.3 = 1.5 m^{3}s

For uniformly discharing side weir = $V_1 = V_2 = V$

Hence V_2 = 1.55 m/s and $A_2 = \frac{1.5}{1.55} = 0.967 \text{ m}^2$

(i) If side contouring is adopted, $y_1 = y_2 = y$

Bed width at section 2 = $B_2 = \frac{A_2}{y_2} = \frac{0.967}{0.58} = 1.667$

The bed width varies from 2 m at section 1 to 1.667 m at section 2, distance 5.17 m downstrean of section 1 lineary.

(ii) If bed controuring is adopted $B_1 = B_2 = B$

Depth of flow at section 2 = $y_2 = \frac{A_2}{B_2} = \frac{0.967}{2} = 0.48$ m

$$= 0.48 \text{ m}$$

ΔZ_2 = change in elevation of Bed at section 2
= 0.58 – 0.48 = 0.1

The nariation of Δ_2 is linear with a value of 0 at section 1 and 0.1 m at Section 2

Problem 6: ***Show that the following equation is applicable to a control sectin where critical depth occurs in a fricionless lateral spillway cannel***

$$\frac{S_0^2 g A_c T_c}{4 \beta q_*^2} = 1$$

Solution: Since the channel is frictionless therefore $S_f = 0$, hence equation 6.7 reduces to

$$\frac{dy}{dx} = \frac{\dfrac{S_0 - 2\beta Q q_*}{gA^2}}{1 - \dfrac{\beta Q^2 T}{gA^3}} \qquad \text{...(a)}$$

At the control section $Q=Q_c$ and at critical depth

$$Q_c = \sqrt{\frac{gA^3}{\beta T}} \text{ or } Q_c^2 = \frac{gA^3}{\beta T} \qquad \text{...(b)}$$

Put value of equation (b) in equation (a) we get

$$\frac{dy}{dx} = \frac{\dfrac{S_0 - 2\beta Q q_*}{gA^2}}{1 - \left(\dfrac{Q}{Qc}\right)^2} \quad \left(\text{if } Q = Q_c, \text{ then } \frac{dy}{dx} = 0\right)$$

$$\frac{\dfrac{S_0 - \beta Q q_*}{gA^2}}{1 - \left(\dfrac{Q}{Qc}\right)^2} = 0$$

$$\frac{S_0 - 2\beta Q q_*}{gA^2} = 0$$

or $$S_0 g A^2 = 2\beta Q q_* \qquad \text{...(c)}$$

Squaring both side equaion (c)

$$S_0^2 g^2 A^4 = 4\beta^2 Q^2 q_*^2$$

As we know $Q_2 = Q_c^2 = \dfrac{gA^3}{\beta \Gamma}$

$$S_0^2 g^2 A^4 = 4\beta^2 \frac{gA^3}{\beta \Gamma} q_*^2$$

$$\Rightarrow \qquad 4\beta q_*^2 = S_0^2 g A_c T_c$$

$$\boxed{\frac{S_0^2 g A_c T}{4\beta q_*^2} = 1}$$

Problem 7: ***Show that for a wide rectangular channel having SVF with a constant inflow rate of q_* the critical flow section is given by***

$$x_c = \frac{8q_*^2}{g\left(s_0 - \frac{g}{c^2}\right)^3}$$

where $\beta = 1$ and Chezys formula is used.

Solution: For a wide rectangular chennel if we use Chezy's equation

$$K = AC\sqrt{R} \qquad \boxed{R = y}$$

$$K = AC\sqrt{y}$$

$$K^2 = A^2C^2y$$

The transitional profile is given by

$$\frac{1}{x_t} = \frac{1}{2}\left(\frac{S_0 T}{A} - \frac{gA^2}{\beta k^2}\right)$$

For rectangular channel $A = By$, $T = B$

$$\frac{1}{x_t} = \frac{1}{2}\left(\frac{S_0 B}{By} - \frac{gB^2y^2}{\beta B^2C^2y^2 \cdot y}\right)$$

$$\frac{1}{x_t} = \frac{1}{2}\left(\frac{S_0}{y} - \frac{g}{BC^2 y}\right)$$

$$\frac{1}{x_t} = \frac{1}{2y}\left(\frac{\beta C^2 S_0 - g}{\beta C^2}\right)$$

$$\frac{1}{x_t} = \frac{2y\beta C^2}{\beta C^2 S_0 - g}$$

At critical section $x_t = x_c, y = y_c$

$$x_C = \frac{2\beta C^2 y_c}{\left(\beta C^2 S_0 - g\right)}$$

$$y_C = \frac{\left(\beta C^2 S_0 - g\right)}{2\beta C^2} x_C \qquad \ldots(1)$$

We know that $q_* = \dfrac{Q}{x} = \dfrac{Q_C}{x_C}$

$$Q_C = \sqrt{\frac{gA_C^3}{\beta\Gamma_C}} = \sqrt{\frac{gB^2 y_C^3}{\beta B}} = \sqrt{\frac{gB^2 y_C^3}{\beta}}$$

The equation of the critical depth line is given by

$$Q_C = q_* x_C = \sqrt{\frac{gB^2 y_C^3}{\beta}}$$

$$x_C^2 = \frac{gB^2 y_C^3}{\beta q_*^2}$$

or

$$y_C^3 = \frac{x_C^2 \beta q_*^2}{gB^2} \qquad (\beta = 1)$$

$\therefore$

$$y_C^3 = \frac{x_C^2 q_*^2}{gB^2} \qquad \ldots(2)$$

Cubing equation (1) and equating with equation (2) and putting $\beta = 1$ we get

$$\frac{\left(C^2 S_0 - g\right)^3}{8C^8} x_C^3 = \frac{x_C^2 q_*^2}{gB^2}$$

$$\Rightarrow \qquad x_C = \frac{8C^8 q_*^2}{gB^2\left(C^2 S_0 - g\right)^3}$$

Dividing numerator and denominator by $C^8 B$ we get

$$\boxed{x_C = \frac{8q_*^2}{g\left(S_0 - \frac{q}{C^2}\right)^3}}$$

Problem 8: ***A wide rectangular channel of length L, having a uniform lateral inflow rate has a discharge of q_e per unit width at channel exit. If Darcy Weisbach friction factor f is used for representing friction effects, show that:***

$$(\beta = 1) \qquad y_c = \frac{256 q_e^2}{gL^2\left(8s_0 - f\right)^2}$$

and

$$x_c = \frac{4096 q_e^2}{gL^2\left(8s_0 - f\right)^3}$$

Solution: As we know for wide rectangular channel, $R = y$ and $B = 1$

Also we know $$\frac{1}{x_t} = \frac{1}{2}\left(\frac{S_0 T}{A} - \frac{gA^2}{\beta K^2}\right) \qquad \ldots\text{(i)}$$

$$K_2 = A^2 C^2 R,\ A = By,\ T = B$$

$$\frac{1}{x_t} = \frac{1}{2}\left(\frac{S_0 T}{A} - \frac{gA^2}{A^2 c^2 R}\right)$$

$$\frac{1}{x_t} = \frac{1}{2}\left(\frac{S_0}{y} - \frac{g}{c^2 y}\right)$$

For critical condition

$$\frac{1}{x_c} = \frac{1}{2y_c}\left(S_0 - \frac{g}{c^2}\right) \qquad \ldots\text{(ii)}$$

As we know $C^2 = 8g/f$ put in (ii)

$$\frac{1}{x_c} = \frac{1}{2y}\left(S_0 - \frac{gf}{8g}\right)$$

or
$$x_c = \frac{16y_c}{(8S_0 - f)} \qquad \text{...(iii)}$$

Also
$$Q_c = q_e x_c = \sqrt{\frac{gL^2 y_c^3}{\beta}}$$

$$x_c^2 = \frac{gL^2 y_c^3}{\beta q_e^2} = \frac{gL^2 y_c^3}{q_e^2} \qquad \text{...(iv)}$$

Equating (iii) and (iv) equation, we get

$$\frac{gL^2 y_c^3}{q_e^2} = \left[\frac{16y_c}{(8S_0 - f)}\right]^2$$

$$\Rightarrow \qquad y_c = \frac{256q_e^2}{gL^2(8S_0 - f)^2} \quad \textbf{Proved}$$

Also from equation (iv) we get

$$y_c = \frac{q_e^{2/3} x_c^{2/3}}{g^{1/3} L^{2/3}} \qquad \text{(v)}$$

Put equation (v) in equation (iii) we get

$$x_c = \frac{16q_e^{2/3} x_c^{2/3}}{g^{1/3} L^{2/3} (8S_0 - f)}$$

$$x_c^3 = \frac{4096q_e^2 x_c^2}{gL^2 (8S_0 - f)^3}$$

or
$$x_c = \frac{4096q_e^2}{gL^2 (8S_0 - f)^3} \quad \textbf{Proved.}$$

Review Questions

1. Show that the following equation is applicable to a control section where critical depth occurs in a frictionless lateral spillway channel

$$\frac{S_0^2 g A_c T_c}{4\beta q_*^2} = 1$$

2. Show that for a wide rectangular channel having SVF with a constant inflow rate of q_* the critical flow section is given by

$$x_C = \frac{8q_*^2}{g\left(S_0 - \frac{q}{C^2}\right)^3}$$

where $\beta = 1$ and Chezy's formula is used what would be corresponding value of y_C

3. A 1.5 m wide rectangular channel conveys a discharge of 1.7m^3/s at a depth of 0.6 m. A uniformly discharging side weir with crest at 0.42 m above the bed at the commencement of side weir is proposed to divert a flow of 0.3 m^3/s laterally:

Design the length of side weir and other geometry of channel at the weir.

4. Derive Demarchi equation for side Weirs. Also write assumptions.

5. Derive Mostkow equation for Bottom racks. Also write assumptions.

6. What are SVF. Derive differential equation of SVF with increasing discharge. Also write assumption.

7. Derive differential equation of SVF with decreasing discharge. Also write assumption.

8. Write short notes on:

(a) Numerical methods for SVF computation.

9. A horizontal frictionless rectangular channel lateral spillway of length L has a free overfall. Show that equation of flow profile is

$$\frac{dx^2}{dy} - \frac{x^2}{y} = -\frac{gB^2 y^2}{q_*^2}$$

10. Obtain an expression to determine critical depth and its location for lateral spillway channel of rectangular section. Spillway channel of rectangular section. Use Chezy's formula with C = constant.

11. Find out the expression to determine critical depth and its location for lateral spillway channel of rectangular channel use Manning equation.

Unit 7 Flow in Channel of Non-Linear Alignment and Non-Prismatic Channel

7.1 Introduction

This chapter describes how the flow takes place in bend or how flow takes place in channel of Non-linear Alignment. Moreover it also deals with design consideration of subcritical flow as well as supercritical flow.

Further moving ahead the chapter describes some of the aspects of flow through non-prismatic channel section which includes, flow through sudden transitions flow through constriction, flow through obstruction and flow through channel junction.

7.2 Flow in Channel of Non-Linear Alignment

In channels of non linear Alignment, flow behave differently according to the state of flow. Sub critical flow shows smooth water surface and slight super elevation whereas super critical flow exhibit characteristics cross wave disturbance patterns on the surface and as a result exaggerates super elevation.

In the study of sub critical flow, spiral currents are of primary interest. These currents are a friction phenomenon, hence their analysis requires use of the Reynolds number as parameter where as in the study of supercritical flow, the formation of cross waves is of major concern, these waves represent the gravity effect of the free surface upon the flow; hence their analysis rely on the use of Froude number as a basic parameter.

It should be kept in mind that the presence of curves or bends in alignment is unavoidable in the design of open channels because of the complex nature of flow around curved path. The streamlines of the flow results in the formation of spiral currents and cross waves. Furthermore, the centrifugal force acting on the flow around the bend produces superelevation (rise in water surface at the outer bank and lowering at the inner bank). Also velocity distribution is very irregular and coefficient of α and β is for greater than one.

Now we will discuss some of the major terminology which is of major concern in the study of flow in channel of non-linear Alignment:

7.2.1 Spiral flow

The term spiral flow means moment of water particles along a helical path in the direction of flow. Under such condition there are transverse velocity component on the cross-section, in addition to the major velocity component normal to the channel cross-section. These transverse components will create so called secondary flow in the plane of cross-section.

Spiral flow in curved channel was first observed by Thomson in 1876. It is believed that the spiral flow phenomenon is due to :

1. Friction on the channel walls which causes higher velocities near the centre of channel than near the walls.
2. Centrifugal force, which deflects the particles of water from straight line motion.
3. A vertical velocity distribution which exists in the approach channel and thus indicates motion in the flow.

It should be kept in mind that looking downstream, a channel curve to the right causes counter clockwise spiral, whereas a channel curve to the left causes a clockwise spiral. In order to show the magnitude and effect of spiral flow in different curves under varying condition of flow the term known as strength of a spiral flow is used. The term is defined as the percentage ratio of mean kinetic energy of the lateral motion to the total kinetic energy of the flow at given cross section. It should be noted that kinetic energy depends of square of velocity if channel cross section is represented by *xy* plane the strength of spiral flow at this section is given by

$$S_{xy} = \frac{V_{xy}^2}{V^2} \times 100 \tag{7.1}$$

where V_{xy} is the mean velocity projected on *xy* plane and V is the mean velocity in this section.

7.2.2. Superelevation

It is actually the different between the water surface levels at the inner and outer banks of the bends. It can be obtained if a suitable assumption is made for the distribution of velocity across the width. The expression for superelevation can be obtained by assumption of theoretical free vortex distribution of velocity around bend. By the law of free vortex equation the following expression can be written

$$V_x = \frac{c}{r} \tag{7.2}$$

where V_x is the forward filemental velocity in the curve at a radial distance r from the centre of curvature and c is called circulation constant in a free vortex motion.

Let E be the specific energy at any section and y the depth of flow at a distance r from the centre of curvature; then

$$y = E - \frac{Vx^2}{2g} \tag{7.3}$$

The average forward velocity is

$$V_x = \frac{\int_{r_i}^{r_0} \left(\frac{c}{r}\right) dr}{r_0 - r_i} = \frac{c}{(r_0 - r_i)} \frac{\ln r_0}{r_i} \tag{7.4}$$

and average depth of flow is

$$y_{avg} = \frac{\int_{r_i}^{r_0} y dr}{r_0 - r_i} = \frac{\int_{r_i}^{r_0} \left(E - \frac{C^2}{2gr^2}\right) dr}{(r_0 - r_i)}$$

$$= E - \frac{C^2}{2gr_0r_i} \tag{7.5}$$

In the above equation r_0 and r_i are the outer and inner radii of curve. Now discharge Q is given by

$$Q = V_x y_{avg} (r_0 - r_i) = \mathrm{C}\left(E - \frac{C^2}{2gr_0r_i}\right)\ln\frac{r_0}{r_i} \tag{7.6}$$

If Q, r_0, r_i and E are given the constant c can be find out from above expression.

The superelevation Δh of the water surface can be given by

$$\Delta h = \frac{C^2}{2gr_0^2r_i^2}\left(r_0^2 - r_i^2\right) \tag{7.7}$$

7.2.3 Cross Waves

Cross waves are generally found in supercritical flow in channel of non linear alignment or bends and channels with non prismatic section. These cross waves forming disturbance pattern upto considerable distance downstream, are caused by turning effect of the curved walls, which does not act equally on all streamlines in the channel section. In a channel with bends, the outer wall, which turn inward to the flow, will produce an oblique jump and a corresponding positive disturbance line or positive wave front. The inner wall, which turns away from the flow will develop oblique expansion wave and negative expansion line or negative wave front. The disturbance lines thus produced by both inner and outer will be reflected back and forth between walls and will interface with each other resulting in the disturbance pattern of cross waves.

Following are the some of the means to reduce the formation of cross waves:

1. Provision of easement transition curve at the beginning and end of the bend.
2. Provision of transverse slope to the bed towards the inside wall to counter the effect of normal acceleration.
3. Provisions of diagonal silts at the entrance and exist.

7.2.4 Energy Loss in Bends or Channel of Non-linear Alignement

The energy loss in a channel bend is larger than in a corresponding length of straight channel on account of following reasons.

(i) Changed velocity disturbance over the depth and across the width.

(ii) Presence of secondary circulation

(iii) Possibility of flow separation

The magnitude of loss is influenced by several parameters if C_L is the energy loss coefficient corresponding to the additional loss of energy in a bend E_L and is defined as

$$C_L = \frac{E_L}{\frac{V^2}{2g}} \tag{7.8}$$

where C_l is the function of $(B/y, F_1, R_e, B/R_c, \theta)$

7.3 Design Consideration of Subcritical Flow

In case of open channel flow curves are undesirable as they will increase frictional loss and lead to danger of serious erosion due to spiral flow. So the basic aim of the design of subcritical flow is to reduce the spiral flow. In order to have proper proportioning of the curve a ratio of $r_0/b = 3$ is recommended, because it will give the least radius at which the effect due to spiral flow is minimized. In erodible channels, the effect of spiral flow will develop a configuration in the bed. The size of the configuration is minimum if r_0/b is 3 or larger where as in Alluvial bend, the configuration may be represented by the empirical equation

$$y = 6.35D\left(\sqrt{0.437 - \frac{x^2}{'T^2}} - 0.433\right)\left(1 + \frac{xk}{r_0}\right) \tag{7.9}$$

where y is ordinate or depth in feet, x is the abscissa in ft, D is the hydraulic depth in feet r_0 = radius of curvature on concave side of the channel in feet, k is coefficient equal to 17.52.

From the above equation the following remarks should be noted

1. For r_0 less than 40 times the square root of the water area, no further deepening of the channel seem to result from the increased curvature. Hence in such a cases the value of r_0 used in the equation should be $40\sqrt{A}$. Consequently, bends are constructive and stable when r_0 is greater than $40\sqrt{A}$ whereas sharper bends are destructive, tending to shift the channel.
2. For r_0 greater than $110\sqrt{A}$ the equation is invalid.
3. The equation may be applied to curved channels not occupying the entire width of water way. In such cases k = 26.28 and value of y computed should be increased by 14%.
4. The equation generally gives the width of the channel at hydraulic depth about 20% greater than actual width.
5. When the channel is neither on the curve nor on a strength reach. The maximum reach is about 14.5% less than computed value.

7.4 Design Consideration for Supercritical Flow

The main objective of design of curved channels for supercritical flow is to eliminate super elevation and cross wave.

The scientist knapp has suggest following method to achieve this objective

1. Banking
2. Multiple curved vanes
3. Eastment curves
4. Diagonal sills

1. **Banking:** In case of banking, the bottom cross slope is provided that will supply lateral force to counter act the centrifugal action of the flow. The required cross slope s_t can be computed by equating the gravity component along the cross slope to the centrifugal force determined by the radius r of the curve and velocity V

$$S_t = \frac{V^2}{gr} \tag{7.10}$$

2. **Multiple curved vanes:** The super elevation can be reduced by providing concentric waves which divide the channel width into series of narrow curved channels.
3. **Easement curves:** The disturbances in the simple curved channel may be reduced by employing a compound curve as compound curve will offer a very desirable solution for most curved channels of supercritical flow.
4. **Diagonal sills:** It is installed on the bottom of the channel near the end of the curve. The optimum sill angle is 30°. This method can be used as remedial measure in existing simple curves. The major disadvantage of this method are higher maintenance cost and possible cavitations at high velocity flow.

7.5 FLOW IN NON PRISMATIC CHANNEL

A channel in which the cross section shape and size and also bottom slope are not constant, but vary is termed as non prismatic channel. All natural channel generally have varying cross section and consequently are non prismatic. The flow through non prismatic channels are generally rapidly flow in nature. In this chapter characteristic of flow under conditions in which rapidly varies flow takes place are discussed. We will discussed one by one the following phenomenon:

1. Characteristics of flow through sudden transition
   (a) Subcritical flow through sudden transmission
   (b) Contraction in supercritical flow
   (c) Expansion in super critical flow
2. Characteristics of flow through constriction
   (a) Subcritical flow through constriction
   (b) Backwater effect due to constriction
   (c) Flow through culvert
3. Characteristics of flow through obstructions:
   (a) Bridge piers
   (b) Pile trestles
   (c) Trash racks
   (d) underflow gates
4. Characteristics of flow through channel junction

7.6. Characteristics of Flow through Transition

Transitions with the change of cross sectional dimensions occurring in a relatively short distance will induce rapidly varied flow. Such transitions include sudden contractions and expansion or both.

Experimental investigations on flow through sudden transition will be discussed in the next three articles.

7.6.1 Subcritical Flow through Sudden Transitions

For subcritical flow passing through sudden transitions, experiments on various designs were made by Formica. Formica had calculated values of the energy coefficient α at different sections of the channel for various design of transitions. The values are apparently very close to unity immediately after the sudden contractions but are generally higher than unity after the sudden expansions. This indicates that flow in a sudden expansion is irregularly diffused. Further head losses for various designs of transitions at different discharge have been computed. It can be seen that in general the sudden contraction have higher losses than the sudden expansions. In sudden contraction the flow is first contracted and then expanded. A process of conversion from potential to kinetic energy immediately by a process of reconversion from kinetic to potential energy. As a result much less energy than sudden expansion is recovered. However, the energy loss in sudden the sharp edged corners of the entrance of the reduced channel. Again the energy loss in a sudden expansion can be reduced by gradually enlarging the channel section or decreasing the angle of divergence.

7.6.2 Contractions in Supercritical Flow

When supercritical flow is introduced through a contraction with symmetrical converging walls, cross waves similar to those developed in a curved channel will appear. The cross wave in contraction, however are symmetrical with respect to centre line of the channel. From an experimental and analytical investigation, Ippen & Dawson have found that straight contractions are always better than curved contractions. For the design of curved contractions, the cross wave pattern may be determined experimentally by model test.

In supercritical flow through a straight contraction a symmetrical shock waves are developed at point A and A' at the entrance. These waves extend across the channel at wave angle $\beta 1$, intersect at B on the centre line of the channel and finally after some modification reach the opposite wall at C & C'. In the regions ABC and A'BC', the flow proceeds through new field characterized by Froude number. At the end of contraction, negative disturbance are created at point D and D'. In a good contraction design it is possible to minimize the downstream disturbances. This can be done by directing the shock waves to the opposite walls at D and D'.

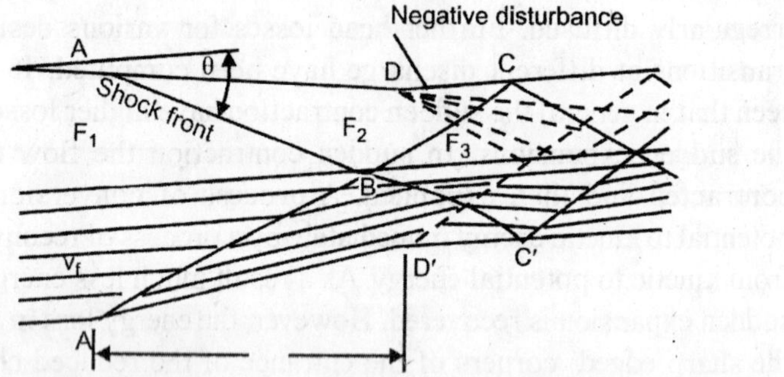

Fig. 7.1 Design of straight wall contractions

7.6.3 Expansion in Super Critical Flow

Channel expansion in supercritical flow occurs frequently at places where flow emerges at high velocity from a closed conduit, sluice gate, spillway, if such expansion is made to diverge too rapidly. The major part of flow will fail to follow the boundaries. Studied indicate that separation of flow take place as shown in figure. The separation surface shown by the dashed lines act as solid boundaries within

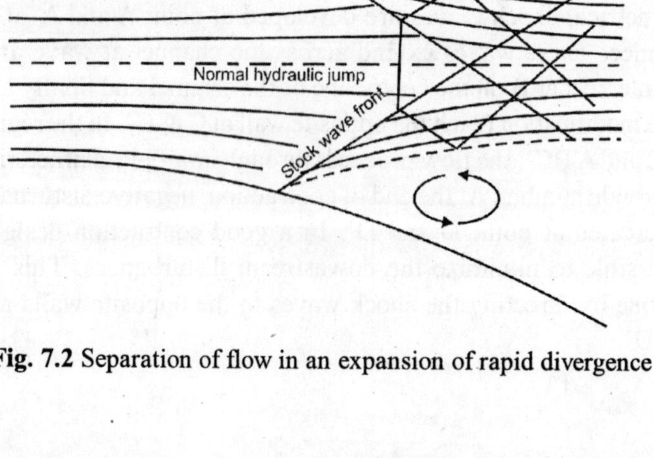

Fig. 7.2 Separation of flow in an expansion of rapid divergence

which the flow has the characteristics. Similar to those in a channel of decreasing width. If the divergence of the expansion is too gradual, waste of structural material will result. Furthermore, if local disturbances of great wave height are produced by improper boundary geometry, either at the expansion or farther downstream, the walls way fail to confine the flow. Satisfactory design for the expansion is therefore of practical importance.

7.7 Characteristics of Flow Through Constrictions

A constriction in an open channel constitutes a reach of sudden reduction in the channel cross-section. The effect of constriction on the flow depends mainly on the boundary geometry, the discharge, and state of flow. The flow through a constriction may be subcritical or supercritical when the flow is subcritical, the construction will induce a pronounced backwater effect that extend a long distance upstream (Fig. a & b). When the flow is supercritical, the constriction will disturb only the water surface that is adjacent to the upstream side of the constriction and will not extend the effect farther upstream. (Fig. c). If the upstream water surface is dammed upto a depth greater than critical depth, the surface will for in an S_1 profile, extending upstream only for a short distance and then ending with hydraulic jump (Fig. d).

7.8 Flow through Culverts

Culvert is a small bridge used for carrying water from one side to another side in the embankment of road or railway. A culvert may have one, two or three spans. Mostly culverts having one or single span are commonly used. In case of road culverts the maximum span is limited to 5m where culverts upto 6m span are used in railway.

Following four types of culverts are commonly used in practice:

a) Arch culvert

b) Slab culvert

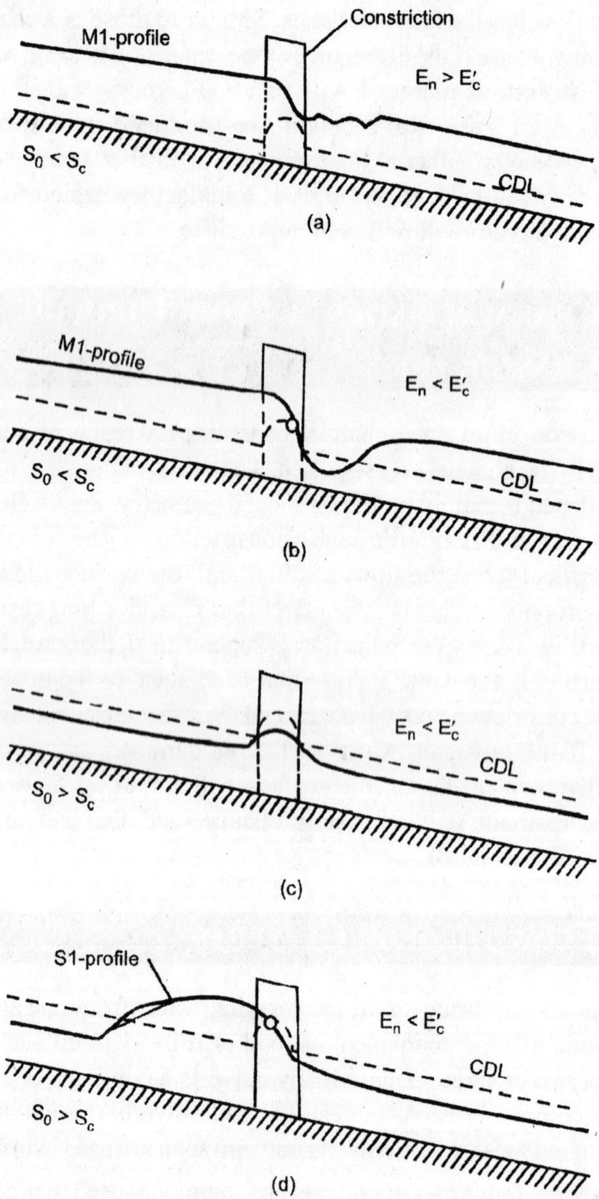

Fig. 7.3: Constriction in uniform-flow channel (a, b) in subcritical flow; (c, d) in supercritical flow.

c) Pipe culvert

d) Box culvert

e) Steel Girder culvert

a) Arch Culvert: Brick or stone masonry arch culverts were very popular in previous times, but nowadays these are very uncommon. These culverts mainly consists of foundation abutments, wing walls, arch and parapet if the bed soil is good, there is no necessity of providing floor and curtain walls in the bed. But if the soil is poor and there is likelihood of scouring, floor and curtain walls are provided. The parapet is provided on both sides.

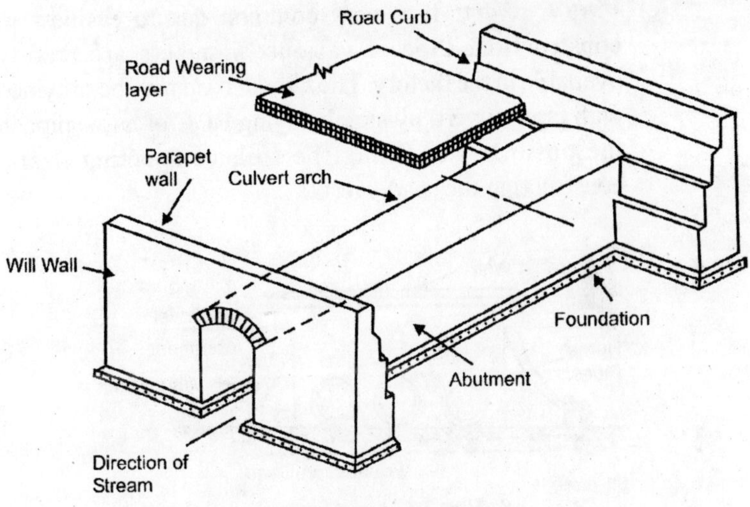

Fig.7.4 Masonry Arch Culvert

b) Slab Culvert: Masonry culvert with RCC slab are very common these days. Due to difficulty in centring, shuttering and less life and possibilities of crack, more dead load etc., the masonry arches culverts have been replaced by simple RCC slab construction.

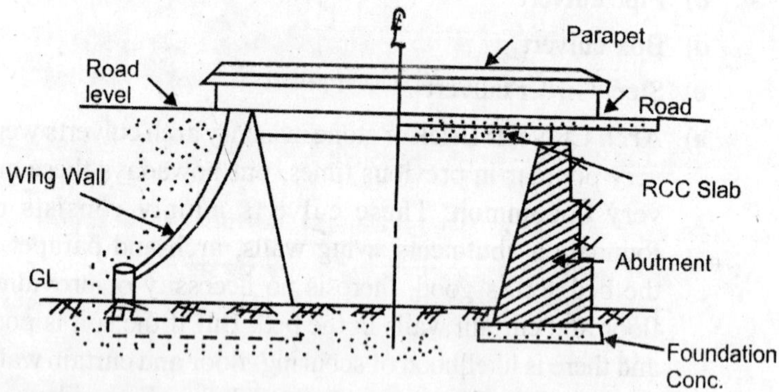

Fig. 7.5 Slab Culvert

c) **Pipe Culvert:** It is very common due to easiness in construction. Pipe of various diameters are readily available in the factory. The *kutcha* road may be provided with pipe culvert, by simply laying RCC or hume pipe in the position and filling the soil, compacting it and constructing the road over it.

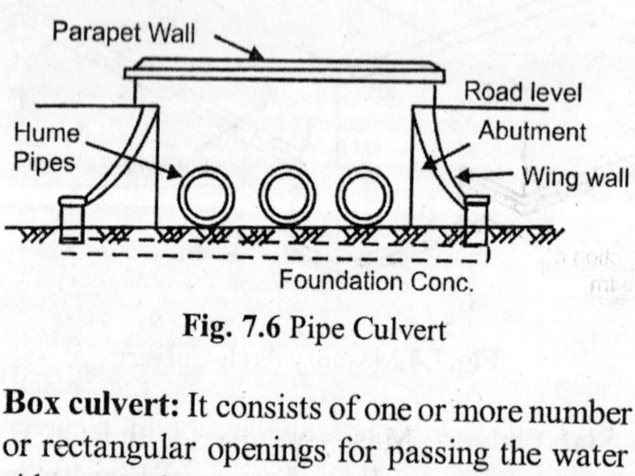

Fig. 7.6 Pipe Culvert

d) **Box culvert:** It consists of one or more number of square or rectangular openings for passing the water from one side to another. In soft soils where there is possibility of scouring and bearing capacity of soil is poor. These culverts are commonly used.

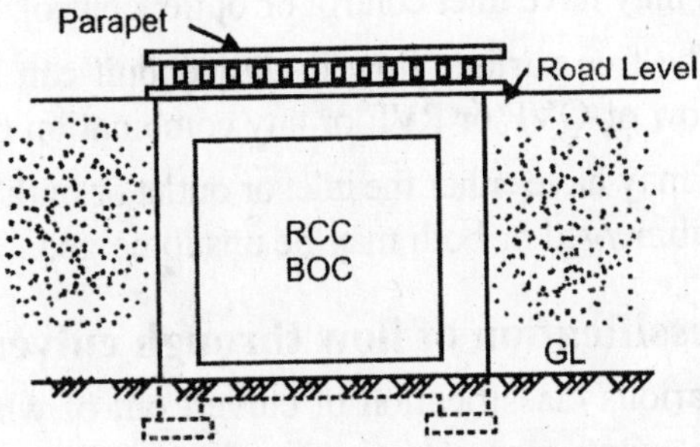

Fig. 7.7 Box Culvert

e) **Steel Girder Culvert:** This type of culvert only provided in railways. Two main girders are laid just below the rails. Wooden sleepers are provided between these girders and the rails. Therefore sometimes these culverts are also known as open deck culvert.

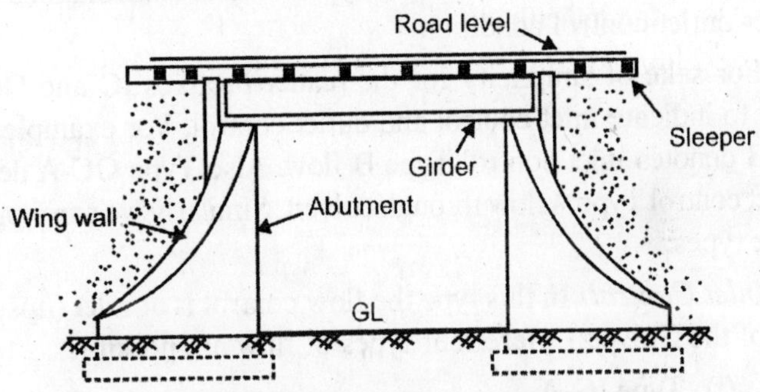

Fig. 7.8 Steel Girder Culvert

7.8.1 Characteristics of flow through Culverts

The characteristics of flow are very complicated because the flow is controlled by many variables including the inlet geometry, slope, size, roughness, approach and tailwater condition, etc.

Following are some of the important characteristics

(i) The flow through culvert can be subcritical or supercritical.

(ii) It can be closed conduit flow or an open channel flow or both.

(iii) It may have inlet control or outlet control.

(iv) The free surface flow in the conduit can have uniform flow or GVF or RVF or any combination of all.

(v) If may have either the inlet or outlet or both inlet or outlet submerged or both may be unsubmerged.

7.8.2 Classification of flow through culvert

There are various classification of culvert out of which currently most widely used classification is that of Federal Highway Administration (FHWA) of US Department of transport as given in Hydraulic Design Series-5 (HDS-5). So we will discuss HDS-5 classification as well as chow classification (1956).

7.8.2.1 Hydraulic Design Series-5 Classification:

Depending upon the control of flow at the inlet and outlet, the flow in a culvert is classified into four types at inlet control and 5 types under outlet control condition.

For sake of simplicity for the reader prefixes IC and OC are used to indicate inlet control and outlet control. For example Type 1C-B denotes inlet control Type B flow. Also Type OC-A denotes outlet control Type A flow through culvert. Similarly we can represent other types:

Inlet Control: In this type the flow control is at inlet (upstream end of the Culvert). Here four types of flow are possible.

(i) Type IC-A

(ii) Type IC-B

(iii) Type IC-C

(iv) Type IC-D

Type IC-A: In this case both the Inlet and Outlet are unsubmerged and channel is steep. The inlet acts like a weir and critical depth is formed just downstream of the Inlet edge. The free surface flow will be GVF of S_2 type in the initial portion with the normal depth occurring in the downstream part depending upon the length and other hydraulic characteristics. The tailwater level is too low and does not influence the flow at the outlet.

Type IC-B: This type of flow is similar to type IC-A but with the additional feature of the tailwater causing the submergence of the outlet. The flow at the inlet is still like a weir with critical depth at the upstream end. The S_2 curve formed will have a hydraulic jump at an appropriate location to cause full conduit pressure flow in the downstream part of the conduit.

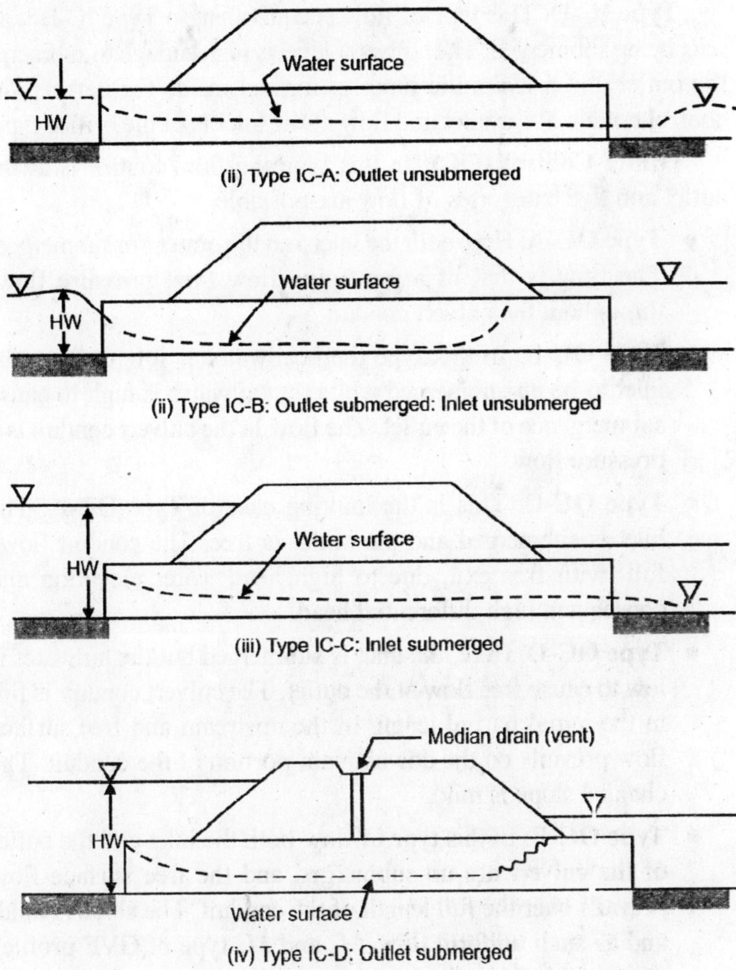

Fig. 7.9 Type of inlet control in culverts

Type IC-C: In this type the inlet is submerged with HW> 1.2d. The tailwater level is too low and free surface flow take place as in type IC-A. Critical depth at the upstream end section, S_2 profile and possibility of occurrence of normal depth at the downstream section are the characteristics features of the water surface profile in the conduit.

Type IC-D: This type of flow is equivalent to Type IC-B with inlet being submerged. The tailwater level is high enough to submerge the outlet and a hydraulic jump is formed inside the barrel. The control remain at the inlet and flow at the inlet is of the orifice type.

Outlet Control (OC): In this type the flow control is at the outlet and five categories of flow are possible.

- **Type OC**-A: Here both the inlet and the outlet are submerged. The flow is that of a pure pipe flow (viz. pressure flow) throughout the culvert conduit.
- **Type OC**-B: In this type the headwater is low causing the inlet to be un-submerged while the tailwater is high to cause submergence of the outlet. The flow in the culvert conduit is a pressure flow.
- **Type OC**-C: This is the limiting case of Type OC-A. The inlet is submerged and the outlet is free. The conduit flows full, with free exit, due to high head water elevation and consequent high differential head.
- **Type OC**-D: Here, the inlet is submerged but the tailwater is low to cause free flow at the outlet. The culvert conduit is full in the initial partial length in the upstream and free surface flow prevails on the downstream portion of the conduit. The channel slope is mild.
- **Type OC**-E: In this type of flow both the inlet and the outlet of the culvert are un-submerged and the free surface flow prevails over the full length of the conduit. The slope is mild, and as such uniform flow, M_2 and M_1 type of GVF profiles are possible depending upon the hydraulic properties and the tailwater elevation.

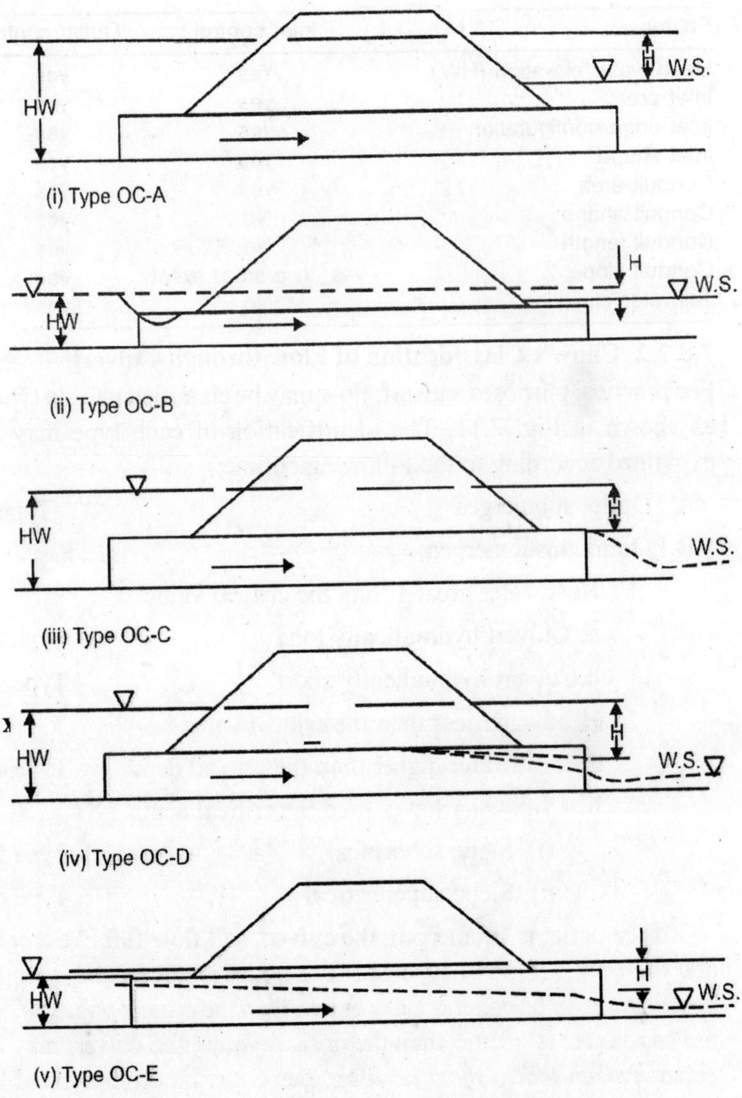

Fig. 7.10 Type of outlet control in culvert

Table 7.1 Factors affecting Culvert Flow

Factor	Inlet control	Outlet control
Head water elevation (HW)	Yes	yes
Inlet area	yes	yes
Inlet edge configuration	yes	yes
Inlet shape	yes	yes
Conduit area	yes	yes
Conduit shape	No	yes
Conduit length	No	yes
Conduit slope	yes, to a small exent	yes
Tailwater elevation	No	yes

7.8.2.2 Chow's Classification of Flow through Culvert

For practical purposes, culvert, flow may be classified into six types, as shown in Fig. 7.11. The identification of each type may be explained according to the following outline:

A. Outlet submerged — Type 1

B. Outlet unsubmerged

1. Headwater greater than the critical value
   - a. Culvert hydraulically long — Type 2
   - b. Culvert hydraulically short — Type 3
2. Headwater less than the critical value
   - a. Tailwater higher than the critical depth — Type 4
   - b. Tailwater lower than the critical depth
     - i) Slope subcritical — Type 5
     - ii) Slope supercritical — Type 6

If the outlet is submerged, the culvert will flow full like a pipe, and the flow will be of type 1. If the outlet is not submerged, the headwater may be either greater or less than the critical value. When the headwater is greater than the critical value, the culvert may be either *hydraulically short or long*; these can be differentiated by means of the charts in Figs. 7.12 and 7.13. The flow is of type 2 if the culvert is hydraulically long and of type 3 if it is hydraulically short. When the headwater is less than the critical value, the tailwater may be either higher or lower than the critical depth of the flow at the culvert outlet. For higher tailwater, the flow is of type 4. For lower tailwater, the flow is of type 5 if the culvert slope is subcritical and of type 6 if the slope is supercritical.

In the above classification, there is an exception in that type 1 flow can occur with tailwater slightly higher than the critical depth or with tailwater higher than the top of the outlet if the bed slope is very steep. The first two types of flow are pipe flow, and the other types are open channel flow. For type 3 flow, the culvert acts like an orifice. The coefficient of discharge varies approximately from 0.45 to 0.75. For type 4, 5 and 6 flows, the entrance is not sealed by water and it acts like a weir. The discharge coefficient varies approximately from 0.75 to 0.95, depending on the entrance geometry and headwater condition. As shown in Fig. 7.11, type 4 flow is subcritical throughout the barred length. Type 5 flow is subcritical and, hence, the control section is at the outlet type 6 flow is supercritical and hence, the control section is at the entrance.

Hydraulically short or long culvert: Culvert usually with a square edge at the top of the entrance, will not flow full even if the entrance is below headwater level when the outlet is not submerged under these conditions, the flow entering the culvert will contract to a depth less than height of the culvert barrel in a manner very similar to the contraction of flow in the form of jet under a sluice gate. This high velocity jet will continue through the barrel length, but its velocity will be reduced slowly as head is gradually lost by friction. If the culvert is not sufficiently long to allow the expanding depth of flow below the contraction to rise and fill the barrel, the culvert will never flow full. Such a culvert is considered hydraulically short. Otherwise, the culvert is hydraulically long, for it will flow full like a pipe.

Whether a culvert hydraulically long or short cannot be determined by length of barrel along. It depends on other characteristics such as slope, size, entrance geometry, headwater, entrance and outlet condition, etc.

7.8.3 Design of Culvert

The design of a culvert is influenced by cost, hydraulic efficiency, purpose, and the topography at the proposed culvert site. Thus

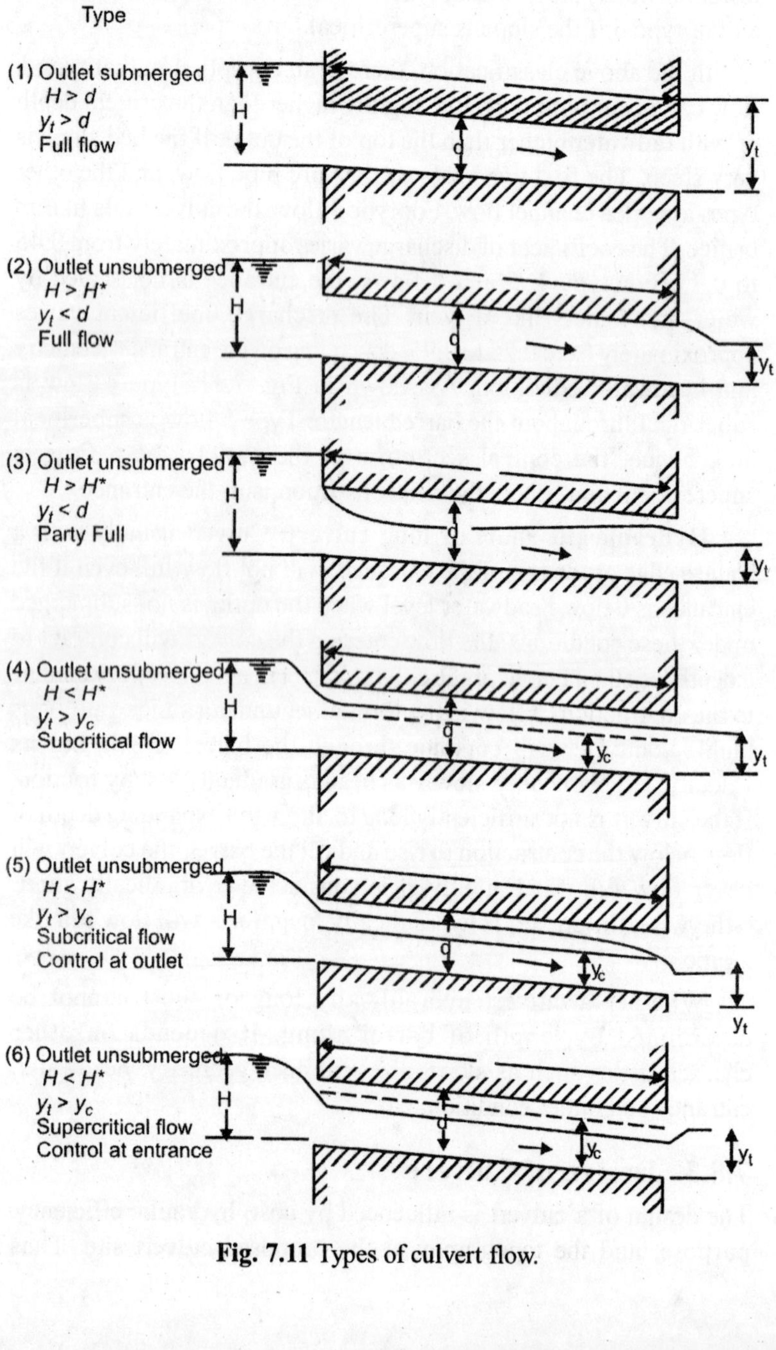

Fig. 7.11 Types of culvert flow.

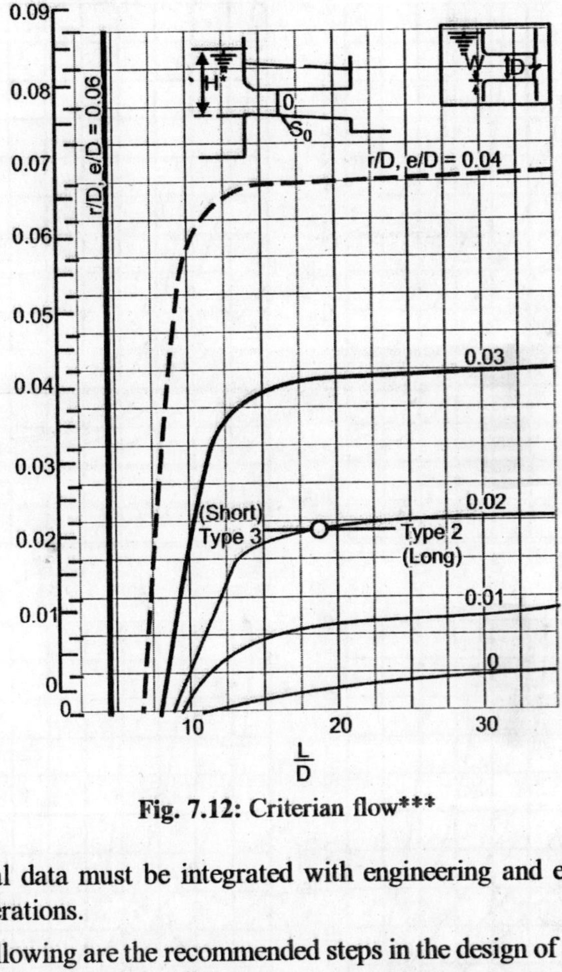

Fig. 7.12: Criterian flow***

physical data must be integrated with engineering and economic considerations.

Following are the recommended steps in the design of a culvert in order to ensure that all design aspects are taken into account.

Step 1: Determine and Analyze Site Characteristics: Site characteristics include the generalized shape of the roadway embankment, bottom elevations and cross sections along the stream bed, the approximate length of the culvert, and the allowable headwater elevation. In determining the allowable headwater elevation, roadway elevations and the elevation of upstream property should be considered. The consequences of exceeding the allowable headwater elevation should bee

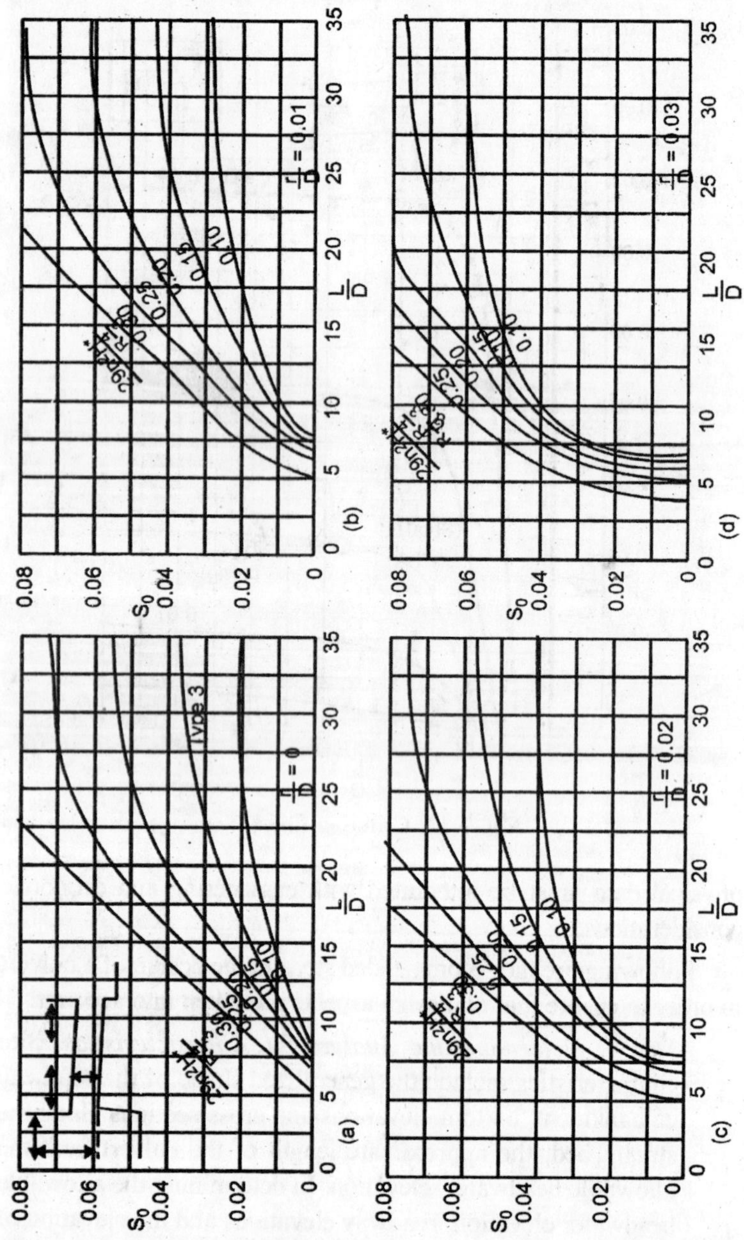

Fig. 7.13: Criteria for hydraulically short and long culverts with rough barrels of corrugated pipes. (U.S. Geological survey)

evaluated and kept in mind throughout the design process. Culvert design is actually a trial-and-error procedure because the length of the barrel cannot be accurately determined until the size is known, and the size cannot be precisely determined until the length is known. In most cases, however, a reasonable estimate of length will be accurate enough to determine the culvert size.

Step 2: Perform Hydrologic Analysis: Delineate the drainage area above the culvert site. Develop flow estimates for the design frequencies. The probable accuracy of the estimate should be kept in mind as the design proceeds.

Step 3: Perform Outlet Control Calculations and Select Culvert: These calculations are performed before inlet control calculations in order to select the smallest feasible barrel which can be used without the required headwater elevation in outlet control exceeding the allowable headwater elevation. The full flow outlet control performance curve for a given culvert (size, inlet edge, shape, material) defines its maximum performance. Therefore, the inlet improvements beyond the beveled edge or changes in inlet invert elevation will not reduce the required outlet control headwater elevation. This makes the outlet control performance curve an ideal limit for improved inlet design. The results of these calculations should be the outlet control performance curve. In addition to considering the allowable headwater elevation, the velocity of flow at the exit to the culvert should be checked to determine if downstream erosion problems will be created.

Step 4: Perform Inlet Control Calculations for Conventional and Beveled Edge Culvert Inlets: Perform the inlet control calculations to develop the inlet control performance curve to determine if the culvert design selected will be on inlet or outlet control for the design and check flood frequencies. A drop may be incorporated upstream of the culvert to increase the flow through the culvert.

Step 5: Perform Throat Control Calculations for Side: And Slope-Tapered Inlet: The same concepts are involved here as with conventional or beveled edge culvert design.

Step 6: Analyze the Effect of a Drop on Inlet Control Section Performance: The purpose of this step is to determine if having a drop before the inlet of the culvert would increase he capacity of the culvert and if a drop can be justified from a cost perspective and site characteristics.

Step 7: Design side-and/or Slope Tapered Inlet: Side and slope tapered inlets can be used to significantly increase the capacity of many culvert designs. Develop performance curves based on side- and/r slope tapered inlets and determine from a cost perspective and site characteristics if such a design would be justified.

Step 8: Complete File Documentation: Complete a documentation file for the final design selected.

7.8.3.1 Engineering Design Criteria

The design of a culvert should take into account many different engineering and technical aspects at the culvert site and adjacent areas. The following design criteria should be considered for all culvert designs as applicable

Engineering aspect

- Flood frequency
- Velocity limitation
- Buoyancy protection

Site criteria

- Length and slope
- Debris control

Design limitations

- Headwater
- Tailwater conditions
- Storage

Design options

- Culvert inlets
- Inlets with headwalls
- Wingwalls and aprons
- Improved inlets
- Material selection
- Culvert skews
- Culvert sizes

Related designs

- Weep holes
- Outlet protection
- Erosion and sediment control
- Environmental considerations
- Safety considerations
- Loading requirements

Some culvert designs are relatively simple, involving a straight-forward determination of culvert size and length. Other designs are more complex where structural, hydraulic, environmental, or other considerations must be evaluated and provided for in the final design. The design engineering must incorporate personal experience and judgment to determine which criteria must be evaluated and how to design the final culvert installation.

7.9 CHARACTERISTICS OF FLOW THROUGH OBSTRUCTIONS

An obstruction in open channel flow presents a phenomenon very similar to that of a constriction, since both have the effect of contracting the cross-sectional area of the flow. However, the constriction reduces the cross-section to single opening, whereas the obstruction creates at least two openings. The degree of contraction offered by a constriction is usually higher than that by an obstruction. The types of obstruction commonly encountered in engineering problems include, bridge piers, piles trestles, trash racks, piers and abutments on top of overflow spillway, etc.

The flow through an obstruction may be subcritical or supercritical. Now we will discuss various examples of obstructions.

7.9.1 Flow between Bridge Piers

For flow between bridge piers, Yarnell's has made extensive study. The following points are extracted from the results of Yarnell's investigation on bridge piers of variouos shapes and sizes.

1. The height of the backwater due to bridge pier varies directly will depth of the unobstructed channel.
2. For flow of low velocity and least turbulence, the more efficient shapes are lens-shaped nose and tail, lens-shaped nose and semicircular tail, semicircular nose and lens-shaped tail, convex nose and tail, and semicircular nose and tail. However , the data are not sufficient to differentiate among these shapes for high degree of contraction.
3. Twin cylinder piers either with or without connecting diaphragms, piers with 90° triangular noses and tails, and piers with recessed webs are less efficient hydraulically than those just mentioned, and piers with square noses and tails are least efficient.
4. Application of better to the ends of piers slightly increases their hydraulic efficiency, that is, raises the K_N and K_A values.
5. Increasing the length of a pier from 4 times the width to 13 times the width has comparatively little effect on its hydraulic efficiency. In some cases the efficiency is thus increased and in other cases decreased. The optimum length-width ratio probably varies with velocity and is generally between 4 and 7. On the average, the values of K_N and K_A will increase about 3 to 5% for the increase of the ratio from 4 to 13.
6. Placing the piers at an angle with the current has an insignificant effect on the amount of backwater if the angle is les than 10°. Placing the piers at 20° or more materially increases the amount of backwater; the increase depends upon

the quantity of flow, the depth, and the degree of channel contraction. In general, the values of K_N and K_A will decrease about 7% at 20° (where K_N and K_A are the coefficient depending on the degree of channel contraction and characteristic of obstruction).

7.9.2 Pile Trestles

The erection of a pile trestle or a bridge pier in a stream forces the water to flow through a reduced cross section, and in passing this section and water must acquire a velocity greater than that existing in the unobstructed channel. The increase in velocity can be produced only by elevating the water surface up stream from the trestle where the contracted area of flow exists. Thus as the stream enters the contracted area, a drop in the water surface is noted accompanying he increase in velocity.

The changes in cross section and velocity in passing the trestle piling cause much disturbance in the flow of water. Eddies may be formed along the pilling. The high velocities and resultant eddies may scour out the bed of the stream next to the pile bents to such an extent that the trestle itself may be endangered and even swept away. Differences of opinion often occur as to the amount of obstruction to flow caused by highway or railway pile trestle bridges and lawsuits may result over the amount of damages. The amount of obstruction a pile-trestle bent offers to the flow of water may easily be expressed in the form of a trestle-bent coefficient. The trestle-bent coefficient varies with the quantity of flow. For a given height of backwater, depth of flow, and channel contraction, if the trestle coefficient is increased 5 per cent through an improved setting of the bents, the flow capacity of the trestle is increased 5 per cent. The trestle-bent coefficient is, in reality, an index number of the hydraulic efficiency of a pile-trestle bent. The discharge through pile-trestle opening during floods may be computed with a fair degree of accuracy by means of a backwater formula if the drop-down at the trestle opening is known. The converse also is true; the discharge being known, the drop-down or backwater caused by the trestle may be determined.

7.9.3 Trash racks

Trash racks are a critical element of outlet structure design and serve several important functions. When properly designed, installed, and maintained, trash racks:

a) Keep debris away from the entrance to the outlet works where the debris will not clog the critical portions of the structure
b) Capture debris in such a way that relatively easy removal is possible;
c) Ensure that people and large animals are kept out of confined conveyance and outlet areas.
d) Provided a safety system that prevents humans and animals from being drawn into the outlet and allows them to climb to safety.

The surface area of all trash rack should be maximized and the trash racks should be located a suitable distance from the protected outlet to avoid interference with the hydraulic capacity of the outlet.

The spacing of trash rack bars must be proportioned to the size of the smallest outlet protected. However, where a small orifice is provided, a separate trash rack for that outlet should be used, so that a simpler, sturdier trash rack with more widely spaced members can be used for the other outlets. Spacing of the rack bars should be wide enough to avoid interference, but close enough to provide the level of clogging protection required.

To facilitate removal of accumulated debris and sediment from around the outlet structure, the racks should have hinged connections. If the rack is bolted or set in concrete it will preclude removal of accumulated material and will eventually adversely affect the outlet hydraulics. Since sediment will tend to accumulate around the lowest stage outlet, the inside of the outlet structure for a dry basin should be depressed below the ground level to minimize clogging due to sedimentation. Depressing the outlet bottom to a depth below the ground surface at least equal to the diameter of the outlet is recommended.

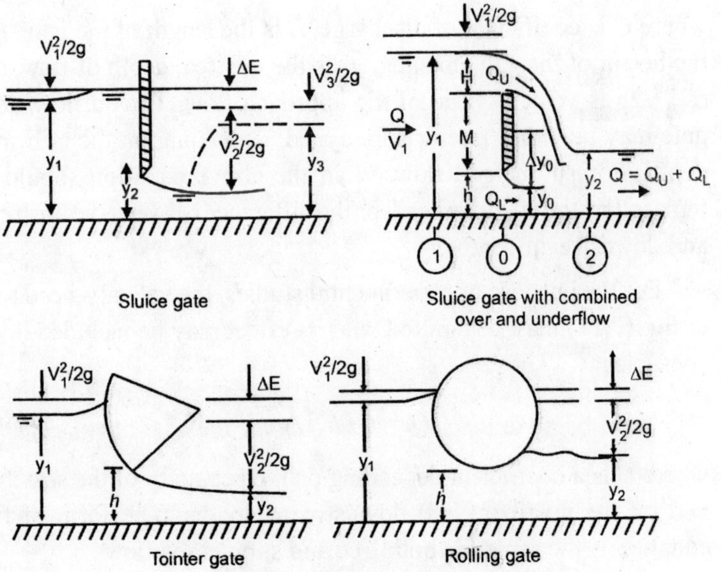

Fig. 7.14 Underflow gates

7.9.4 Under Flow Gates

Under flow gates are the structures from which the water passes underneath the structures. Common examples are the sluice gate, radial gate and rolling gate. In designing such gates the hydraulic engineers is most interested in two measure features; the head-discharge relationship and pressure distribution over the gate surfaces for various positions of the gate and forms of the gate lip. The form of the lip will not only effect the velocity and pressure distribution and energy loss in flow through gate opening, but may also develop very disturbing vibrations that should be avoided during gate operation. As the design of lip varies considerably, independent investigation of the lip for a particular design is usually necessary.

By the energy equation, it can be shown that the discharge through an underflow gate may be expressed as

$$Q = CLh\sqrt{2g\left(y_1 + \alpha\frac{V_1^2}{2g}\right)} \qquad (7.11)$$

where C is coefficient of discharge, L is the length of the gate, h is the height of the gate opening, y_1 is the upstrea, depth of flow, and $\alpha V_1^2/2g$ is velocity head of the approach flow. The outflow of he gate may be either free or submerged, depending on the tailwaver depth. For submerged flow, y_1 in the above equation should be replaced by the effective head, or the difference between the upstream and downstream depths.

For the purpose of experimental studies, the velocity-head term in Eq. (7.11) may be omitted, and its effect may be included in the coefficient C. Thus,

$$Q = CLh\sqrt{2gy_1} \qquad (7.12)$$

where C is a coefficient depending on the geometry of the structure and on the upstream and downstream depths. The form of this equation is the same for both free and submerged flows.

7.10 Characteristics of Flow through Channel Junction

Flow throw a channel junction is a phenomenon that involves numerous variables, such as the number of adjoining channels, the directions and discharges of flow, the rounding of the corner at the junction, etc.

For subcritical flow passing through the channels at a junction Taylor has investigated the specific cases as shown in figure. The channels are horizontal and of equal width.

In case of combining flow, the following assumptions were made by Taylor:

(1) The flow is from channel 1 and 2 into channel 3.

(2) Channels 1 and 3 lie in a straight line.

(3) The flow is parallel to channel walls and velocity is uniformly distributed immediately above and below the junction.

(4) Ordinary wall friction is negligible in comparison with other forces induced.

(5) The depth in channels 1 and 2 are equal immediately above the function.

The experimental data showed clearly that assumption 5 is valid regardless of the angle of intersection of the channels.

Dividing flow cannot be easily analyzed theoritically. The application of momentum principle is difficult because it involves some unknown quantities, such as assumption 5 for combining flow are not available. Basically the division of flow will depend upon the backwater effects of the two branch channels and the dynamic conditions existing at the junction

For supercritical flow passing through the channel junction, the flow condition is more complicated studies of few cases of combining flow have been made when the principle of momentum is applied to such a problem, it is necessary to know the position of jump or depth of flow at the upstream edge of the junction if the flow passing through the junction is supercritical without formation of any hydraulic jump upstream, the flow condition will be complicated by the development of crosswaves. These waves will propagate for a considerable distance downstream before being damped by the channel friction. Hence, side walls higher than those normally provided are required not only near the junction but also for a considerable distance downstream.

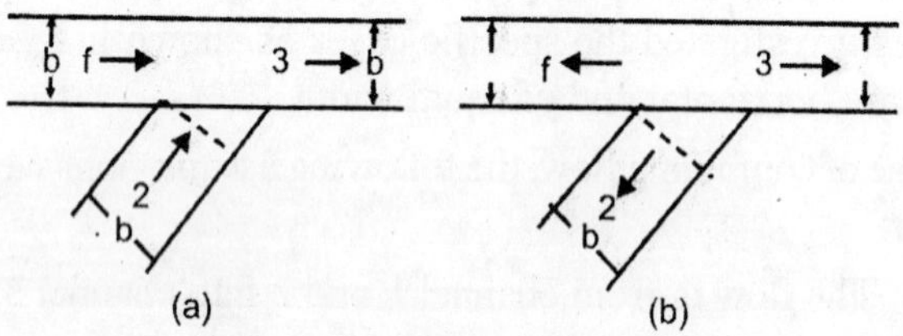

Fig. 7.15 Simple channel junction (a) combining flow; (b) dividing flow

Review Questions

1. How you will determine that the culvert is hydraulically short or long? Give the neat sketches of culvert with outlet un-sumberged condition.
2. Design the design consideration for determining the shape of cross-section at an open channel bend i.e., a channel of non-linear alignment under the supercritical flow condition.
3. Write short notes on the following:
   (*a*) Characteristics of flow through channel function
   (*b*) Characteristics of flow through sudden transitions in subcritical flow.
   (*c*) Flow through an obstruction in open channel flow.
4. Discuss the design procedure for culvert.
5. Give examples of channels of non-linear alignment and discuss design considerations under supercritical flows. What are non prismatic channels.

6. Write short notes on the following:
   (*a*) Spiral flow
   (*b*) Superelevation
   (*c*) Cross waves
   (*d*) Energy loss in bends
7. Discuss design consideration for
   (*a*) Subcritical flow
   (*b*) Supercritical flow.

Objective Type Questions

1. A triangular channel with sides at 60° to the horizontal flows at its normal depth of 0.8 m. carrying a discharge of 0.5 m^3/s. The bed slope of the channel is

   (a) adverse (b) critical

   (c) mild (d) steep

2. In a triangular channel which has equal side slope of 1 vertical to 1 horizontal the ratio between critical depth y_c to specific energy E, is

   (a) 4/5 (b) 5/4

   (c) 1/2 (d) 2/3

3. The sequent depth ratio in a hydraulic jump for med in horizontal rectangular channel is 16.48. The Froude number of super critical stream is

   (a) 4 (b) 8

   (c) 12 (d) 120

4. In a hydraulic jump occurring in a rectangular horizontal channel the depth before and after jump are 0.5 m and 1 m respectively then head loss due to jump will be

   (a) ½ m b) 1/8 m

   (c) 1/16 m d) 1/32 m

5. In a rectangular channel, the value of discharge intensity q is $\sqrt{g}$ m^3/s/m then critical depth of flow will be

(a) 1.0 (b) 2

(c) 0.5 (d) 1.5

6. The value of friction factor $f = 0.02$ then the value of chezy constant will be

   (a) 20 g (b) $20\sqrt{g}$

   (c) $\sqrt{80g}$ (d) $\sqrt{160g}$

7. In a hydraulic efficient triangular channel the depth of flow is $5\sqrt{2}$ then hydraulic radius will be

   (a) 2.5 m (b) 2.0 m

   (c) 1.5 m (d) 5 m

8. In a rectangular channel the Froude's numbers corresponding to alternative depths are 0.5 and 4.0 respectively. The ratio of corresponding alternative depths is

   (a) 1.0 (b) 0.52

   (c) 0.37 (d) 0.46

9. The kinetic energy correction factor is

   (a) applied to continuity equation

   (b) expressed as $\frac{1}{A}\int_A \left(\frac{v}{V}\right)^2 dA$

   (c) expressed as $\frac{1}{A}\int_A \left(\frac{v}{V}\right)^3 dA$

   (d) expressed as $\frac{1}{A}\int_A \left(\frac{v}{V}\right) dA$

10. The momentum correction factor for the velocity distribution shown in figure is

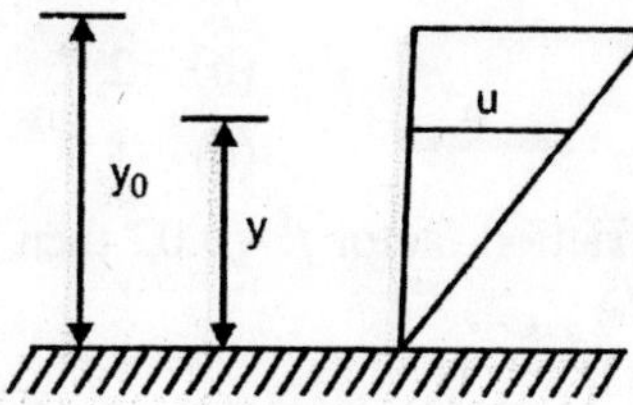

Fig.

(a) 1/3 (b) 1
(c) 4/3 (d) 2

11. Equation of continuity is based on the principle of conservation of
(a) mass (b) energy
(c) momentum (d) none

12. Which of the following has highest coefficient of discharge
(a) sharp edged orifice
(b) venturimeter
(c) Borda Mouthpiece running full
(d) Cipolettti weir

13. In a Sutro weir, the discharge is proportional to
(a) $H^{1/2}$ (b) $H^{3/2}$
(c) $H^{5/2}$ (d) H
Where H is head.

14. The discharge over broad crested weir is maximum when the depth of flow is
(a) $H/3$ (b) $H/2$
(c) $2H/5$ (d) 2H/3

15. The discharge through a Cipolletti weir is given by

(a) $\frac{2}{3}C_d\sqrt{2g}LH^{3/2}$ (b) $\frac{8}{15}C_d\sqrt{2g}\frac{\tan\theta}{2}H^5$

(c) $\frac{2}{3}C_d\sqrt{2g}(L-0.2H)H^{3/2}$ (d) $\frac{2}{3}C_d\sqrt{2g}LH^{3/2}$

Where symbols have their usual meaning.

16. In gradually varied flow, if $\frac{dy}{dx}$ is positive, then $\frac{dE}{dx}$ will be

(a) always zero (b) positive if $y > y_c$

(c) negative if $y > y_c$ (d) always negative

17. Which of the following is the correct representation of the sequence of profiles if the channel slope changes from mild to steep?

(a) M_1, S_1 (b) M_3, S_2

(c) M_2, S_3 (d) M_2, S_2

18. Which of the following equation are used for the derivation of the differential equation for water surface profile in open channel flow?

1 Continuity Equation 2. Energy Equation

3. Momentum equation

Select the correct answer using the code given below:

(a) 1, 2 and 3 (b) 1 and 3

(c) 1 and 2 (d) 2 and 3

19. S_2 profile can occur at

(a) a break in slope from mild to steep

(b) a break in slope from steep to mild

(c) the downstream of a sluice gate on steep slope

(d) a sudden drop in bed in steep slope

20. When a channel bottom slope changes abruptly from mild to steep the resulting surface profile is

(a) M_1 changing to S_1 (b) M_3 changing to S_3

(c) M_2 changing to S_2 (d) M_2 changing to S_1

21. The flow in channels is considered to be in transitional state if the reynold's number is

(a) less than 500 (b) between 500 and 2000

(c) between 2000 and 4000 (d) greater than 4000

22. For shooting flow the Froude number is

(a) zero (b) less than one

(c) one (d) greater than one

23. The Chezy's coefficient

(a) is dimensionless

(b) has the dimension of velocity

(c) has the dimension of discharge

(d) has dimension $L^{1/2} T^{-1}$

24. If f is the friction factor, then the Chezy's coefficient is proportional to

(a) f (b) $\sqrt{f}$

(c) $\frac{1}{f}$ (d) $\frac{1}{\sqrt{f}}$

25. The relationship between Manning's coefficient 'n' and Chezy's coefficient 'C; is given by

(a) $C = \frac{R^{2/3}}{n}$ (b) $C = \frac{R^{1/6}}{n}$

(c) $C = \frac{R^{1/3}}{n}$ (d) $C = \frac{R^{1/4}}{n}$

26. In case of S_2 water surface profile (y = depth of flow, y_0 = normal deptt and y_c = critical depth)

(a) $y_0 > y_c > y$ (b) $y_c > y_0 > y$

(c) $y_c > y > y_0$

27. A 3 m wide rectangular channel flowing at its normal depth of 0.8 carries a discharge of 9.5 m^3/s. The bed slope of channel is

(a) steep (b) critical

(c) mild (d) adverse

28. If the froude number characterizing flow in an open channel is less than unity an increase in channel width causes the water surface elevation to

(a) form ripples (b) remain same

(c) decrease (d) increase

29. While conducting a flow measurement using rectangular notch an error of 2% in head over the notch and error of -3% in length of notch occurred, then the % error in discharge will be

(a) no change (b) 3%

(c) 6% (d) 5%

30. The Froude number for a hydraulic jump F = 5.0 then jump formed is called

(a) Oscillating (b) Undular

(c) Strong (d) Steady

31. For a trapezoidal channel section to be most economical, its hydraulic radius must be equal to

(a) $y/4$ (b) $y/2$

(c) $\dfrac{y}{2\sqrt{2}}$ (d) $\dfrac{y}{3\sqrt{2}}$

32. The critical depth of flow in a most economical triangular channel section for a discharge of 1 m³/s is given by

(a) $\left(\dfrac{1}{9.8}\right)^{1/5} m$ (b) $\left(\dfrac{1}{9.8}\right)^{1/3} m$

(c) $\left(\dfrac{1}{4.9}\right)^{1/5}$ m (d) $\left(\dfrac{1}{4.9}\right)^{1/3}$ m

33. For a given specific energy E, the critical depth y_c for a rectangular channel is given by

(a) $y_c = \dfrac{3}{2}\text{E}$ (b) $y_c = \dfrac{2}{3}\text{E}$

(c) $y_c = \frac{4}{5}E$ (d) $y_c = \frac{3}{4}E$

34. For the same specific forces, the two depths at which a given discharge can occur are called

(a) Alternate depths (b) Normal depths

(c) Critical depths (d) Conjugate depth

35. When the slope of bottom of a channel rises in the direction of flow, it is called

(a) Critical slope (b) Mild slope

(c) Steep slope (d) Adverse slope

36. The hydraulic jump occurs in a rectangular channel and the depths of flow before and after the jumps are 0.5 and 2 m then the critical depth of flow will be

(a) 1.077 m (b) 1.0 m

(c) 1.25 m (d) 0.25 m

37. The Froude number of flow in an open channel is more than 1 then the flow is said to be

(a) Critical (b) Shooting

(c) Streaming (d) Transitional

38. Which one of the following condition is a typical characteristics of critical flow?

(a) $\frac{Q^2T}{gA^3}=1$ (b) $\frac{QT^2}{gA^3}=1$

(c) $\frac{Q^2}{gA^2}=1$ (d) $\frac{Q^2T^2}{gA^2}=1$

39. The critical velocity for a flow of q m³/sec/m of a wide rectangular channel is given by

(a) $\left(\frac{q^2}{g}\right)^{1/3}$ (b) $(qg)^{1/3}$

(c) $\sqrt{qg}$ (d) none

40. For a steep slope profile S_1, the type of flow will be
    (a) Supercritical (b) Critical
    (c) Subcritical (d) None

41. Which of the following Froude number range indicates a weak jump
    (a) 1 to 1.7 (b) 1.7 to 2.5
    (c) 2.5 to 4.5 (d) 4.5 to 9.0

42. The speed of an elementary wave in still liquid at depth y is given by
    (a) $\sqrt{\frac{gy}{3}}$ (b) $\sqrt{\frac{gy}{2}}$
    (c) $\sqrt{gy}$ (d) $\sqrt{2gy}$

43. The height of hydraulic jump is equal to
    (a) initial depth
    (b) sequent depth
    (c) difference in alternating depth
    (d) difference in conjugate depth

44. In a Sutro weir, the rate of flow for all flows above the rectangular base of width w and depth 'a' is proportional to the head.
    (a) Above the crest
    (b) above the rectangular box
    (c) above a datum $a/3$ above the crest
    (d) $\frac{2a}{3}$ above the crest

45. At a section in a channel expansion, the velocity over a one quarter of cross-section is zero and is uniform over the remaining three fourth of the area the kinetic energy correction factor is

(a) 1.78 (b) 1.33
(c) 1.67 (d) 2

46. Piezometric head is the sum of
    (a) pressure head, datum head and velocity head
    (b) datum head and velocity head
    (c) pressure head and velocity head
    (d) pressure head and datum head

47. The difference between total head line and Piezometric head line represents:
    (a) Velocity head
    (b) Pressure head
    (c) Elevation of the bed of the channel
    (d) depth of flow

48. The specific force is constant
    (a) In all frictional channel irrespective of the magnitude of the longitudinal slope.
    (b) In horizontal, frictionless channels of any shape
    (c) In all horizontal channels of any shape
    (d) in any open channel

49. The two alternative depths in a 4 m wide rectangular channel are 3.86 m and 1.0 m respectively. The discharge in the channel in m^3/s is
    (a) 15 (b) 1.5
    (c) 7.76 (d) 31.0

50. In rectangular channel, the alternative depths are 6.0 m and 2.0 m respectively. The specific energy is
    (a) 3.38 (b) 1.33
    (c) 2.33 (d) 3.0

51. A rectangular channel carries a certain flow for which the alternative depths are found to be 3 m and 1 m. The critical depth in m will be

(a) 2.65 (b) 1.65

(c) 0.65 (d) 0.33

52. For triangular channel of side slope 1:1, the Froude number is given by

(a) $\frac{m}{\sqrt{gy}}$ (b) $\frac{v}{\sqrt{2gy}}$

(c) $\frac{v\sqrt{2}}{\sqrt{gy}}$ (d) $\frac{v}{\sqrt{gy}}$

53. A triangular channel has a vertex angle of 90° and carries a discharge of 1.9 m^3/s at a depth of 0.8. The Froude number of the flow is

(a) 0.68 (b) 1.06

(c) 0.75 (d) 1.5

54. A triangular channel of apex channel of 120° carries a discharge of 1573 l/s. The critical depth in m is

(a) 0.6 (b) 0.7

(c) 0.8 (d) 0.632

55. A triangular channel of apex angle 60° has a critical depth of 0.25 m. The discharge in l/s is

(a) 60 (b) 640

(c) 160 (d) 40

56. At critical depth

(a) The discharge is minimum for given specific energy

(b) The discharge is maximum for given specific force

(c) The discharge is maximum for given specific energy

(d) The discharge is minimum for given specific force

57. If F_0 = Froude number of flow in rectangular channel at a depth of flow = y_0. Then y_c/y_0 =

(a) $\sqrt{F_0}$ (b) $F_0^{2/3}$

(c) $F_0^{1/3}$ (d) $F_0^{3/2}$

58. Supercritical flow at Froude number of $F_0 = 2$ occurs at a depth of 0.63 m in a rectangular channel. The critical depth in m is

(a) 0.857 (b) 0.735
(c) 1.00 (d) 0.5

59. The Froude number of flow in a rectangular channel is 0.73. If the depth of flow is 1.5 m the specific energy in meter is

(a) 1.9 (b) 1.5
(c) 1.73 (d) 0.73

60. For Triangular channel, the first hydraulic exponent M is

(a) 2.0 (b) 3.0
(c) 5.0 (d) 5.33

61. The conveyance of a triangular channel will side slope 1 horizontal : 1 vertical is expressed as $k = cy^{8/3}$ where c equal to

(a) $2^{8/3}$ (b) $1/n$
(c) $1/2n$ (d) $\frac{2\sqrt{2}}{n}$

62. In a wide rectangular channel, if the normal depth is increased by 20%, the discharge would increased by:

(a) 20% (b) 15.5%
(c) 35.5% (d) 41.3%

63. In circular channel the ratio of maximum discharge to the pipe full discharge is about

(a) 1.5 (b) 0.94
(c) 1.08 (d) 1.00

64. For hydraulically efficient rectangular channel, B/y_0 is equal to

(a) 1.0 (b) 2.0
(c) 0.5 (d) $1/\sqrt{3}$

65. A triangular channel is hydraulically efficient when vertex angle θ is
    (a) 90° (b) 120°
    (c) 60° (d) 30°
66. For hydraulically efficient Triangular channel with depth of flow y the hydraulic radius R is equal
    (a) $2\sqrt{2}y$ (b) $y/2$
    (c) $\sqrt{2}\,y$ (d) $\frac{y}{2\sqrt{2}}$
67. In hydraulically efficient trapezoidal channel section, the ratio of bed width to depth is
    (a) 1.155 (b) 0.867
    (c) 0.707 (d) 0.5
68. The following types of GVF profiles do not exist:
    (a) C_2, H_2, A_1 (b) A_2, H_1, C_2
    (c) H_1 A_1, C_2 (d) C_1, A_1, H_1
69. The total number of possible types of GVF profiles are
    (a) 9 (b) 11
    (c) 12 (d) 15
70. In a channel the value of dE/dx equal to
    (a) $S_0 - S_f$ (b) $S_f - S_0$
    (c) $S_0 - S_f - dy/dx$ (d) $S_0 (1 - F^2)$
71. In M_3 profile, the inequality between various depth are
    (a) $y_0 > y_c > y$ (b) $y > y_0 > y_c$
    (c) $y_c > y_0 > y$ (d) $y > y_c > y_0$
72. In GVF profile as depth $y \to y_c$
    (a) $\frac{dy}{dx} \to 0$ (b) $\frac{dy}{dx} \to \infty$
    (c) $\frac{dy}{dx} \to S_0$ (d) $\frac{dy}{dx} \to a$ finite value

73. $\frac{dy}{dx}$ is negative in the following GVF profiles:

(a) M_1, S_2, A_2 (b) M_2, A_2, S_3

(c) A_3, A_2, M_2 (d) M_2, A_2, H_2, A_2

74. The standard step Method aims to solve:

(a) The differential equation of GVF

(b) differential energy equation of GVF

(c) Bernoulli equation

(d) Momentum equation

75. The direct step method of calculating the GVF profile uses the relation

(a) $\Delta E = \Delta x\,(S_0 - S_f)$ (b) $\Delta x = \dfrac{\Delta E}{\left(S_0 - \bar{S}_f\right)}$

(c) $\Delta x = \Delta E\left(S_0 - \bar{S}_f\right)$ (d) $\Delta y = \Delta x \dfrac{\left(S_0 - S_f\right)}{\left(1 - F^2\right)}$

76. The direct step method

(a) is best suited for natural channels

(b) is accurate for all step sizes

(c) is most accurate for calculating supercritical flow profiles

(d) None of these

77. If the Froude no of hydraulic jump is 5.5, it is

(a) Oscillating jump (b) Weak jump

(c) Strong jump (d) Steady jump

78. Seventy percent of initial energy is lost in a jump taking place in horizontal rectangular channel. The Froude number of flow at toe is

(a) 4 (b) 9

(c) 20 (d) 15

79. The sequent depth in a hydraulic jump formed in horizontal rectangular channel is 16.48. The Froude number of the

supercritical stream is

(a) 8.0 (b) 4.0

(c) 20 (d) 12

80. The Froude no. of a subcritical stream at the end of the jump in horizontal rectangular channel is 0.22. The sequent depth ratio of the jump is:

(a) 11.25 (b) 15.25

(c) 8.35 (d) 6.5

81. The discharge Q in a triangular Weir varies as:

(a) $H^{0.5}$ (b) $H^{1.5}$

(c) H^2 (d) $H^{2.5}$

82. In a Triangular notch there is a+2% error in the observation of the head. The error in the computed discharge is

(a) +2% (b) +5%

(c) -5% (d) +2.5%

83. A nominal 90° triangular notch was found to have 2% error in the vertex angle. While discharging under a constant head, the error in the estimated discharge is

(a) π% (b) π/4%

(c) π/2% (d) 2%

84. The discharge equation of a parshall flume is expressed as $Q=$

(a) $K\sqrt{H_a}$ (b) $KH_a^{3/2}$

(c) KH_a (d) $K\left(H_1^{3/2}-H_a^{3/2}\right)$

85. The basic differential equation of SVF with increasing discharge is based on

(a) Continuity equation (b) Energy equation

(c) Momentum equation (d) Manning's equation

86. The basic differential equation of SVF with decreasing discharge is based on

(a) continuity equation (b) Energy equation

(c) Momentum equation (d) Manning's equation

87. The flow profile in a side spillway channel can be determined by using
    (a) Standard step method
    (b) Standard Runge-Kutta method
    (c) De-March equation
    (d) Mustokow equation
88. The De Marchi varied flow function is
    (a) used in SVF over bottom rack
    (b) used in SVF in lateral spillway channel
    (c) Meant for side weirs in frictionless rectangular channels
    (d) Meant for subcritical flows only
89. The De Marchi coefficient for discharge C_M for side weir is
    (a) independent of Froude number
    (b) same as that of normal weir
    (c) essentially a function of inlet Froude number F_1
    (d) approaches unity as $F_1 \to 0$
90. In gradually varied unsteady, open channel flow, $dQ/dx = 0.1$, if top width of channel is 10 m, dA/dt is
    (a) 0.1 (b) 0.01
    (c) –0.1 (d) –0.01
91. A positive surge of height 0.5m was found to occur in a rectangular channel with a depth of 2m. The celerity of surge in m/s is
    (a) ± 4.43 (b) ±2.25
    (c) ±1.25 (d) ±5.25
92. A trapezoidal channel with B = 0.06m, m = 1 and depth of flow = 2m has a positive surge of height 0.8 m. The celerity of surge in m/s is
    (a) ±4.43 (b) ±9.5
    (c) ±3.5 (d) none of these

93. A stone thrown into a shallow pond produced a way of amplitude 2cm and velocity of 1.8m/s. The depth of pond in m is:

(a) 1.8 (b) 0.03

(c) 0.4 (d) 0.02

94. In negative surge:

(a) wave velocity v_w is constant.

(b) celerity is always negative.

(c) water surface is a uniformly progressive wave.

(d) The celerity varies with depth.

95. A canal has velocity of 2.5 m/s and depth of flow 1.63m. A negative wave formed due to decrease in the discharge at an upstream control moves at this depth with a celerity of:

(a) 6.5m/s (b) –6.5m/s

(c) 1.5m/s (d) –4m/s

96. In a channel bed slope changes from mild to steep slope. The resulting GVF profiles are:

(a) (M_1, S_2) (b) (M_1,S_3)

(c) (M_2, S_2) (d) (M_2,S_1)

97. The flow will be in the supercritical state in the following types GVF profiles

(a) All S curve (b) M_2

(c) A_3,M_3,S_2 (d) S_2,M_2,S_3

98. The Keifer and Chu varied flow function are useful for GVF computations in

(a) All types of channels

(b) Channels with closing top

(c) Circular channel only

(d) Rectangular Channel only.

99. The hydraulic jump is a phenomenon:

(a) In which the water surface connects the alternative depths.

(b) Which occurs only in frictionless channel.

(c) Which occurs only in rectangular channel.

(d) None of these.

100. The Transitional profile in a lateral spillway channel is

(a) Independent of the roughness of the channel.

(b) Independent of rate of lateral flow.

(c) Independent of channel Geometry

(d) Independent of bottom slope of channel.

Answers of Objective Questions

1. (c)	2. (b)	3. (c)	4. (c)	5. (a)	6.(b)	7. (a)	8. (c)
9.(c)	10.(c)	11. (a)	12. (b)	13. (d)	14. (d)	15. (a)	16. (b)
17. (d)	18. (c)	19. (a)	20. (c)	21. (b)	22. (d)	23. (d)	24. (d)
25. (b)	26. (c)	27. (a)	28. (d)	29. (a)	30.(d)	31. (b)	32. (c)
33. (b)	34. (d)	35. (d)	36. (a)	37. (b)	38. (a)	39. (b)	40. (c)
41. (b)	42. (c)	43. (d)	44. (c)	45. (a)	46. (d)	47. (a)	48. (b)
49. (d)	50. (c)	51. (b)	52. (c)	53. (d)	54. (b)	55. (d)	56. (c)
57. (b)	58. (c)	59. (a)	60. (c)	61. (c)	62. (c)	63. (c)	64. (b)
65. (a)	66. (d)	67. (a)	68. (c)	69. (c)	70. (a)	71. (a)	72. (b)
73. (d)	74. (c)	75. (c)	76. (d)	77. (d)	78. (b)	79. (d)	80. (a)
81. (d)	82. (b)	83. (a)	84. (c)	85. (b)	86. (c)	87. (b)	88. (c)
89. (c)	90. (c)	91. (d)	92. (d)	93. (c)	94. (d)	95. (a)	96. (c)
97. (c)	98. (c)	99. (d)	100. (b)				

Previous Year Questions Paper

B.Tech.
(SEM. VII)
THEORY EXAMINATION
2011-12
Code CE043
OPEN CHANNEL FLOW

Time : 3 Hours *Total Marks : 100*

Note:—Attempt **all** questions. Assume any data suitably if required.

1. Attempt any **four** of the following: **(5×4=20)**

(a) Show that for narrow deep rectangular channels, hydraulic mean depth (R) is approximately equal to B/2.

(b) Show that for channels with larger slopes, the pressure variation with depth is given by $p = \gamma y \cos^2 \theta$.

(c) Show that triangular channel section is most efficient when its central angle is 90°.

(d) For a rectangular channel, show that Q is maximum at the critical depth when specific force P is constant.

(e) Determine the normal depth in a rectangular channel with apex angle 90°. when it carries a discharge of 1.5 m³/s at a slope of 0.0001. Take Manning's *n* as 0.015.

(f) A discharge of 40.0 m³/s flows in a trapezoidal channel with bottom width 4.0 m and side slope 2 horizontal : 1

vertical. If the normal depth at bottom slope of 0.0016 is 2.0 m, determine the values of n and C. Is the flow is supercritical or subcritical ?

2. Attempt any **two** of the following : (10×2=20)

(a) Using the Chezy's equation, show that general equation for hydraulic exponent N is $N = \frac{y}{A}\left(3T - R\frac{dP}{dy}\right)$.

(b) Sketch the flow profile in the following cases :

(i) Mild slope to sleep slope

(ii) Mild slope to horizontal slope

(iii) Mild slope to adverse slope

(iv) Steep slope to horizontal slope

(c) Prove $\frac{dy}{dx} = S_0 \frac{1-(K_n/K)^2}{1-(Z_n/Z)^2}$, where all symbols have usual meaning.

3. Attempt any **one** of the following : (20×1=20)

(a) A trapezoidal channel having b = 6 m. side slope z = 2. S_0 = 0.0016 and n = 0.025 carries a discharge of 11.30 m^3/s. Compute the back water profile by Graphical Method created by dam which backs up the water to a depth of 1.50 m immediately behind the dam. The upstream end of the profile is assumed at depth equal to 1% greater than normal depth. Take α = 1.10.

(b) A wide rectangular channel carrying 5.0 m^3/s has a bottom slope 0.00266 and Manning's n of 0.015. If the channel is followed by a sudden drop, determine how far upstream from the drop depth of flow will be 2.40 m. Use direct step method.

4. Attempt any **two** of the following : (10×2=20)

(a) A hydraulic jump occurs in a rectangular channel and depths of flow before and after jump are 0.5 m and 2.0 m respectively. Calculate the critical depth and energy loss in the jump.

(b) For hydraulic jump in a rectangular channel show that

$$F_2^2 = \frac{8F_1^2}{\left(\sqrt{1+8F_1^2}-1\right)^3}$$

(c) What is surge in open channel ? Prove celerity of wave 'c' is given by $C \approx \sqrt{gy_1}$, where all symbols have usual meanings.

5. Attempt any **two** of the following : **(10 × 2 = 20)**

(a) How will you determine that the culvert is hydraulically short and hydraulically long ? Give the neat sketches of culvert with outlet unsubmerged condition.

(b) What do you understand by bottom racks and classify it into different categories? Discuss various types of flow that can occur over bottom racks and draw its profile also.

(c) A quadratic weir is designed for installation in a rectangular channel of width 30 cm. The rectangular base of the weir occupies the full width of channel and is 6 cm in height. The crest of the base weir coincides with the channel bed.

(j) Determine the discharge through the weir when the depth of flow in the channel is 15 cm.

(ii) What would be the depth of flow upstream of the weir when discharge in the channel is 25 litres/s {assume $C_d = 0.62$} ?

B. Tech

(Sem. VIII) Examination 2008-09

Open Channel Flow

Time : 3 Hours] *[Total Marks: 100*

1. Attempt any two parts of the following:

   (a) Discuss energy and momentum principles for prismatic and non-prismatic channels. Explain specific energy specific force.

   (b) Give examples with neat sketches for a uniform, gradually varied, rapidly varied, spatially varied or unsteady flow.

   (c) A concrete lined trapezoidal channel with uniform flow carries a discharge of 30 cumecs. The base width is 5m and the side slopes are equal at 1:2 (vertical Horizontal), Manning's n can be taken as 0.015 and the bed slope = 0.001. What is normal depth of the flow?

   (d) In a rectangular two-dimensional open channel flow the water depth is 0.9 m and the specific energy 2.4 m. For the same flow rate what depths may be expected for a specific energy of 3 m?

   (e) A trapezoidal channel is to have a longitudinal slope of 1 in 1000 and is to be lined with brickwork. It has to carry a discharge of 20 cumecs and the slope stability requires the side slopes to be limited to an angle of 30°. What should be the bed width for the most efficient section?

2. Attempt any **two** parts of the following:

   (a) State assumptions and obtain and the governing equation for GVF (gradually varied flow) as given below:

$$\frac{dy}{dx} = S_0 \frac{\left[1-\left(\frac{y_n}{y}\right)^3\right]}{\left[1-\left(\frac{y_c}{y}\right)^3\right]};$$

where $\frac{dy}{dx}$ = slope of the water surface with reference to the channel bed: S_0 = bed slope ; y = depth of water; y_n = normal depth, and y_c = critical depth.

(b) A rectangular channel, 6m wide, carries a discharge of 36.16 m³/s, with a bed slope. $S_0 = 0.0016$, and Manning's $b = 0.015$. Compute the water surface slopes at sections where depths of flow are 2.5 m and 1.8 m, respectively. Assume following data to be applicable to the channel under consideration.

The normal depth, $y_n = 2.00$ m,

The critical depth, $y_c = 1.55$ m

Uniform flow velocity, $V_n = 3.01$ m/s,

$\cos\theta = 1$, and

Velocity coefficient $\alpha = 1.0$

(θ is bottom-slope angle; α is the energy coefficient.)

(c) A 1.25 m diameter circular concrete pipe is laid at a longitudinal slope of 1 in 2000. The normal depth of flow for certain discharge is equal to 0.8 times the diameter. If the flow depth at a section under gradually varied conditions is 0.75, what would be the nature of the flow profile?

3. Attempt any **two** parts of the following:

(a) Discuss evaluation of the hydraulic jump elements in rectangular and non-rectangular channels on horizontal and sloping beds. Give examples with neat sketches of hydraulic jump in gradually and suddenly expanding channels. Discuss celerity of the wave and open channel positive and negative surges.

(b) A hydraulic jump is formed in a 5 m wide rectangular channel carrying a discharge of 20 m³/s. The pre jump depth is 0.5 m. Find the post jump depth, post jump, Froude number, and length of the jump.

(c) A 1 m wide rectangular tilting flume ends in a free overfall where the depth of the flow is measured as 0.3 m. The flow depth upstream of the depth was not measured but was observed to be considerably large than the brink depth. What is discharge in the channel? The discharge was then reduced till the flow in the channel becomes just supercritical and the brink depth was measured as 0.05 m. Compute the discharge under this condition.

4. Obtain any **two** parts of the following:

(a) Obtain the differential equation of spatially varied flow (SVF)? Classify flow profiles for SVF. Discuss forward Euler's method to plot the flow profiles. Sketch for flows over bottom racks.

(b) A rectangular channel has bed width of 3m, bed slope of 1 in 1000, and Manning's roughness of 0.015. Design a side weir such that it becomes effective when the discharge in the channel is 1 m^3/s of the flow when the discharge is 1.5 m^3/s.

(c) A 3 m wide rectangular channel has bed width of 5m, bed slope of 0.1, and may be assumed to be frictionless. It is 40 m long and carries water away from a spillway whose crest is parallel to the channel. The discharge per unit length of the spillway is 0.5 $m^3/s/m$. The channel outlet is submerged such that the water depth above the bed is 2m. Plot the water surface profile. Assume the momentum correction factor to be 1.

5. Attempt any one part of the following 20

(a) What is design procedure for a box culvert? What are hydraulically short or long culverts? Discuss inlet and outlet controls.

(b) Give examples of channels of non linear alignment and discuss design considerations under supercritical flows. What are non-prismatic channels?

B. Tech

(Sem. VIII) Examination 2007-08

Open Channel Flow

Time : 3 Hours] *[Total Marks: 100*

Note:(1) Attempt all the questions.

(2) All questions carry equal marks.

(3) Assume any data suitable, if not given

1. Attempt any four part of the following: $5 \times 4 = 20$

(a) Explain the following:

(i) Velocity distribution is open channel.

(ii) Pressure distribution in open channel.

(iii) Specific energy and specific force.

(iv) Prismatic channel and non-prismatic channel.

(b) Derive an expression for determining the discharge passing through an open channel and state the assumptions made there in.

(c) Show that for a trapezoidal channel of given area of flow, the condition of maximum flow requires that its hydraulic radius in equal to half the depth of flow and its sides make an 60° angle with the horizontal.

(d) An irrigation channel carries a discharge of 46 cumec. Determine the critical depth of flow in each case

$2 \times 2.5 = 5$

(i) A triangular channel with each of the sides slopes of 1.5 H:IV

(ii) A trapezoidal channel width of 10m and side slope of 1H:IV on one side and $\frac{1}{2}$ H:IV on other side.

(e) For a given discharge, show that the minimum specific force in open channel flow occurs at the critical depth, when Froude's Number is unity. 5

(f) A 3.6 m wide rectangular channel carries water to a depth

1.8 m. In order to measure the discharge, the channel width is reduced to 2.4 m and a hump of 0.30m height is provided in the bottom. Calculate the discharge if water surface in the contracted section drops by 0.15 m. Assume to losses. 5

2. Attempt any two parts of the following $10 \times 2 = 20$

(a) Derive an expression for the variation of depth along length of bed of the channel for gradually varied flow in an open channel flow. State clearly all the assumptions made there in. 10

(b) Sketch the possibly GVF profiles in the following serial arrangement of channels and controls. The flow is from left to right. 10

Give explanation with every sketches

(i) Steep-Horizontal-Mild slopes

(ii) Sluice gate - Adverse-Horizontal

(iii) Mild-Sluice gate - Steep

(iv) Free intake-steep-sluice gate

(c) A rectangular channel 10m wide carries a discharge of 30 m^3/s. It is laid at a slope of 1.0×10^{-4}. If at a section in this channel the depth is 1.6 m, how for (upstream or down stream) from the section will the depth be 2.0 m? Take Manning's '*n*' as 0.015.

3. Attempt any **two** parts of the following: $10 \times 2 = 20$

(a) Determine from first principle the conditions required for the formation of such a jump in the case of a rectangular channel of constant width and calculate the loss of head in terms of depth just and after the jump. 10

(b) Differentiate between Broad-crested weir and Narrow-crested weir. Find the condition for maximum discharge over a broad-crested weir and hence derive an expression for maximum discharge over a broad-crested weir 10

(c) The depth and velocity of flow in a rectangular channel are 1m and 1.5 m/s respectively. If the rate of inflow at the upstream end is suddenly doubled, what will be the height and absolute velocity of the resulting serve and the celerity of the wave? 10

4. Attempt any **two** parts of the following 10×2=20

(a) Show that for a wide rectangular channel having SVF with a constant in flow rate of 'q' the critical flow section is given by 10

$$x_c = \frac{8q^2}{g\left(s_0 - g/c^2\right)^3}$$

when $B = 1.0$ as assumed and Chezy's formula with C = Constant is used. What would be the corresponding value of Y_c?

(b) A rectangular channel 2.0 m wide carries a discharge of 3.5 m^3/s at a Froude number of 0.30. A2.0 m long parallel longitudinal bar bottom rack having $E = 0.2$ is provided at a section. Super critical flow is known to occur over the rack Estimate the discharge diverted out. 10

(c) An axis symmetric radial flow emanates from a source on to a horizontal plane. Show that the basic differential equation of SVF with decreasing discharges can be expressed in the case as 10

$$\frac{dy}{dx} = \frac{\left(\frac{y^2}{gr} - s_f\right)}{\left(1 - \frac{v^2}{gy}\right)}$$

5. Attempt any two parts of the following: 10×2=20

(a) A 120° band with B = 10m, and Rc = 12 m is designed to connect two straight channels of the same width carrying 15 cumec at a normal depth of 7.5 m. Determine the specific energy at the sector containing maximum surface. Depression in the band Assume $\alpha = 1.0$ and $n = 0.01$. 10

(b) Design a straight contraction to reduce the width of a rectangular channel from 6 to 3 m. The channel carries a discharge of 20 cumec with an approach velocity of 2.5 m/s. Assume $y_3/y\ \pi = 3.00$ 10

(c) A culvert under a road must carry 4.3 m^3/s 10

(i) The culvert length will be 30 m and the slope will be 3×10^{-3}. If the maximum permissible head water level of 3.6 m above the culvert invert, what size of corrugated pipe culvert ($n = 0.025$) would you select? The outlet will discharge freely. Neglect velocity of approach. Assume square-edged entrance with Ke = 0.5, ed = 0.65.

B. Tech

(Sem. VIII) Examination 2006-07

Open Channel Flow

Time : 3 Hours] *[Total Marks: 100*

Note:(1) Attempt all the questions. All questions carry equal marks.

(2) Use illustrations, whereever needed.

(3) Assume missing data suitably, if any, and state the assumptions made.

1. Attempt any four part of the following: $5 \times 4 = 20$

(a) Differentiate between the following with neat sketches:

(i) Rapidly varied flow and spatiallty varied flow?

(ii) Prismatic and non-prismatic channels.

(b) Figure - 1, show a free overfall in a horizontal frictionless rectangular channel. Assuming the flow to be horizontal at section 'O' and the pressure at the brink section '*e*' to be atmospheric throughout the depth, show that

$$\frac{y_0}{y_e} = \frac{2f_0^2}{(2f_0^2 + 1)}$$

where $f_0^2 = \frac{q^2}{gy_0^3}$ and q = discharge per unit width.

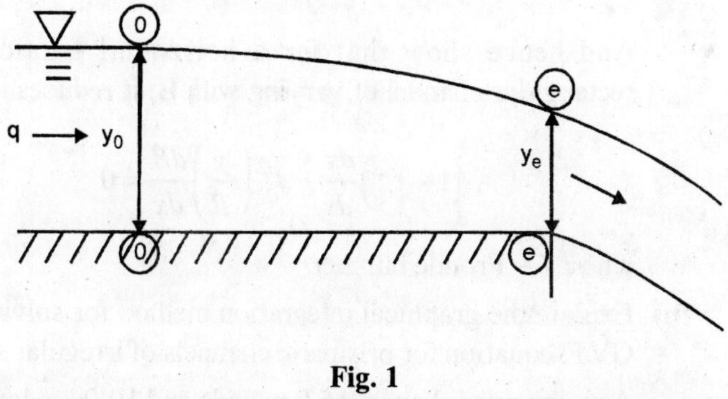

Fig. 1

(c) For a rectangular channel of width B = 2.0 m, calculate the critical depth and the corresponding specific energy for a discharge of 6.0 m^3/s.

(d) A rectangular channel section is to have critical flow and at the same time the wetted perimeter is to be minimum. Show that these two conditions to occur simultaneously, the width of the channel must be equal to 8/9 times the minimum specific energy head.

(e) Prove that for a most economical section of a trapezoidal channel:

(i) Half of top width = length of one of sloping sides.

(ii) Hydraulic mean depth = half of depth of flow.

(f) A triangular channel with an apex angle of 60° carries a flow of 1.2 m^3s at a depth of 0.80 m. If the bed slope is 0.009, find the roughness coefficient of the channel.

2. Attempt any **two** parts of the following: 10 × 2 = 20

(a) Derive the differential equation of G.V.F. in a rectangular channel of variable width *B* with stating assumptions, in the following form

$$\frac{dy}{dx} = \frac{s_o - s_f + \left(\frac{Q^2 y}{gA^3}\frac{dB}{dx}\right)}{1 - \frac{Q^2 B}{gA^3}}$$

And hence show that for a horizontal frictionless rectangular channel of varying with B, it reduces to

$$\left(1 - f^2\right)\frac{dy}{dx} - f^2\left(\frac{y}{B}\right)\frac{dB}{dx} = 0$$

where f = Froude number.

(b) Explain the graphical integration method for solving the G.V.F. equation for prismatic channels of irregular shape. A rectangular channel 15.0 m wide and 10.0 km long has a slope of 10^{-4} and connects two lakes. The water depths at the upstream and downstream ends of the channel are 1.50 and 2.00 m. respectively. If $n = 0.015$ determine the type of G.V.F. profile.

(c) Sketch the possible water surface profile in case of flow from a steep slope to a wild slope?

A rectangular channel 5.0 m wide carries water at a depth of 1.5 m. The bed slope is 10^{-4} and n= 0.016. The channel ends in an abrupt drop. Find using Chow's method, how far upstream of the fall the depth of flow would be 1.4 m.

Given: first hydraulic exponent 'M' = 3.0 and second hydraulic exponent 'M = 3.0'.

3. Attempt any **two** parts of the following: 10 × 2 = 20

(a) Derive the differential equation of spatially varied flow (S.V.F.) with decreasing discharge. Clearly state the assumption made.

(b) Horizontal rectangular lateral spillway has a free overfall outlet. The inflow is uniformally distributed along the channel with a rate of q^* per unit lengthy of the channel. The channel length is 'L'. Prove that the equation of flow is:

$$\frac{dy^2}{dx} - \frac{x^2}{y} = \frac{-gb^2y^2}{q^{*2}}$$

Hence, show that for the condition of critical depth 'y_c' occurring at the outlet solution will be

$$\left(\frac{x}{L}\right)^2 = \frac{3}{2}\left(\frac{y}{y_c}\right) - \frac{1}{2}\left[\frac{y}{y_c}\right]^2$$

(c) Write short notes on any **two** of the following:

(i) Classification of different types of flows over bottom vacks (with neat sketch)

(ii) Classification of flow profiles for S.V.F. with lateral inflows (Li's Classification)

(iii) Standard fourth-order Range-Kutta method for surface profile plotting.

4. Attempt any four parts of the following: $5 \times 4 = 20$

(a) Show that in a hydraulic jump formed in a horizontal, frictionless rectangular channel. The energy loss relative to the critical depth 'y_c' can be expressed as

$$\left(\frac{E_L}{y_c}\right)^3 = \frac{(a-1)^9}{32(a+1)a^4}$$

where a = sequent depth ratio $= \frac{y_2}{y_1}$

(b) Give the classification of hydraulic jumps formed in horizontal rectangular channels with neat sketches, stating the relative energy loss.

(c) A hydraulic jump takes place in a horizontal triangular channel having side slopes of 1.5 H: IV. The depths before and after the jump are 0.40 m and 1.50 m respectively. Estimate the flow rate and Froude number at the beginning and end of the jump.

(d) Explain the use of hydraulic jump as an energy dissipator.

(e) A trapezoidal channel is 7.0 m wide at the bottom and the side slopes are 1 H:IV carries a discharge of 20 cumes. Draw the specific force diagram for this channel.

(f) What is a submerged jump? Hence, detail the submergence factor.

5. Attempt any **two** parts of the following: $10\times2 = 20$

(a) How you will determine that the culvert is hydraulically short or hydraulically long? Give the neat sketches of culvert with outlet unsubmerged condition.

(b) Discussed the design consideration for determining the shape of cross section at an open channel bend i.e. a channel of non-linear alignment under the supercritical flow condition.

(c) Write short notes on any **two** of the following:

(i) Characteristics of flow through channel junction.

(ii) Characteristics of flow through sudden transitions in subcritical flow.

(iii) Flow through an obstruction in open channel flow.

Glossary

Abutment Part of the valley side against which the dam is constructed. Artificial abutments are sometimes constructed to take the thrust of an arch where there is no suitable natural abutment.

Afflux Rise of water level above normal level (i.e. natural Hòod level) on the upstream side of a culvert or of an obstruction in a channel. In the US, it is commonly referred to as maximum backwater.

Alternate depth In open channel flow, for a given flow rate and channel geometry, the relationship between the specific energy and flow depth indicates that, for a given specific energy, there is no real solution (i.e. no possible flow), one solution (i.e. critical flow) or two solutions for the flow depth. In (he latter case, the two flow depths arc called alternate depths. One corresponds to a subcritical flow and the second to a supercritical flow.

Apron The area at the downstream end of a weir to protect against erosion and scouring by water.

Aqueduct A conduit for conveying a large quantity of flowing waters. The conduit may include canals, siphons, pipelines etc

Arch dam Dam in plan dependent on arch action lor its strength.

Arched dam Gravity dam which is curved in plan. Alternatives include 'curved-gravity dam' and "arch-gravity dam'.

Archimedes Greek mathematician and physicist. He lived between BC 290-280 and BC 212 (or 211). He spent most of his life in Syracuse (Sicily, Italy) where he played a major role in the defence of the city against the Romans. His treaty 'On Floating Bodies' is the first-known work on hydrostatics, in which he outlined the concept of buoyancy.

Backwater In a tranquil flow motion (i.e. subcritical flow) the longitudinal flow profile is controlled by the downstream flow conditions: e.g. an obstacle, a structure, a change of cross-section. Any down-stream control structure (e.g. bridge piers, weir) induces a backwater effect. More generally the term backwater calculations or backwater profile refer to the calculation of the longitudinal flow profile. The term is commonly used for both supercritical and subcritical flow motion.

Barrel For a culvert, central section where the cross-section is minimum. Another term is the throat.

Bakhmeteff Boris Alexandrovitch Bakhmeteff (1880—1951) was *a* Russian hydraulician. In 1912, he developed the concept of specific energy and energy diagram for open channel flows.

Barrage Hrench word for dam or weir, commonly used to described large dam structure in Hnglish.

Bed-form Channel bed irregularity that is related to the flow conditions. Characteristic bed-forms include ripples, dunes and antidunes (e.g. Plates 3, 4 and 22).

Bed-load Sediment material transported by rolling, sliding and saltation motion along the bed.

Belanger Jean-BaptisteCh. Belanger (1789—1874) was a French hydraulician and Professor at the bcole Nationals Superieure des Ponts-et-Chaussees (Paris). He suggested first the application of the momen-tum principle to hydraulic jump flow (Belanger, 1828). In the same book, he presented the first 'back-water' calculation for open channel flow.

Belanger equation Momentum equation applied across a hydraulic jump in a horizontal channel (named after J.B.C, Belanger).

Bernoulli Daniel Bernoulli (1700-17S2) was a Swiss mathematician, physicist and botanist who developed the Bernoulli equation in his *Hydrodynamica, de viribus et motibus fluidorum* textbook (first draft in 1733, first publication in 1738, Strasbourg).

Bidone Giorgio Bidone (1781 1839) was an Italian hydraulician. His experimental investigations on the hydraulic jump were published between 1 820 and 1826.

Biesel Francis Biesel (1920-1993) was a French hydraulic engineer and a pioneer of computational hydraulics.

Borda Jean-Charles de Borda (1733-1799) was a French mathematician and military engineer. He achieved the rank of Capitaine dc Vaisseau and participated to the US War of Independence with the

French Navy. He investigated the flow through orifices and developed the Borda mouthpiece.

Borda mouthpiece A horizontal re-entrant tube in the side of a tank with a length such that the issuing **jet** is not affected by the presence of the walls,

Bore A surge of tidal origin is usually termed a bore (e.g. the Mascaret in the Seine river, France).

Bottom outlet Opening near the bottom of a dam for draining the reservoir and eventually flushing out reservoir sediments.

Boundary layer Flow region next to a solid boundary where the flow field is affected by the presence of the boundary and where friction plays an essential part. A boundary layer flow is characterized by a range of velocities across the boundary layer region from zero at the boundary to the free-stream velocity at the outer edge of the boundary layer.

Boussinesq Joseph Valentin Boussincsq (1842 1929) was a French hydrodynamicist and Professor at the Sorbonne University (Paris). His treatise *Essai sur la theorie des caux coitrautes* (1877) remains an outstanding contribution in hydraulics literature.

Bomsinesq coefficient Momentum correction coefficient named after J.V Boussinesq who first proposed it (Boussincsq, 1877).

Boussuiesq-Favre wave An undular surge (see Undular surge).

Bresse Jacques Antoine Charles Bresse (1822 1883) was a French applied mathematician and hydraulician. He was Professor at the Ecole Nationale Superieure des Ponts-et-Chaussees, Paris as successor of J.B.C. Belanger. His contribution to gradually varied flows in open channel hydraulics is considerable (Bresse, 1860).

Broad-crested weir A weir with a flat long crest is called a broad-crested weir when the crest length over the upstream head is greater than 1.5 to 3. If the crest is long enough, the pressure distribution along Ihe crest is hydrostatic, the flow depth equals the critical flow depth $d_c = (q^2/g)^{1/3}$ and the weir can be used as a critical depth meter.

Cavitation Formation of vapour bubbles and vapour pockets within a homogeneous liquid caused by excessive stress (Franc *etal.,* 1995). Cavitation may occur in low-pressure regions where the liquid has been accelerated (e.g. turbines, marine propellers, baffle blocks of dissipation basin). Cavitation modifies the hydraulic characteristics of a system, and it is characterized by damaging erosion, additional noise, vibrations and energy dissipation.

Chezy Antoine Chezy (1717—1798) (or Antoine de Chezy) was a French engineer and member of the French 'Corps dcs Ponts-ct-Chaussccs[1]. lie designed canals for the water supply of the city of Paris. In 1768 he proposed a resistance formula for open channel flows called the Chezy equation. In 1798, he became Director of the Ecole Nationale Superieure des Ponts-et-Chaussees after teaching there for many years.

Chezy coefficient Resistance coefficient for open channel flows first introduced by the Frenchman A. Chezy. Although it was thought to be a constant, the coefficient is a function of the relative roughness and Reynolds number.

Choke In open channel flow, a channel contraction might obstruct the flow and induce the appearance of critical flow conditions (i.e. control section). Such a constriction is sometimes called a 'choke'.

Choking flow Critical flow in a channel contraction. The term is used for both open channel flow and compressible flow.

Conjugate depth In open channel flow, another name for sequent depth.

Control Considering an open channel, subcritical flows are controlled by the downstream conditions. This is called a 'downstream flow control'. Conversely, supercritical flows are controlled only by the upstream flow conditions (i.e. 'upstream flow control').

Contivl section In an open channel, cross-section where critical flow conditions take place, The concept of 'control'and 'control section' are used with the same meaning.

Control surface Is the boundary of a control volume.

Control volume Refers to a region in space and is used in the analysis of situations where flow occurs into and out of the space.

Crest of spillway Upper part of a spillway. The term 'crest of dam' refers to the upper part of an uncontrolled overflow.

Critical depth It is the flow depth tor which the mean specific energy is minimum.

Critical flow conditions In open channel flows, the flow conditions such as the specific energy (of the mean flow) is minimum are called the critical flow conditions. With commonly used Froude number definitions, the critical flow conditions occur for $Fr = 1$. If the flow is critical, small changes in specific energy cause large changes in flow depth. In practice, critical flow over a long reach of channel is unstable.

Culvert Covered channel of relatively short length installed to drain water through an embankment (e.g. highway, railroad, dam).

Froude William Froude (1810 1879) was a English naval architect and hydrodynamicist who invented the dynamometer and used it for the testing of model ships in towing tanks. He was assisted by his son Robert Edmund Froude who, after the death of his father, continued some of his work. In 1868, he used Reech's Law of Similarity to study the resistance of model ships,

Froude number The Froude number is proportional to the square root of the ratio of the inertial forces over the weight of fluid. The Froude number is used generally for scaling free-surface flows, open channels and hydraulic structures. Although the dimension less number was named after William Froude, several French researchers used it before, Dupuit (1848) and Bresse (I860) highlighted the significance of the number to differentiate the open channel flow regimes, Bazin (1865a) confirmed experimentally the findings. Ferdinand Reech introduced the dimensionless number for testing ships and propellers in 1852. The number is called the Reech Froude number in France.

Gate Valve or system for controlling the passage of a fluid. In open channels the two most common types of gates are the underflow gate and the overflow gate.

Hydraulic diameter It is defined as the equivalent pipe diameter; i.e. four times the cross-sectional area divided by the wetted perimeter. The concept was first expressed by the Frenchman P.L.G. du Buat (1779).

Hydraulic jump Transition from a rapid (supercritical flow) to a slow flow motion (subcritical flow) (e.g. Plates 8 and 25). Although the hydraulic jump was described by Leonardo da Vinci, the first experimental investigations were published by Giorgio Bidone in 1820. The present theory of the jump was developed by Belanger (1828) and it has been verified experimentally numerous researchers (e.g. Bakhmeteff and Matzke, 1936).

Ideal fluid Frictionless and incompressible fluid. An ideal fluid has zero viscosity: i.e. it cannot sustain shear stress at any point.

Inflow (1) Upstream flow; (2) incoming flow.

Inlet (1) Upstream opening of a culvert, pipe or channel; (2) a tidal inlet is a narrow water passage between peninsulas or islands.

Intake Any structure in a reservoir through which water can be drawn into a waterway or pipe. By extension, upstream end of a channel.

Interface Surface forming a common boundary of two phases (e.g. gas-liquid interface) or two fluids.

Irrotational flow It is defined as a zero vorticity flow. Fluid particles within a region have no rotation. If a frictionless fluid has no rotation at rest, any later motion of the fluid will be irrotational. In irrotational flow each element of the moving fluid undergoes no net rotation, with respect to chosen co-ordinate axes, from one instant to another.

Kennedy Professor John Fisher Kennedy (1933-1991) was a Professor in Hydraulics at the University of Iowa. He succeeded Hunter Rouse as Head of the Iowa Institute of Hydraulic Research.

Laminar flow This flow is characterized by fluid particles moving along smooth paths in laminas or layers, with one layer gliding smoothly over an adjacent layer. Laminar flows are governed by Newton's Law of Viscosity which relates the shear stress to the rate of angular deformation: $\tau = \mu \partial V/\partial y$.

Manning Robert Manning (1816-1897) was the Chief Engineer of the Office of Public Works, Ireland. In 1889, he presented two formulae (Manning, 1890). One was to become the so-called 'Gauckler-Manning formula' but Robert Manning did prefer to use the second formula that he gave in his paper. It must be noted that the Gauckler-Manning formula was proposed first by the Frenchman P.G. Gauckler (Gauckler, 1867).

Meandering channel Alluvial stream characterized by a series of alternating bends (i.e. meanders) as a result of alluvial processes.

Navier Louis Marie Henri Navier (1785-1835) was a French engineer who primarily designed bridges but also extended Euler's equations of motion (Navier, 1823).

Navier-Stokes equation Momentum equation applied to a small control volume of incompressible fluid. It is usually written in vector notation. The equation was first derived by L. Navier in 1822 and S.D. Poisson in 1829 by a different method. It was derived later in a more modern manner by A.J.C Barre de Saint-Venant in 1843 and G.G. Stokes in 1845.

Outflow Downstream flow.

Outlet (1) Downstream opening of a pipe, culvert or canal; (2) artificial or natural escape channel.

Pascal Blaise Pascal (1623-1662) was a French mathematician, physicist and philosopher. He developed the modern theory of probability. Between 1646 and 1648, he formulated the concept of pressure and showed that the pressure in a fluid is transmitted through the fluid in all directions. He measured also the air pressure both in Paris and on the top of a mountain over-looking Clermont-Ferrand (France).

Side-channel spillway A side-channel spillway consists of an open spillway (along the side of a channel) discharging into a channel running along the foot ofthe spillway and carrying the flow away in a direction parallel to the spillway crest (e.g. Arizona-side spillway ofthe Hoover dam, USA).

Slope (1) Side of a hill; (2) inclined face of a canal (e.g. trapezoidal channel); (3) inclination of the chan-nel bottom from the horizontal.

Sluice gate Underflow gate with a vertical sharp edge for stopping or regulating flow.

Specific energy Quantity proportional to the energy per unit mass, measured with the channel bottom as the elevation datum, and expressed in metres of water. The concept ofspecific energy, first developed by B.A. Bakhmeteff in 19 12, is commonly used in open channel flows.

Spillway Opening built into a dam or the side of a reservoir to release (to spill) excess flood waters.

Steady flow Occurs when conditions at any point of the fluid do not change with the time:

$$\frac{\partial V}{\partial t}=0 \quad \frac{\partial \rho}{\partial t}=0 \quad \frac{\partial P}{\partial t}=0 \quad \frac{\partial T}{\partial t}=0$$

Stilling basin Structure for dissipating the energy ofthe flow downstream of a spillway, outlet work, chute or canal structure. In many cases, a hydraulic jump is used as the energy dissipator within the stilling basin.

Streamline It is the line drawn so that the velocity vector is always tangential to it (i,c, no flow across a streamline). When the streamlines converge the velocity increases. The concept of streamline was first introduced by the Frenchman J.C. de Borda.

Supercritical flow In open channel, when the flow depth is less than the critical flow depth, the flow is supercritical and the Froude number is larger than one. Supercritical flows are controlled from upstream.

Surge A surge in an open channel is a sudden change of flow depth (i.e. abrupt increase or decrease in depth). An abrupt increase in flow depth is called a positive surge while a sudden decrease in depth is termed a negative surge, A positive surge is also called (improperly) a 'moving hydraulic jump' or a 'hydraulic bore'.

Surge wave Results from a sudden change in flow that increases (or decreases) the depth.

Tailwater level Downstream free-surface elevation.

Total head The total head is proportional to the total energy per unit mass and per gravity unit. It is expressed in metres of water, *Trashrack* Screen comprising metal or reinforced concrete bars located at the intake of a waterway to prevent the progress of Moating or submerged debris.

Turbulence Flow motion characterized by its unpredictable behaviour, strong mixing properties and a broad spectrum oflenglh scales (Lesieur, 1994).

Turbulent flow In turbulent flows the fluid particles move in very irregular paths, causing an exchange of momentum from one portion of the fluid to another. Turbulent flows have great mixing potential and involve a wide range of eddy length scales.

Undular hydraulic jump Hydraulic jump characterized by steady stationary free-surface undulations downstream of the jump and by the absence of a formed roller. The undulations can extend far down-stream of the jump with decaying wave lengths, and the undular jump occupies a significant length of the channel. It is usually observed for $1 < Fr_1 <$ 1.5-3 (Chanson, 1995a), The first significant study of undular jump flow can be attributed to Fawer (1937) and undular jump flows should be called Fawer jump in homage to Fawer's work.

Undular surge Positive surge characterized by a train of secondary waves (or undulations) following the surge front. Undular surges arc sometimes called Boussincsq Favrc waves in homage to the contribu-tions of J.B. Boussinesq and H. Favre.

Upstream flow conditions Flow conditions measured immediately upstream of ihe investigated control volume.

Vena contracta Minimum cross-sectional area of the flow (e.g. jet or nappe) discharging through an orifice, sluice gale or weir.

Venturi meter In closed pipes, smooth constriction followed by a smooth expansion. The pressure difference between the upstream location and the throat is proportional to the velocity-square. It is named after the Italian physicist Giovanni Battista Venturi (1746-1 822).

Weak jump A weak hydraulic jump is characterized by a marked roller, no free-surface undulation and low energy loss. It is usually observed after the disappearance of undular hydraulic jump with increasing upstream Froude numbers.

Weir Low river dam used to raise the upstream water level (e.g. Plate 8). Measuring weirs are built across a stream for the purpose of measuring the flow.

Weisbach Julius Weisbach (1 806-1 87 1) was a German applied mathematician and hydraulician.

Wetted perimeter Considering a cross-section (usually selected normal to the flow direction), the wetted perimeter is the length of wetted contact between the flowing stream and the solid boundaries. For example, in a circular pipe flowing full, the wetted perimeter equals the circle perimeter.

Wetted surface In open channel, the term 'wetted surface' refers to the surface area in contact with the flowing liquid.